Digital Pathology

Implementation in Clinical Practice With AI Applications

Digital Pathology

Implementation in Clinical Practice With AI Applications

Edited by

Meera Hameed

Department of Pathology and Laboratory Medicine, Memorial Sloan Kettering Cancer Center, New York, NY, United States

Matthew G. Hanna

Department of Pathology and Laboratory Medicine, Memorial Sloan Kettering Cancer Center, New York, NY, United States

ELSEVIER

ACADEMIC PRESS

An imprint of Elsevier

Academic Press is an imprint of Elsevier
125 London Wall, London EC2Y 5AS, United Kingdom
525 B Street, Suite 1650, San Diego, CA 92101, United States
50 Hampshire Street, 5th Floor, Cambridge, MA 02139, United States

ISBN: 978-0-443-13809-6

For Information on all Academic Press publications
visit our website at https://www.elsevier.com/books-and-journals

Publisher: Stacy Masucci
Acquisitions Editor: Patricia Osborn
Editorial Project Manager: Tracy Tufaga
Production Project Manager: Gomathi Sugumar
Cover Designer: Vicky Pearson Esser

Typeset by MPS Limited, Chennai, India

Contents

List of contributors xiii
Preface xv
Acknowledgment xix

CHAPTER 1 Basics of digital pathology 1
Matthew G. Hanna
Introduction 1
Digital imaging basics 2
Digital pathology timeline 3
Digital pathology ecosystem 6
Digital photomicroscopy 6
Gross imaging 7
Video streaming 8
Whole slide scanning 9
Monitors/display devices 11
Input devices for digital pathology 13
Mouse and keyboard interactions 13
Touchscreen displays 13
3D input devices 13
Voice and gesture recognition 14
Augmented reality and virtual reality 14
Microscope attachments for mobile devices 14
Interoperability and standards in digital pathology 14
Applications 15
Challenges/limitations 19
References 21

CHAPTER 2 The seven S's for success 23
Matthew G. Hanna
Introduction 23
Technology adoption 23
Sponsorship from leadership 25
Space allocation 26
Assessment of current space utilization 26
Ergonomics and user experience 27
Staffing resources 29
Scanner considerations 31
Imaging mode 32

Slide throughput/continuous load ... 33
Scan speeds ... 33
Slide size/format ... 33
Magnification/resolution ... 34
Flexibility in scan area ... 34
Live viewing needs ... 34
Multi z-plane support ... 35
Rescan rates ... 35
Whole slide image file format ... 36
Lab space/countertop weight load ... 36
User experience ... 36
Storage ... 36
Software ... 37
Service and maintenance ... 38
Honorable mention ... 40
Standards ... 40
Conclusion ... 43
References ... 43

CHAPTER 3 Building a business case for digital pathology ... 47
Orly Ardon
Introduction ... 47
Understanding the costs associated with digital pathology ... 49
Acquisition and ongoing costs ... 49
Digital scanners cost ... 52
Other scanner-related costs ... 53
Total cost of infrastructure ... 58
Financial benefits of digital pathology ... 59
Additional revenue ... 61
Cost avoidance and savings ... 62
Hybrid analog and digital pathology workflows ... 63
Creating a business plan ... 64
Digital pathology cost calculators ... 69
Understanding financial indicators ... 69
Discussion ... 71
References ... 72

CHAPTER 4 Preanalytic factors in digital pathology implementation ... 79
Peter Ntiamoah, George Ayob, David Beckles, Rene Serrette and Grecia Castro
Introduction ... 79

Basic histology workflow ... 80
Accession ... 80
Specimen grossing and fixation ... 80
Tissue processing ... 83
Tissue embedding ... 85
Microtomy ... 85
Staining ... 86
Digital pathology in histology laboratory ... 88
Artifacts ... 90
Immunohistochemistry and special stains ... 91
Logistical challenges ... 92
Adoption of whole slide imaging workflow ... 94
References ... 96

CHAPTER 5 Quality management in digital pathology: analytic and postanalytic phases ... 103
Orly Ardon
Introduction ... 103
Digital pathology process: preanalytical, analytical, and postanalytical phases ... 104
Ensuring image quality: multicheckpoint process ... 104
Analytic step ... 106
Scanner level quality checkpoint ... 106
Scanner quality variability ... 110
Barcode quality considerations ... 111
Postanalytic step quality factors ... 112
Whole slide image viewer software ... 114
Display monitors ... 115
Automated digital slide quality tools ... 117
Image quality and computational pathology tools ... 119
Published quality error rates ... 119
Digital pathology quality management systems ... 121
Establishing quality and performance metrics ... 122
The cost of quality ... 124
Conclusions ... 125
References ... 126

CHAPTER 6 Use cases in surgical pathology ... 135
Meera Hameed
Introduction ... 135
Telepathology ... 136

Diagnostic accuracy and feasibility 136
Intraoperative consultation 137
Primary diagnosis 137
Tumor detection and classification 139
Metastases detection in lymph nodes 143
Quantitative biomarker detection 144
Prediction of genetic mutation on hematoxylin and eosin whole slide images 147
Image-based prognosis and outcome prediction 149
Summary 151
References 152

CHAPTER 7 Use cases in cytology 171
Brie Kezlarian-Sachs
Introduction 171
Whole slide imaging in cytopathology 171
Whole slide images for primary diagnosis 173
AI applications for primary diagnosis of cytology images 174
Practical considerations for the development and deployment of AI solutions for primary diagnosis 174
Digital imaging for education 175
Telecytology 176
Telecytology configurations 176
Rapid onsite evaluation 178
AI applications in rapid onsite evaluation telecytology 178
Telecytology for primary diagnosis and expert consultation 179
References 180

CHAPTER 8 Regulatory aspects of digital pathology 187
S. Joseph Sirintrapun
Introduction 187
Food and Drug Administration regulations in digital pathology whole slide imaging devices 187
History of whole slide imaging Food and Drug Administration regulation 188
Food and Drug Administration perspective on whole slide imaging components 189
Background and history of Clinical Laboratory Improvement Amendment and remote whole slide imaging sign-out 191

A brief history of Clinical Laboratory Improvement Amendment 191
Defining remote 192
Defining what constitutes a primary diagnosis in digital pathology 192
Remote sign-out and CLIA'88 192
Ramifications and requirements for the enforcement discretion "Waiver" 194
Continued enforcement discretion "Waiver" 196
Accreditation 197
CAP inspectors will look for the sampling records (documentation) for training and system validation 197
Additional considerations around remote primary sign-out through digital pathology 205
Risks of using digital pathology for remote sign-out 205
Standard operating procedures to cover for remote sign-out procedures 205
HIPAA/Privacy-related considerations 207
Role of hospital IT when implementing remote sign-out 207
Validation of digital pathology platforms prior to engaging in remote sign-out, particularly during an emergency 208
Validating/verifying monitors or laptops from remote locations, particularly if the digital pathology system is already validated on-premises 209
A minimal required specifications for hardware and software for use in remote sign-out 210
Role of the Food and Drug Administration on remote whole slide imaging sign-out 210
Medicolegal considerations 210
Diagnostic procedures remotely reportable and billable from a non-CLIA-licensed facility under the waiver 211
Billing 212
Legislation potentially affecting digital pathology regulation 214
What has happened so far with VALID? 214
Food and Drug Administration moving forward with rulemaking on diagnostic testing regulation 215
An editorial on VALID and the feature of Food and Drug Administration overreach 215
Concluding thoughts on legislation regulating laboratory-developed tests 216
References 216

CHAPTER 9 Standards in digital pathology **219**
Rajesh C. Dash and William R. Jeck
Standards and interoperability in digital pathology 219
What is a standard and why is it important? 219
Who develops standards? 220
Types of standards in healthcare that support digital pathology 220
Standards for digital pathology in more detail 223
Enhancing interoperability 231
Closing remarks 239
References 240

CHAPTER 10 The basics of machine learning and artificial intelligence **243**
Stanley Cohen
Introduction 243
"Shallow" learning 244
Deep learning 246
Convolutional neural networks 248
Recurrent neural networks 250
Generative models: autoencoders and generative adversarial networks 251
Reinforcement learning 253
ChatGPT: everything, everywhere, all at once 253
Things to come 254
Summary and overview 255
References for further reading 256

CHAPTER 11 Academic industry partnerships **257**
Liron Pantanowitz
Introduction 257
Digital pathology clinical trials 257
Academic industry partnership benefits 258
Risks associated with academic industry partnership 258
Academia versus industry priorities 258
Contracts 259
Innovation cycle 260
The pathologist's role 262
Disclosure and conflict of interest 262
Intellectual property 262
Conclusion 263

Disclosures ... 263
References ... 263

CHAPTER 12 Ethical aspects of using artificial intelligence in digital and computational pathology ... 267
Chhavi Chauhan
Introduction ... 267
The shift from pathology to digital and computational pathology ... 268
Ethical concerns in leveraging artificial intelligence in digital pathology and computational pathology ... 268
Bias and fairness ... 269
Addressing bias and equity in healthcare ... 269
Ethical concerns pertaining to patient privacy and choice ... 270
Ethical concerns pertaining to ownership and control of patient data ... 272
Regulation of artificial intelligence and machine learning and oversight ... 272
Successful artificial intelligence case studies in digital pathology and computational biology ... 273
Predicting patient outcomes with multimodal data integration ... 273
Stratifying prostate cancer risk ... 273
Augmenting pathologists in prostate cancer detection ... 273
Achieving ethical use of artificial intelligence in digital pathology ... 274
Conclusions ... 274
Disclosures ... 274
Funding ... 274
References ... 275

CHAPTER 13 Next-generation imaging in pathology ... 277
Richard Torres
Why next-gen imaging matters ... 277
Looking beyond whole slide imaging ... 278
What makes a new imaging tech relevant to pathology? ... 280
What is the new tech and how does it work? ... 281
Stimulated Raman histology ... 284
Photoacoustic microscopy ... 288
Microscopy with ultraviolet surface excitation and fluorescence imitating brightfield illumination ... 290

Optical coherence tomography/optical coherence microscopy 294
Confocal microscopy 296
Selective plane illumination microscopy (lightsheet) 298
Multiphoton microscopy 302
A note on 3D imaging 305
Superresolution 308
A framework for validation is evolving 310
Making sense of it all 310
References 311

Index 323

List of contributors

Orly Ardon
Department of Pathology and Laboratory Medicine, Memorial Sloan Kettering Cancer Center, New York, NY, United States

George Ayob
Department of Pathology and Laboratory Medicine, Memorial Sloan Kettering Cancer Center, New York, NY, United States

David Beckles
Department of Pathology and Laboratory Medicine, Memorial Sloan Kettering Cancer Center, New York, NY, United States

Grecia Castro
Department of Pathology and Laboratory Medicine, Memorial Sloan Kettering Cancer Center, New York, NY, United States

Chhavi Chauhan
American Society for Investigative Pathology, Rockville, MD, United States

Stanley Cohen
Department of Pathology, Rutgers-NJMS, Newark, NJ, United States; Perelman School of Medicine at the University of Pennsylvania, Philadelphia, PA, United States; Kimmel School of Medicine, Jefferson University, Philadelphia, PA, United States

Rajesh C. Dash
Department of Pathology, Duke University Health System, Durham, NC, United States

Meera Hameed
Department of Pathology and Laboratory Medicine, Memorial Sloan Kettering Cancer Center, New York, NY, United States

Matthew G. Hanna
Department of Pathology and Laboratory Medicine, Memorial Sloan Kettering Cancer Center, New York, NY, United States

William R. Jeck
Department of Pathology, Duke University Health System, Durham, NC, United States

Brie Kezlarian-Sachs
Department of Pathology and Laboratory Medicine, Memorial Sloan Kettering Cancer Center, New York, NY, United States

Peter Ntiamoah
Department of Pathology and Laboratory Medicine, Memorial Sloan Kettering Cancer Center, New York, NY, United States

Liron Pantanowitz
Department of Pathology, University of Pittsburgh Medical Center, Pittsburgh, PA, United States

Rene Serrette
Department of Pathology and Laboratory Medicine, Memorial Sloan Kettering Cancer Center, New York, NY, United States

S. Joseph Sirintrapun
Department of Pathology, Mass General Brigham, Massachusetts General Hospital, Brigham and Women's Hospital, Boston, MA, United States; Harvard Medical School, Boston, MA, United States

Richard Torres
Applikate Technologies Inc., Fairfield, CT, United States; Laboratory Medicine, Yale School of Medicine, New Haven, CT, United States

Preface

In the last several years the discipline of pathology has made enormous strides in technological advancement, leading to an explosion of scientific knowledge aided by "disruptive technologies," such as "next-generation sequencing" in molecular diagnostics, which has fundamentally changed cancer medicine to precision medicine, offering a wide range of treatment options to patients to ensure better quality of care and unimagined prolongation of life even in the late stages of disease. Visual inspection of tissue sections of disease processes under a microscope by skilled pathologists was considered "disruptive technology." In 1868 Dr. Alfred Stille, a Physician and Professor of Medicine from the University of Pennsylvania, stated, "The use of a microscope in pathological investigations is an art so difficult that none but those who especially cultivated it (and for a long time) can depend on it for correctness. On this account it can never become employed in ordinary practice." One hundred and fifty years later, this very special skill is well known in practice and is the cornerstone for diagnosing diseases such as cancer. Currently, we are in the process of embarking on to another disruptive technology, digital pathology, which enables pathologists to make the same diagnosis through the integration of information technology to transmit microscopic images from glass slides and associated metadata on a digital screen with the ability to view, share, and transmit diagnoses from anywhere and at any time.

The adoption of digital pathology necessitates multiple steps. Instead of bringing glass slides to the pathologist, H&E images are available as whole slide images (WSIs) on a screen for review and diagnosis. The adoption of this technology has been slower than initially thought, mostly related to the incorporation of additional steps into the routine analog laboratory workflow. The implementation steps include appropriate infrastructure, including space, specialized equipment, and expertise (hardware, software, network, and storage) with accompanying costs; interfacing into the laboratory information system; integration with routine histopathological laboratory workflows; hiring and training of laboratory personnel; and establishment of a quality management system, as well as necessary steps for validation and regulatory compliance. This must all be done without interruption of current practice or compromising patient care. This process may appear complex and perhaps raise doubts in return on investment; however, the near-future and long-term benefits are enormous and practice changing. The plethora of use cases include the availability of WSIs at any time, anywhere for primary diagnosis, remote pathology consultation practically anywhere in the world, the instantaneous availability of patients' prior pathology images, digital annotations that can be shared immediately, simultaneous comparisons of multiple images including immunohistochemical stains, remote intraoperative consultations, and digital and more accurate assessments for macrodissection for tumor purity (for molecular diagnostic assays). There are obvious additional advantages such as facilitating multidisciplinary tumor boards within and outside institutions including global consultations, trainee education, objective quantitation of biomarkers, and the ability to explore petabytes of data within WSIs and the development of new multimodal fusion models. The digital footprint provides the scaffolding to layer on computational pathology atop and further disrupts conventional pathology practices. Finally, digital pathology provides the ability to reconstruct 3D images that can provide information about patterns and distributions of lesions within the tissue.

In this book we address practical solutions for the successful implementation of digital pathology in pathology laboratories. The book is divided into several chapters that address the roadmap and blueprint laboratories should consider, which include preanalytic, analytic, and postanalytic processes, including hardware, software, space, staffing, quality control, laboratory information system applications, clinical validation, use cases, and future applications, including computer-assisted diagnosis and decision support tools. This book provides, for example, experience from a large surgical pathology laboratory as "early adopters" of digital pathology. While early adoption invariably leads to "growing pains," it also provides the opportunity to innovate the technology and improve outcomes. The content of this book reflects our collective experience laid out in a pragmatic manner for the end user to apply within their laboratory system. We hope that the end user will be able to utilize the contents chapter by chapter to apply and solve problems at the preanalytic, analytic, and postanalytic phases of their implementation.

Our target audience include pathology trainees, practicing pathologists, laboratory directors, machine learning scientists, as well as researchers, administrators, and policy makers, for whom the core principles and understanding of digital pathology implementation in the clinical laboratory and its future potential for practice change are essential to learn. This book will take the audience through the basics of anatomic pathology workflows to adopt scanning glass slides to view digital images, use cases in surgical pathology and cytology to understand the basics of machine learning, and apply artificial intelligence (AI), which lies in the future of pathology and medicine. Regulatory frameworks and industry partnerships are discussed as separate chapters to broaden the understanding of current regulations and industry collaboration opportunities. Additionally, basic concepts and applications of AI will be discussed in addition to important considerations encompassing AI ethics, an area not well explored and very important to be aware of in the domain of translational research.

Our goal is to highlight the digital transformation occurring in our field, which is the beginning of an evolutionary milestone with the potential to change the future practice of pathology. We hope that our comprehensive and practical approach for implementing digital pathology will be useful to readers and that descriptions of the basic concepts and applications of AI will provide food for thought toward moving the field forward. The advancement of pathology is dependent on the wisdom, expertise, and knowledge of pathologists to steer the ability of AI algorithms toward a common goal of discovery, efficiency, and improvement of patient care and well-being. We thank all the readers for sharing their thoughts and wisdom toward a bright future for the practice of pathology and, ultimately, patient care.

Acknowledgment

The book is dedicated to Drs. Victor Reuter and David Klimstra who had the vision and passion to support the development and implementation of digital pathology at our institution, the Warren Alpert Foundation for their generous support to build the infrastructure, and Dr. Stanley Cohen, Chair of the External Advisory Committee (EAC) of our Warren Alpert Center for Digital and Computational Pathology, who with EAC members, Drs. Joel Saltz, Anil Parwani, and Benjamin Mitchell, provided expertise, wisdom, and encouragement to explore the potential of digital and computational pathology as springboards for both artificial intelligence and the development of other new methods to advance pathology.

We would also like to acknowledge Drs. Liron Pantanowitz and Anil Parwani for pointing the way toward digital pathology integration in pathology departments and laying a foundation to move the field forward. We are also indebted to Dr. Ronald Weinstein whose pioneering work in telepathology which inspired the use of digital pathology to enable pathologists at distant locations to interact with one another, first via linked robotic microscopy and then via digital pathology. Last but not the least, we also wish to acknowledge the senior administration at MSKCC for their unwavering support.

CHAPTER

Basics of digital pathology

1

Matthew G. Hanna

Department of Pathology and Laboratory Medicine, Memorial Sloan Kettering Cancer Center, New York, NY, United States

Introduction

Pathology is a specialized field of medicine focused on the study of diseases. Pathologists work in clinical pathology laboratories alongside allied professionals, such as technologists, to conduct diagnostic testing for patients. Traditionally, laboratory testing and diagnostic work have relied on manual processes and the use of conventional microscopes for certain testing on patient specimens, such as histopathology, cytopathology, hematopathology, and microbiology.

However, with the advancement of digital technology, the practice of pathology has evolved to incorporate digital imaging. Digital pathology encompasses various approaches, including static images, live streaming of images, and whole slide imaging (WSI). WSI allows the digitization of glass slides, reducing the requirement for pathologists to use traditional microscopes.

The digital pathology ecosystem comprises three major components: information systems, digital pathology systems (DPS), and system tools (Hanna & Pantanowitz, 2019). An ideal ecosystem is one where these components are interoperable to facilitate a seamless digital workflow. The information systems enable the DPS to interface with various enterprise resources, such as the hospital information system (HIS), electronic medical record (EMR), laboratory information system (LIS), and radiology picture archiving and communication system (PACS). The DPS serves as a central digital workspace where pixel data flows from image acquisition devices, such as whole slide scanners, to the end user's display, like a computer workstation or tablet. It consists of two interconnected subsystems: the image acquisition device responsible for acquiring and managing digital images and the display cockpit for viewing and sharing images. Additionally, system tools, including image analysis algorithms, interact with the DPS to analyze and manipulate digital images for tasks like computer-assisted diagnosis.

The applications of digital pathology are vast and impactful in both clinical and nonclinical settings (Hanna et al., 2020; Luo & Hassell, 2016). Clinically, digital images can be used for primary pathology diagnoses, remote intraoperative consultations (IOC), telepathology for second opinions, multidisciplinary diagnostic management

Digital Pathology. DOI: https://doi.org/10.1016/B978-0-443-13809-6.00003-8

team meetings, archiving, quality assurance, and decision support tools. Non-clinical uses include proficiency testing, research, biobanking, marketing, and education. Ongoing innovation in imaging and computers has furthered the development of digital pathology, leading to more global connectivity among pathology laboratories and the implementation of novel image-related applications. Moreover, it enables the integration of cost-efficient technology in daily practice and participation in computational pathology or big data analytics, all contributing to the improvement of patient care.

Digital imaging basics

Pixels, or "picture elements," are the basic unit of an image in digital pathology. Pixels represent the individual units of information in a digital image of a tissue specimen. Each pixel corresponds to an area (usually very small and individually inconspicuous) on the digitized slide. Pixels are typically square-shaped, where the individual pixels are arranged in a grid-like pattern. Each pixel corresponds to a discrete unit of color and intensity on the screen, ultimately reflecting the underlying specimen on the glass slide from which it was captured. The square shape ensures that the pixels are evenly spaced, allowing for a uniform representation of the image. Each pixel contains data (binary, 0 and 1), storing values such as brightness and color. High-resolution pathology images contain a larger number of pixels, allowing pathologists to examine intricate details of the tissues and make more accurate diagnoses. The bit depth in digital pathology refers to the number of bits used to represent the color and intensity variations in each pixel. Higher bit depth images provide a broader range of colors and shades, which is essential in pathology for distinguishing different tissue types and cellular features. Pathologists may require higher-bit-depth images, such as 24 or 32-bit, for precise analysis and identification of subtle pathological changes.

Resolution is critical in pathology as it determines the level of detail visible in an image. High-resolution pathology images are crucial for accurate diagnosis, especially in cases where pathologists need to identify specific cellular structures or abnormalities. These are typically measured as microns per pixel (mpp) as the resolution of the digital image. The equivalent resolution of the digital scan can be calculated when dividing 10 by the mpp resolution. For example, a 0.25 microns/pixel scan would by 10/0.25 = "40 ×" equivalent magnification. Pathology images, particularly whole slide images (WSIs), can be quite large in file size due to their high resolution and bit depth. Managing large file sizes is essential in digital pathology for efficient storage, retrieval, and sharing of slides between pathologists, collaborators, and external medical institutions. Image compression is vital in digital pathology to manage the large file sizes of whole slide images. Compression techniques that balance file size reduction with minimal loss of image quality are often used to make storage and transmission more manageable. Lossless compression is often preferred to retain the original image quality, ensuring that pathologists can examine the slides without any loss of critical information. Lossy compression is generally avoided in diagnostic pathology to prevent the loss of essential details that may affect diagnostic accuracy. Some examples of the imaging compression algorithms used in digital pathology are based on the lossless (e.g., no loss in data during compression) tag image file format (TIFF). TIFF images are commonly used for preserving high-quality images that retain all the necessary information and can be used for diagnosis and other clinical and nonclinical activities. Other formats, such as JPEG or JPEG2000, may be used for sharing images with reduced file sizes while maintaining acceptable image quality. File formats in digital pathology range from statis images with JPG, PNG, and TIFF to WSI whole slide images, where a wide variety of file formats (extensions) are available, including JPEG, JPEG2000, TIFF, SVS, BigTIFF, MRXS, BIF, VSI, iSyntax, NDPI, RTS, and DICOM.

For any imaging workflow, there is a photoelectric conversion process in the sensor that captures light and stores the binary digital pixel data. Sensors are crucial components of cameras, including gross camera workstations or whole slide scanners. These sensors capture the light passing through the tissue specimens and convert it into digital data. Two predominant sensors are used: charge-coupled device (CCD) and complementary metal-oxide semiconductor (CMOS). While the process in each sensor varies, the underlying technology uses a photoelectric conversion where the light passing through the lens generates electric charges. The voltage represents the intensity of light at each pixel, and the binary data is digitally stored. High-quality sensors are essential for accurately capturing the colors, textures, and structures of the tissues, ensuring that pathologists can make precise diagnoses based on the digital images.

The basic concepts of digital imaging and their correlation to pathology are fundamental for pathologists and other medical professionals working with digital tools in this medical discipline. Proper image resolution, bit depth, file size management, and the use of appropriate file formats are crucial to ensuring accurate and efficient diagnostic processes and research in pathology.

Digital pathology timeline

Over the centuries, pathology has undergone a remarkable transformation, transitioning from pure macroscopic examinations to utilizing microscopy. Early pathologists, predominantly physicians, conducted autopsies to gain insights into diseases (Rosai, 1997). The late 1500s saw a pivotal moment when Zacharias Jansen invented the first compound microscope. Cell theory and cellular pathology relied on the invention of the microscope. The earliest microscopes suffered from various optical problems; however, by 1665, the English scientist Robert Hooke (1635–1702), who was a gifted illustrator, coined the term "cell" while studying bark from cork oak trees under a microscope. While looking at the tree bark, Hooke observed hexagonal and box-shaped structures, which he called "cells" as they reminded him of the cells, or rooms, in monasteries. This discovery led to the development of the classical cell theory, where the reflectance of light (e.g., candle) could be coupled with magnification of samples under the microscope. Subsequently, in the mid-17th century, Anton van Leeuwenhoek employed microscopes to explore cellular details, particularly in microbiology. Rudolf Virchow further advanced cellular pathology through the 18th century. By the late 19th century, significant progress was made in enhancing the microscope's lenses and optics, leading to higher resolution capabilities that profoundly impacted the study of pathology. The introduction of new light sources, such as halogen, fluorescent, and light-emitting diode (LED), further improved the resolution of light microscopes (Table 1.1).

The advent of sensors and cameras opened new possibilities in pathology, enabling pathologists to capture and photograph postmortem bodies, gross pathology specimens, microbiology specimens, and eventually glass slides. Cameras coupled with microscopes facilitated photomicroscopy, including brightfield and fluorescence microscopy, as well as the recording of ultrastructural findings using electron microscopes. The transition from analog to digital cameras allowed pathologists to use digital images for clinical work, research, and education. Initially, digital microscopy images were static snapshots, representing only a limited portion of the cellular material on a glass slide, confined to a single field of view and plane of focus. Overcoming these limitations, advancements led to the integration of digital cameras with microscopes, streaming live video microscopy, and even remotely controlled robotic telemicroscopy. The first digital pathology system was developed by a team of researchers led by Dr. Ronald S. Weinstein in the mid-1980s. Dr. Weinstein, a pathologist, and his team at the University of Arizona created the first practical and functional system for the digital capture and analysis of pathology slides. Pathologists could now remotely review tissue by controlling microscope stages, switching objectives, and adjusting focal planes using digital tools on a computer workstation (Weinstein et al., 2019). In the late 1990s, efforts emerged to digitize entire glass slides, marking the inception of whole slide

Table 1.1 Digital pathology timeline: key milestones.

Key milestones for digital pathology	
1960s	First telepathology experiments.
1986	Public demonstration of remote telepathology diagnosis.
1990s	Virtual microscopy hybrid dynamic-robotic telepathology/static image telepathology.
1996	First digital pathology system to stitch whole slide image.
2000s	Technology advancement to make low and high throughput whole slide scanners commercially available.
2010s–2022	Several FDA approved digital pathology systems (i.e., full systems, standalone viewers, standalone display devices, and decision support systems).

imaging (WSI). Renato Ferreira at the University of Maryland Institute for Advanced Computer Studies and Joel Saltz from Johns Hopkins University developed the first software system to support whole slide images. Initially, the process involved a robotic microscope and software to stitch together individual static images in a tile-based mosaic pattern. However, this approach had drawbacks, such as lengthy image acquisition times and the generation of massive digital files, with individual slides reaching up to 210 GB of data (Pantanowitz et al., 2018). Advancements in digital cameras, objective lenses, robotics, and computers paved the way for new whole slide scanners capable of automatically scanning multiple glass slides within minutes, producing high-resolution digital slides with manageable file sizes. By the early 2000s, various commercial whole slide scanners were available, ranging from large instruments handling over hundreds of slides to lower-throughput devices with niche applications (e.g., fluorescence). More recently, whole slide scanners have integrated real-time robotic, hybrid technologies, allowing both whole slide imaging and robotic telepathology, further evolving the tools available for pathologists to practice with.

In the last 10 years, the field of digital pathology has seen significant advancements, and the U.S. Food and Drug Administration (FDA) has played a crucial role in facilitating its adoption in clinical practice. One notable development during this period is the authorization of digital pathology systems for primary diagnosis. In 2017, the FDA granted the first marketing clearance for a whole slide scanning system for primary diagnosis in certain pathology applications. This milestone allowed pathologists to use an FDA-cleared device for reviewing whole slide images for making diagnostic decisions in surgical pathology. Subsequent to this initial authorization, the FDA has continued to evaluate and approve additional digital pathology solutions for primary diagnostic use. This trend has accelerated the adoption of digital pathology in various pathology labs. With digital pathology systems now authorized for primary diagnosis, pathologists can perform remote consultations, participate in multidisciplinary disease management team meetings, and collaborate with experts from different locations seamlessly.

Moreover, the FDA has also been involved in approving digital pathology algorithms and image analysis tools (Cornish, 2020). These algorithms can assist pathologists in automating certain tasks, such as tumor detection, classification of pathology (e.g., tumor grading), rare event counting (e.g., mitoses), and immunohistochemical stain quantification. By harnessing the power of artificial intelligence and machine learning, these tools aim to enhance the accuracy and efficiency of pathology diagnoses. As digital pathology continues to evolve, the FDA remains actively engaged in evaluating new technologies and applications. The ongoing support and authorization of digital pathology systems have significantly contributed to the transformation of pathology practices, driving improvements in patient care through faster and more accurate diagnoses, streamlined workflows, and better access to pathology expertise across healthcare networks. With the FDA's continued focus on ensuring patient safety and the advancement of medical technology, the future of digital pathology holds great promise for revolutionizing diagnostic medicine.

Digital pathology ecosystem

The digital pathology ecosystem comprises three major components: hospital information systems, digital pathology systems (DPS), and system tools. The ideal setup enables seamless interoperability, supporting an integrated digital workflow between various hospital information systems (HIS), electronic medical records (EMR), laboratory information systems (LIS), and picture archiving and communication systems (PACS). At the core of this ecosystem is the DPS, which plays a central role in facilitating the flow of pixel data from image acquisition hardware to viewer applications through a digital pathology workflow.

The image acquisition process is enabled by specialized hardware, such as camera sensors that capture color information from pixel data on glass slides. Through the use of Health Level 7 (HL7) or other health information exchange standards, the unique identifiers embedded in the decoded barcodes are used to link to the file storage location of the whole slide image (WSI). The resulting whole slide images are accompanied by associated metadata from the LIS, which includes patient health information and specimen information (e.g., specimen type, block, and stain data). The retrieved WSI files can be stored locally or on remote servers, depending on the intended use case and scale. Digital images can be used for various applications. Internal use cases allow the images to be used within the healthcare institution, such as in multidisciplinary diagnostic management team conferences or external sharing for consultations with experts. Enhanced sharing capabilities lead to improved communication and value-added reporting, ultimately contributing to the enhancement of patient care.

Downstream tools in the digital pathology ecosystem further enhance the pathologist's workflow by integrating advanced functionalities, such as image analysis and computer-assisted diagnosis. These tools leverage augmented intelligence (AI) and sophisticated models to aid pathologists in analyzing and interpreting WSIs. Overall, the digital pathology ecosystem relies on a well-coordinated and integrated approach involving information systems, DPS, and decision support tools. By facilitating seamless data flow, efficient image exchange, and enhanced collaboration, this ecosystem empowers pathologists with formidable tools to improve diagnostic accuracy, research capabilities, and most importantly, patient care.

Digital photomicroscopy

The advent of integrated microscopy (e.g., camera attached to a microscope) has paved the way for the transformation of traditional pathology into the realm of digital technology. Integrated microscopes serve as a crucial gateway, linking the analog world of microscopy to the digital landscape of pathology. These systems ingeniously combine the capabilities of digital cameras with the conventional microscope, creating a seamless bridge between the two domains. At its core, an integrated microscope is the result of coupling a digital camera to a traditional microscope through a c-mount interface. This integration marks a significant step in capturing the dynamic essence of microscopy within the digital framework. Initially, these systems focused on capturing static images originating from the microscope's objective lens. However, the early iterations were confined to capturing a solitary field of view (FOV) per image, limiting their scope.

The process of capturing images from integrated microscopes typically involves a desktop software application that facilitates the capture, storage, and sharing of digital images. These images can be effortlessly disseminated via various mechanisms, ranging from simple email attachments to dedicated telepathology websites. The digital age further extends its reach through smartphone-compatible microscope attachment adapters. These adapters enable users to position their device's camera over the microscope's eyepiece (e.g., ocular), enabling the capture of microscopy through mobile device-specialized software. This opens avenues for sharing images through electronic means, such as text message, email, and social media platforms.

Cameras associated with integrated microscopes come in two varieties: dedicated static capture and live streaming. The cutting-edge software available today enhances the synergy between integrated cameras and image stitching algorithms. This real-time integration enables the dynamic capture of user-defined FOVs, creating a seamless collage of images as the operator moves the glass slide on their conventional brightfield microscope. This operator-driven approach offers a flexible array of multiple z-planes and FOV digitization, yielding images with smaller file sizes. While this method is influenced by operator-dependent factors such as FOV, magnification, and blur, it also provides the versatility needed for varying digital representations.

The generated image files from integrated microscopes possess interoperability with a range of existing platforms and software designed for image management, annotation, and sharing. Despite being characterized by lower throughput, integrated microscopes offer a cost-effective entry point into the world of digital pathology. This amalgamation of traditional and digital microscopy not only shapes the present landscape of pathology but also foretells the potential avenues of its future evolution.

Gross imaging

Gross pathology workstations represent an essential cornerstone in modern diagnostic pathology. These specialized workstations provide a dedicated environment for the examination, documentation, and imaging of gross specimens. They seamlessly integrate digital technology with traditional macroscopic analysis, enabling the capture of high-resolution specimen images. These images can be incorporated into the LIS and associated with the patient's case. The utility of gross pathology workstations extends beyond immediate visualization; they facilitate collaboration, enhance documentation, and support educational endeavors. By capturing detailed images of gross specimens, pathologists can document findings, collaborate remotely with colleagues, and archive visual records for future reference. These workstations are pivotal in the shift toward digital pathology, promoting efficiency, accuracy, and comprehensive case management. As a result, gross pathology workstations have emerged as indispensable tools that bridge the gap between traditional macroscopic examination and the digital age, elevating the quality of diagnostic processes and fostering advancements in the field of pathology.

Video streaming

Video streaming technology has emerged as a transformative tool in the realm of pathology, enabling collaborative workflows for pathologists to communicate and educate with. This technology enables real-time transmission of live video feeds, allowing pathologists to remotely participate in consultations, procedures, and educational sessions. The application of video streaming in pathology has profound implications for both clinical practice and academic pursuits.

Video streaming technology has transcended geographical boundaries, enabling pathologists to participate in remote consultations, consensus opinions, educational activities, and multidisciplinary discussions. By streaming live video of glass slides or specimens, experts can provide real-time insights, second opinions, and guidance to colleagues located anywhere in the world with internet connectivity. This not only enhances diagnostic accuracy but also facilitates timely decision-making, particularly in challenging cases that require input from diverse perspectives. One prime example includes intraoperative consultations and guidance, which are critical in surgical pathology. Video streaming allows pathologists to virtually join procedural consults for frozen sections or cytology evaluations and provide immediate feedback. This real-time interaction supports surgical teams in making informed decisions during procedures, reducing the need for delays and reoperations. Video streaming technology thus empowers pathologists to contribute actively to patient care beyond the confines of the pathology laboratory.

This technology has also allowed for significant advancement in pathology education by providing dynamic, interactive learning experiences. Educational institutions can stream live pathology sessions, including lectures and slide reviews, to students in various locations. This fosters engagement, enables real-time review, and offers students valuable exposure to practical scenarios. Additionally, archived video content, if stored, facilitates on-demand learning, allowing trainees to revisit material and hone their skills at their own pace.

While video streaming technology offers immense benefits, it also presents challenges. Maintaining high-quality video feeds, addressing network connectivity issues, and ensuring data security are essential considerations. Furthermore, securing patient privacy and compliance with regulations such as HIPAA are paramount when streaming medical information. Nonetheless, video streaming technology has propelled pathology into a new era of collaboration, education, and clinical decision-making. By harnessing its potential, pathologists can move beyond physical limitations, facilitate expert consultations, and cultivate a dynamic learning environment. As technology continues to evolve, video streaming promises to redefine the boundaries of pathology practice and education, enhancing patient care and professional development alike.

Whole slide scanning

The introduction of whole slide scanners has propelled pathology into a new era of digital transformation. Revolutionizing the way glass slides are transformed into high-resolution digital images. Operating as automated robotic microscopes, these sophisticated devices capture each field of view (FOV) of a glass slide and seamlessly stitch them together to create comprehensive whole slide images. Leveraging image viewing software, users can navigate through these digital slides, aptly referred to as "virtual slides," with capabilities encompassing panning, zooming, and dynamic exploration. Beyond replicating the physical slide, whole slide imaging technology unlocks a range of possibilities that were previously unattainable with traditional microscopes, including image overlays, telepathology consultations, and intricate image analysis.

Integrated with digital cameras, these scanners utilize sensors to capture each FOV. However, no single sensor is sufficient to capture an entire slide in one instance. Consequently, whole slide scanners adopt tile- or line-based scanning methodologies. In tile-based scanning, contiguous squares (tiles) of a slide are imaged and later merged into a single mosaic image. Alternatively, line scanning involves capturing consecutive strips of a slide, which are combined to generate the whole slide image. Powerful computer software algorithms stitch together these individual images into a cohesive whole. Recent advancements in technology have significantly enhanced scan speed through innovations like real-time focus and tissue detection algorithms.

Whole slide scanners employ varying objective lenses to replicate microscope magnifications. Objective lenses ranging from 1.25 × to 100 × can be used, catering to different diagnostic needs. For routine surgical pathology, medium magnification objectives (e.g., 20 ×) are generally adequate. However, higher magnifications (e.g., 40 × or more) are crucial when identifying minute details like microorganisms. When comparing scanners, the preferred vendor-neutral parameter for image resolution is micron/pixel rather than scan magnification (e.g., 0.5 mm/pixel vs 20× or 0.25 mm/pixel vs 40×). A lower micron/pixel value enhances the depth of information obtainable from the image.

In scenarios involving slides with intricate cellular material, such as cytology cases, visualization along both horizontal (X and Y) and vertical (Z) axes becomes essential. Z-stack scanning addresses this by capturing the same slide at multiple focal planes along the Z-axis and merging these images into a composite multiplane image. However, Z-stack scanning comes with extended processing times and larger digital file sizes. Pathologists using such files may experience performance issues if their computer and network are not optimal.

In practice, low-throughput scanners cater to lower to medium scanning requirements, often fulfilling specialized needs that higher-throughput counterparts cannot readily provide. Despite sharing similar mechanics and engineering principles with their high-throughput counterparts, low-throughput scanners occupy smaller laboratory footprints and are generally more cost-effective. Pathology laboratories have the option of employing either low-volume scanners, designed to handle up to approximately 100 slides, or high-capacity systems catering to 200–1000 slides. The latter, often deployed in large-scale settings requiring higher throughput, features continuous loading capabilities, allowing slides to be loaded and removed while scanning is ongoing. On the other hand, high-throughput whole slide scanners are designed for laboratories with significant scanning demands. Equipped to accommodate hundreds of slides, these scanners are commonly found in large-volume core labs, offering manual and automated slide input bays with capacities ranging from 100 to 1000 slides. Some models even offer continuous loading capabilities, allowing uninterrupted scanning even during slide loading. The expansion of scanning capabilities to larger slide formats, such as 2×3 or 8×6 in., broadens their utility. These scanners maintain high image resolution, often reaching 20× equivalent magnification with varying microns per pixel for enhanced detail. Scanning speed is influenced by numerous factors, encompassing imaging acquisition, sensor size, scan area, automation, motorized stage speed, and more (Fig. 1.1).

Moreover, the emergence of real-time dynamic-robotic imaging devices is notably shaping the landscape of telepathology and intraoperative diagnosis. With a focus on low throughput, these scanners offer real-time remote-control capabilities, enabling pathologists to remotely view and operate microscopes during procedures. The ability to pan, zoom, and focus in real time facilitates remote evaluations, enhancing the collaborative nature of modern pathology.

In summary, the advent of whole slide scanners represents a pivotal advancement in pathology, merging traditional practices with cutting-edge automation. By offering dynamic imaging capabilities, these scanners empower pathologists with detailed insights, flexible scanning options, and remote collaboration possibilities that transcend physical limitations. As the field of pathology continues to evolve, whole slide scanning remains a cornerstone technology, ushering in a new era of precision and efficiency in diagnostic practices.

Scanner Category	Image	Function	Size	Cost
High Throughput	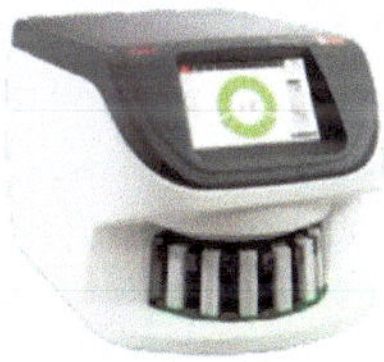	High-Volume Lab	Large	$$$$
Low		Low-Volume Lab/ Niche Use Case	Medium	$$$
Hybrid/ Real-Time Robotic	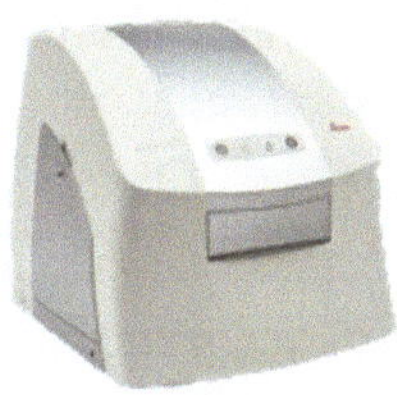	Telepathology	Medium	$$
Integrated Mobile		Global Health	Small	$

FIGURE 1.1

Whole slide scanner comparison.

Monitors/display devices

Accurate interpretation of digital pathology images is paramount for ensuring precise diagnoses. In place of oculars on microscopes, monitors/display devices play a tremendous role in this process, serving as the primary interface through which pathologists view and interact with digital slides. The display devices are instrumental in translating the visual information captured by whole slide scanners into clinically meaningful insights (Abel et al., 2020). Monitors utilized in digital pathology must meet stringent standards for color accuracy, resolution, brightness, and contrast. High-resolution displays with fine pixel pitch ensure the faithful representation of intricate cellular structures. Additionally, a wide color gamut and color accuracy are crucial for reproducing the diverse staining patterns present in pathology specimens with high fidelity. Uniformity across the display surface is also imperative to eliminate variations that might hinder accurate interpretation. Display devices may appropriately have quality assurance protocols whereby regular calibration and adherence to standardized color profiles are warranted. Calibration ensures that the monitor's performance aligns with established reference standards, maintaining consistent color and luminance over time. Color profiles, such as Digital Imaging and Communications in Medicine (DICOM) Part 14 Grayscale Standard Display Function, aid in achieving uniform display characteristics across various monitors, essential for harmonized diagnostic practices. Furthermore, in modern pathology workflows, integrating various imaging modalities, such as radiology, static clinical images, and microscopy, is increasingly common. Monitors with multimodality capabilities can display a diverse range of medical images while maintaining the requisite color accuracy and detail. This integration streamlines diagnostic assessments that involve correlating information from different sources.

Ergonomics and optimal viewing conditions are important to prevent visual fatigue and, more importantly, prevent errors in rendering diagnoses (Stathonikos et al., 2019). Ambient lighting, monitor positioning, and glare reduction are critical factors to consider. Adjustable monitor stands and ergonomic workstations allow pathologists to customize their viewing environment, minimizing strain during prolonged review of patients' digital slides. Emerging display technologies, such as organic light-emitting diode (OLED) and high dynamic range (HDR) displays, hold promise for further enhancing the accuracy and visual experience of digital pathology interpretation. OLED screens offer improved color rendition and contrast, while HDR displays enable the visualization of subtle details in bright and dark areas simultaneously. Higher resolution displays also allow for more tissue to be reviewed in a single field of view, compared to standard definition monitors. Higher-resolution monitors can allow for more efficient viewing of pathology specimens and aid in the overall review process.

Monitors and display devices form the visual gateway to the visual rendering of diagnoses in pathology. Their adherence to standards, calibration practices, ergonomic considerations, and incorporation of advanced technologies collectively contribute to accurate diagnoses, efficient workflows, and enhanced collaboration among medical professionals. As digital pathology continues to evolve, the role of monitors and display devices remains pivotal in ensuring the highest quality patient care and research outcomes.

Input devices for digital pathology

In the realm of digital pathology, input devices play a crucial role in facilitating interaction between pathologists and the digital images they analyze. These devices are integral to navigating, annotating, and manipulating whole slide images, enhancing diagnostic precision and efficiency. These tools or attachments aid pathologists in reviewing patient specimens. Input devices for digital pathology include traditional computer mice, touchscreen displays, and 6 degrees of freedom (6-DOF) input devices that enable users to manipulate objects in three-dimensional space, among others (Mateos et al., 2016; Molin et al., 2015). These various input devices tailor to different user preferences and task requirements and may be user-dependent in terms of usability or productivity enhancements.

Mouse and keyboard interactions

Traditional input devices like computer mice and keyboards remain essential tools for navigating digital pathology software interfaces. Pathologists can pan, zoom, and rotate whole slide images using mouse interactions, while keyboard shortcuts expedite common actions and annotations. Ergonomic designs and customizable buttons cater to the intricate demands of image analysis. These may be comfortable for many users due to the familiarity of their daily use in routine computer interactions; they may not be an efficient tool to drag across gigapixels of images.

Touchscreen displays

Incorporating touchscreen displays into digital pathology workstations provides a direct and intuitive means of interaction. Pathologists can use gestures to zoom in and out, pan across images, and annotate directly on the screen. Touchscreens enhance the tactile experience, facilitating natural interactions similar to traditional glass slide handling. Stylus and pen tablets offer precision and control, particularly when annotating and marking up images. These input devices simulate the experience of working with traditional glass slides and allow pathologists to draw, highlight, and annotate with accuracy. Pressure sensitivity and customizable buttons enhance the versatility of stylus interactions. These, however, should be ergonomically positioned in order to facilitate heavy workloads of digital slide review.

3D input devices

Advancements in 3D input devices, or input devices with six degrees of freedom, are emerging as potential tools for digital pathology. Devices like 3D mice or haptic controllers offer multiaxis movement and tactile feedback, allowing pathologists to navigate and manipulate images in a more immersive manner. Some input devices are application-ready "off-the-shelf," while others may require additional programming to work in the specified software (e.g., image viewer).

Voice and gesture recognition

Incorporating voice commands and gesture recognition technology can streamline interactions with digital pathology software. Pathologists can navigate, annotate, and manipulate images using voice commands or intuitive hand gestures, reducing reliance on traditional input devices and further enhancing workflow efficiency.

Augmented reality and virtual reality

As augmented reality (AR) and virtual reality (VR) technologies advance, they present opportunities for pathologists to immerse themselves in digital images in new and dynamic ways. AR headsets can overlay digital images onto the pathologist's view, while VR environments can simulate microscopic exploration in a three-dimensional space, enhancing education and collaborative analysis.

Microscope attachments for mobile devices

Using a conventional brightfield microscope, separate attachments enable pathologists to capture images using smartphones devices equipped with cameras. These attachments align the device's camera with the microscope's eyepiece, allowing for rapid capture of images directly from the microscope's field of view. This mobility extends the reach of pathology professionals, enabling remote consultations and quick image acquisition during intraoperative assessments.

Input devices are pivotal in shaping the interaction between pathologists and digital pathology images. These devices empower pathologists with diverse tools for image manipulation, annotation, and analysis, ultimately contributing to accurate diagnoses, efficient workflows, and enhanced collaboration in the digital pathology landscape.

Interoperability and standards in digital pathology

The evolving landscape of digital pathology is marked by the emergence of diverse proprietary whole slide imaging (WSI) file formats and viewer software, often tethered to specific vendors. This situation, while reflective of innovation, presents challenges for seamless interoperability. Viewer software, available as stand-alone desktop applications or web-based platforms, can be confined to a vendor's format, hindering cross-compatibility. Open-source solutions such as

OpenSlide and Open Microscopy Environment have aimed to bridge this gap by offering compatibility with multiple WSI formats. Conversely, certain vendors have gone beyond software to develop image management systems supporting elevated clinical integration.

A focal point for interoperability progress is the DICOM standard, initially prevalent in radiology. This standard encompasses both file formats and network protocols, offering a framework for exchanging medical images and patient data. Through the DICOM network protocol, imaging devices communicate in a client-server model, facilitating secure data transfer over networks. Notably, this has found substantial success in radiology, allowing integration across multiple vendors. DICOM's influence is now extending to digital pathology, with initiatives like DICOM Working Group 26 concentrating on addressing pathology-specific needs. DICOM supplements 122, 145, and 222 outline provisions for pathology specimen identification, WSI storage and retrieval, and digital annotations. This extension of DICOM standards into pathology holds potential for streamlined workflows, improved efficiency, and enhanced communication across specialties. These are further detailed in Chapter 10, Standards section.

The picture archiving and communication system (PACS) stands as a pivotal conduit for visual data management. In hospital imaging systems, PACS enables concurrent image viewing across multiple locations, fostering integration in workflows and collaborative practices. Its application expands beyond radiology to accommodate various medical domains, including pathology. However, the inherent differences in file sizes between radiology and WSI necessitate exploration of image compression techniques to ensure efficient data storage without compromising vital pixel information. Research endeavors in this direction strive to maintain diagnostic quality while optimizing storage capacities.

Interoperability and standards in digital pathology are useful to navigate the intricate web of proprietary formats and software. While challenges persist, initiatives like DICOM's expansion into pathology and the development of open-source solutions signal progress toward a more harmonized digital ecosystem. As these efforts continue, the overarching goal remains steadfast: to cultivate a seamless exchange of medical images and data while preserving the integrity of diagnostic information, ultimately elevating the quality of patient care and research in the realm of digital pathology.

Applications

Digital pathology systems can be used for myriads of applications. Many of these will be covered in more detail in Chapters 6 and 7. The below applications provide a high-level overview for available digital pathology use cases (Luo & Hassell, 2016) (Table 1.2).

Table 1.2 Applications of digital pathology.

Applications in digital pathology	
Primary diagnosis	Rendering a diagnosis using only whole slide images in lieu of reviewing patient pathology specimens on a microscope with glass slides.
Second opinion (consultation)	Pathologists sending or receiving patient specimens via whole slide images to collaborate and share cases.
Telepathology	Category of technology where pathologists may share cases over the internet.
Quality assurance	Programmatic efforts to ensure consistent and accurate diagnoses.
Archiving and sharing	Retaining digital images for clinical, educational, and research initiatives.
Education/ conferences	Educators can use whole slide images for didactic teaching, multidisciplinary tumor boards, and other conferences.
Image analysis	Software to perform quantitative assessments on digital images.
Publications	Researchers can include whole slide images to support their findings and enhance the comprehensibility of their work.
Marketing and business	Using digital images to showcase the expertise and capabilities of the practice, attracting potential trainees, clients, and collaborators.
Tracking and auditing	Tracking and auditing of cases.
Frozen section	Pathologists can rapidly access digital images of frozen tissue sections, enabling real-time evaluations.
Flow cytometry	Digitization of images, or juxtaposition of anatomic pathology slides with flow cytometry data can be useful.
SPEPs (serum protein electrophoresis)	Electrophoretic patterns from serum protein electrophoresis can be digitized and analyzed using digital platforms.
FISH (fluorescence in situ hybridization)	Digital pathology simplifies the review and interpretation of fluorescence in situ hybridization (FISH) images. Pathologists can analyze FISH signals and patterns in a digital format.
Genomic review	Integrating genomic data with digital images provides a comprehensive view of genetic variations in pathological specimens. Pathologists can correlate genetic findings with cellular morphology.
Gram stains/plate review	Analyze and annotate digital images of microbiological cultures, identifying bacterial species and assessing antimicrobial susceptibilities more efficiently.
Telecytology	Digital pathology to enable remote assessment of cytology specimens.
Cytology adequacies	Assess cellularity, specimen quality, and suitability for further analysis, aiding in clinical decision-making for subsequent diagnostic processes.
Gross images	Document macroscopic features, annotate key areas, and provide accurate gross descriptions, contributing to detailed case records.
Research	Use of digital images are useful for case review, aggregation of patient images, and machine learning.

Primary diagnosis: Digital pathology revolutionizes primary diagnosis by allowing pathologists to review high-resolution whole slide images, replicating the experience of traditional microscopy. This digital format provides enhanced visualization, enabling pathologists to zoom in and out, navigate across the slide, and analyze various magnifications. This comprehensive view assists in identifying intricate cellular structures and anomalies, ultimately contributing to accurate and efficient diagnoses.

Second opinion (consultation): Telepathology facilitates second opinions and consultations by allowing pathologists to share digital images with experts across different locations. This remote collaboration fosters interdisciplinary discussions and knowledge sharing. Pathologists can seek guidance on challenging cases, receive feedback on diagnoses, and collectively arrive at the most accurate assessments, improving patient care outcomes.

Telepathology: Digital platforms enable telepathology, where pathologists can remotely access and review whole slide images. This is particularly advantageous in underserved areas where access to specialized expertise may be limited. Telepathology bridges geographical gaps, providing timely consultations and diagnostic support, ultimately ensuring patients receive the best possible care, regardless of their location.

Quality assurance: Proficiency testing (PT) and quality assurance activities benefit from digital pathology. Pathologists can participate in proficiency testing by analyzing digital images, ensuring consistent and accurate diagnoses. Digital platforms enable standardized assessments, helping laboratories maintain a high level of quality in diagnostic processes.

Archiving and sharing: Digital pathology enables the creation of comprehensive image archives. These digital archives offer long-term storage of images, making it easier to retrieve and reference cases for research, education, and quality control purposes. Furthermore, digital images can be easily shared with colleagues for collaborative discussions, educational initiatives, and interdisciplinary presentations.

Education/conferences: Digital images enhance educational experiences and conferences in pathology. Educators can use high-quality digital images to illustrate lectures, enabling students and participants to visualize pathological conditions in detail. Additionally, digital images can be shared during conferences, enhancing the effectiveness of presentations and discussions.

Image analysis: Integration with image analysis software empowers pathologists to perform quantitative assessments on digital images. This includes measurements, quantification of biomarkers, and automated cell counting. Image analysis tools provide objective data that contribute to evidence-based diagnoses and research findings.

Publications: High-quality digital images serve as valuable visual aids for scientific publications. Researchers can include these images to support their findings and enhance the comprehensibility of their work. Digital images provide a more accurate representation of cellular structures and pathological conditions, bolstering the credibility of research outputs.

Marketing and business: Digital images can be used for marketing and business purposes in pathology practices. These images showcase the expertise and capabilities of the practice, attracting potential clients and collaborators. They provide a visually compelling way to demonstrate the practice's proficiency and technological advancements.

Tracking and auditing: Digital pathology facilitates accurate tracking and auditing of cases. Each digital image is associated with metadata, including timestamps, pathologist identifiers, and relevant clinical data. This information supports case documentation, quality assurance, and audit trails, ensuring transparency and accountability.

Frozen section: Intraoperative assessments benefit from digital pathology, especially during frozen section analysis. Pathologists can rapidly access digital images of frozen tissue sections, enabling real-time evaluations. This expedites decision-making in critical surgical scenarios, as pathologists can remotely provide guidance to surgeons.

Flow cytometry: Digital platforms integrate flow cytometry data with pathology images, offering a comprehensive view of cellular characteristics. By juxtaposing digital pathology images with flow cytometry results, pathologists gain a deeper understanding of cellular interactions and abnormalities, contributing to accurate diagnoses.

SPEPs (serum protein electrophoresis): Electrophoretic patterns from serum protein electrophoresis can be digitized and analyzed using digital platforms. This supports the identification of abnormal protein patterns, aiding in the diagnosis of various conditions such as multiple myeloma and other plasma cell dyscrasias.

FISH (fluorescence in situ hybridization): Digital pathology simplifies the review and interpretation of fluorescence in situ hybridization (FISH) images. Pathologists can analyze FISH signals and patterns in a digital format, enhancing accuracy in detecting genetic abnormalities and aiding in cancer diagnosis.

Genomic review: Integrating genomic data with digital images provides a comprehensive view of genetic variations in pathological specimens. Pathologists can correlate genetic findings with cellular morphology, contributing to personalized medicine approaches and more precise diagnoses.

Gram stains/plate review: Microbiology images, including Gram stains and agar culture plate reviews, benefit from digitization. Pathologists can analyze and annotate digital images of microbiological cultures, identifying bacterial species and assessing antimicrobial susceptibilities more efficiently.

Telecytology: Telecytology leverages digital pathology to enable remote assessment of cytology specimens. Pathologists can review digital cytology images, provide rapid interpretations, and collaborate with colleagues for challenging cases, enhancing patient care in cytology diagnostics.

Cytology adequacies: Digital images assist in determining the adequacy of cytology specimens. Pathologists can assess cellularity, specimen quality, and suitability for further analysis, aiding in clinical decision-making for subsequent diagnostic processes.

Gross images: Digital gross images offer a comprehensive view of specimens before processing. Pathologists can document macroscopic features, annotate key areas, and provide accurate gross descriptions, contributing to detailed case records.

Research: Digital pathology serves as a vast resource for research. Pathologists can access a wide range of digital images for studies, correlations, and analyses. This vast repository of images supports scientific advancements and insights into pathological processes.

Incorporating digital pathology into these diverse applications enhances the precision, collaboration, and efficiency of pathology practices. It transforms traditional diagnostic methods, offers new avenues for research, and empowers pathologists to provide the highest caliber of patient care.

Challenges/limitations

Implementing digital pathology, despite its substantial advantages, is accompanied by an array of challenges and limitations that must be carefully addressed for successful integration and widespread adoption (Jahn et al., 2020). A significant hurdle is the financial investment required for advanced whole slide scanners, robust information technology infrastructure, and efficient storage solutions. This initial cost can be particularly daunting for smaller healthcare institutions and laboratories seeking to embrace digital pathology technology (Table 1.3).

While the generation of high-resolution whole slide images offers exceptional fidelity, it also results in substantial file sizes. This prompts the need for extensive storage capacities and effective archiving strategies. Transferring and accessing these large files through networks can lead to sluggish loading times and network congestion, especially in areas with limited bandwidth and connectivity. The absence of standardized formats for whole slide images presents a challenge in terms of interoperability. Proprietary formats employed by different vendors hinder the seamless sharing and viewing of images across diverse platforms and software. Furthermore, the variability in viewer settings, monitor calibration, color rendering, and resolution settings introduces potential discrepancies in image interpretation and collaborative discussions. Navigating the regulatory landscape

Table 1.3 Limitations of digital pathology.

Limitations of digital pathology
Large file sizes
Lack of standardization
Regulatory requirements
Culture change/management
Reimbursement
Hardware limitations (e.g., specimen polarization)
Cost

is critical when adopting digital pathology. Regulatory agencies often require rigorous validation and verification of digital systems before clinical use. This requires additional effort on the implementing laboratory with documentation and training. Integrating digital pathology into existing workflows is a complex undertaking. Adapting processes to accommodate digital platforms while maintaining the efficiency of clinical operations requires meticulous planning and execution. Additionally, training pathologists and laboratory staff to effectively navigate digital interfaces, interpret whole slide images, and employ image analysis tools is crucial for a seamless transition. Some organizations may be met with culture change as a limitation, where the adoption curve is skewed such that pathologists are not readily adopting a new technology (Fig. 1.2). This is a well-known challenge to any newly introduced technology and will require demonstration of success to prompt others to adopt. Image analysis tools, while providing quantifiable data, may not consistently match the accuracy of manual assessments by experienced pathologists. Developing and validating accurate algorithms for intricate analyses remains a persistent challenge within the field. Furthermore, ensuring the security and privacy of digital images containing sensitive patient information is an ongoing concern, requiring vigilant cybersecurity measures. The variability in slide preparation techniques, staining methods, and scanner calibration introduces potential challenges in image quality and consistency. Standardizing these processes across different laboratories becomes essential to ensuring accurate and reliable diagnoses. Reimbursement remains a significant challenge in the adoption of digital pathology. Ensuring that healthcare providers are adequately reimbursed for digital pathology services requires navigating complex reimbursement

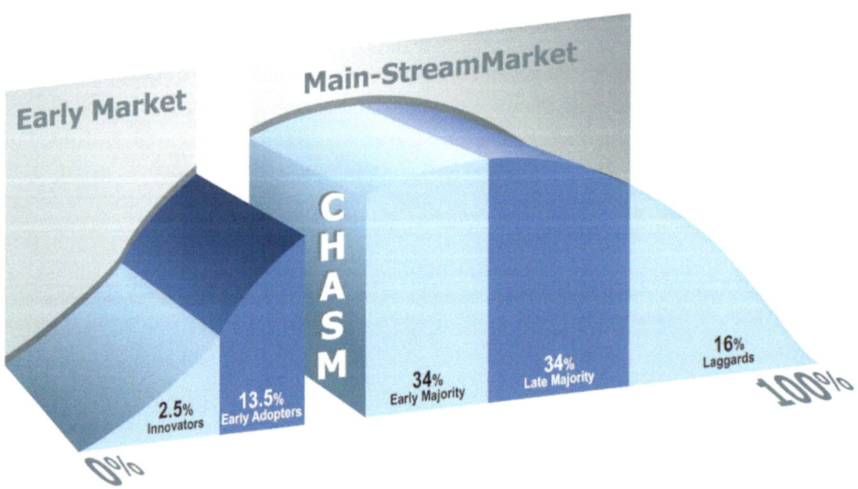

FIGURE 1.2

Technology Adoption Curve, Diffusion of Innovations.

policies, codes, and guidelines. Demonstrating the clinical and economic value of digital pathology to regulatory authorities and payers is essential to secure appropriate reimbursement rates. Hardware limitations of whole slide scanners also need consideration. Some scanners may not be equipped to handle specialized tasks, such as scanning polarized glass slides for specific applications. These limitations can impact the versatility of the technology and hinder its full potential.

In conclusion, while digital pathology holds immense promise, addressing these challenges collectively among pathologists, technology developers, regulatory bodies, healthcare institutions, and payers is imperative. Overcoming these limitations will be instrumental in fully harnessing the potential of digital pathology to enhance patient care, research, and education within the field of pathology.

References

Abel, J. T., Ouillette, P., Williams, C. L., Blau, J., Cheng, J., Yao, K., Lee, W. Y., Cornish, T. C., Balis, U. G. J., & McClintock, D. S. (2020). Display characteristics and their impact on digital pathology: A current review of pathologists' future "microscope". *Journal of Pathology Informatics*, *11*(1)23. Available from https://doi.org/10.4103/jpi.jpi_38_20.

Cornish, T. C. (2020). Clinical application of image analysis in pathology. *Advances in Anatomic Pathology*, *27*(4), 227–235. Available from https://doi.org/10.1097/PAP.0000000000000263, http://www.anatomicpathology.com.

Hanna, M. G., & Pantanowitz, L. (2019). Digital pathology. *Encyclopedia of Biomedical Engineering*, *1-3*, 524–532. Available from https://doi.org/10.1016/B978-0-12-801238-3.99958-6.

Hanna, M. G., Parwani, A., & Sirintrapun, S. J. (2020). Whole slide imaging: Technology and applications. *Advances in Anatomic Pathology*, *27*(4), 251–259. Available from https://doi.org/10.1097/PAP.0000000000000273, http://www.anatomicpathology.com.

Luo, W., & Hassell, L. A. (2016). *Use cases for digital pathology. Digital pathology: Historical perspectives, current concepts future applications* (pp. 5–15). United States: Springer International Publishing. Available from http://doi.org/10.1007/978-3-319-20379-9.

Jahn, S. W., Plass, M., & Moinfar, F. (2020). Digital pathology: Advantages, limitations and emerging perspectives. *Journal of Clinical Medicine*, *9*(11), 1–17. Available from https://doi.org/10.3390/jcm9113697, https://www.mdpi.com/2077-0383/9/11/3697/pdf.

Mateos, E. A., Caballero-Alemán, F., Albarracín-Ferrer, M., Cárceles-Moreno, F., Hernández-Gómez, R., Hernández-Kakauridze, S., Hernández-Sabater, L., Jiménez-Zafra., López-Alacid, A., Moreno-Salmerón, C., Pérez-Ramos, M., Nieto-Olivares, A., Sánchez-Campoy, N., González-Moro, I. M., & Poblet, E. (2016). Research on devices for handling whole slide images on pathology workstations. An ergonomic outlook. *Diagnostic Pathology*, *2*. Available from https://doi.org/10.17629/http://www.diagnosticpathology.eu-2016-2:2322016.

Molin, J., Lundström, C., & Fjeld, M. (2015). A comparative study of input devices for digital slide navigation. *Journal of Pathology Informatics*, *6*(1), 7. Available from https://doi.org/10.4103/2153-3539.151894.

Pantanowitz, L., Sharma, A., Carter, A. B., Kurc, T., Sussman, A., & Saltz, J. (2018). Twenty years of digital pathology: An overview of the road travelled, what is on the horizon, and the emergence of vendor-neutral archives. *Journal of Pathology Informatics*, *9*(1). Available from https://doi.org/10.4103/jpi.jpi_69_18, http://www.jpathinformatics.org/browse.asp?sabs = n.

Rosai, J. (1997). Pathology: A historical opportunity. *American Journal of Pathology*, *151* (1), 3–6.

Stathonikos, N., Nguyen, T. Q., Spoto, C. P., Verdaasdonk, M. A. M., & van Diest, P. J. (2019). Being fully digital: Perspective of a Dutch academic pathology laboratory. *Histopathology*, *75*(5), 621–635. Available from https://doi.org/10.1111/his.13953, http://onlinelibrary.wiley.com/journal/10.1111/(ISSN)1365-2559.

Weinstein, R., Holcomb, M., & Krupinski, E. (2019). Invention and early history of telepathology (1985–2000). *Journal of Pathology Informatics*, *10*(1). Available from https://doi.org/10.4103/jpi.jpi_71_18, http://www.jpathinformatics.org/browse.asp?sabs = n.

CHAPTER

The seven S's for success

2

Matthew G. Hanna

Department of Pathology and Laboratory Medicine, Memorial Sloan Kettering Cancer Center, New York, NY, United States

Introduction

The field of pathology is undergoing a transformative shift with the adoption of digital pathology, mirroring the trajectory of other disruptive technologies. This chapter delves into the intricacies of successfully implementing a digital pathology system, focusing on the critical components we have coined as the "seven S's": (1) sponsorship from leadership, (2) space allocation, (3) staffing resources, (4) scanner considerations, (5) storage solutions, (6) software, and (7) service and maintenance. This chapter will focus on lessons learned and best practices to sustain a seamless and efficient digital pathology workflow. Table 2.1 staks

Technology adoption

The adoption of digital pathology closely follows the technology adoption lifecycle, akin to the diffusion of innovation model. This cycle traverses several stages, each characterized by a distinct group of adopters and their attitudes toward the new technology.

Table 2.1 The seven S's for success implementing digital pathology.

Sponsorship from leadership
Space allocation
Staffing resources
Scanner considerations
Storage solutions
Software
Service and maintenance

Digital Pathology. DOI: https://doi.org/10.1016/B978-0-443-13809-6.00004-X

The journey begins with the innovators, who are the earliest to embrace or invent the technology. They are willing to take risks, and require little convincing to adopt novel ideas. Following them are the early adopters, often key opinion leaders, who readily recognize the need for change and readily adopt new technologies. They do not need much persuasion to adopt and can validate the value of the technology. As the technology gains traction, it moves on to the early majority, a larger group that demands evidence of its effectiveness before adopting it. Success stories, visualizations of the technology's value, and demonstrations from early adopters serve as compelling reasons for their adoption. The late majority comes next, characterized by a more skeptical outlook. They adopt only after the technology has been validated by the early majority, where sufficient success has been adequately shown. Lastly, there are the laggards, or skeptics, the most conservative group, who prefer to maintain the traditional processes. They are hesitant to change and require significant evidence, statistics, appeals, and peer pressure before embracing the new technology. These distinct groups form a bell curve representing the technology adoption curve (Rogers, 2003). Figure 2.1

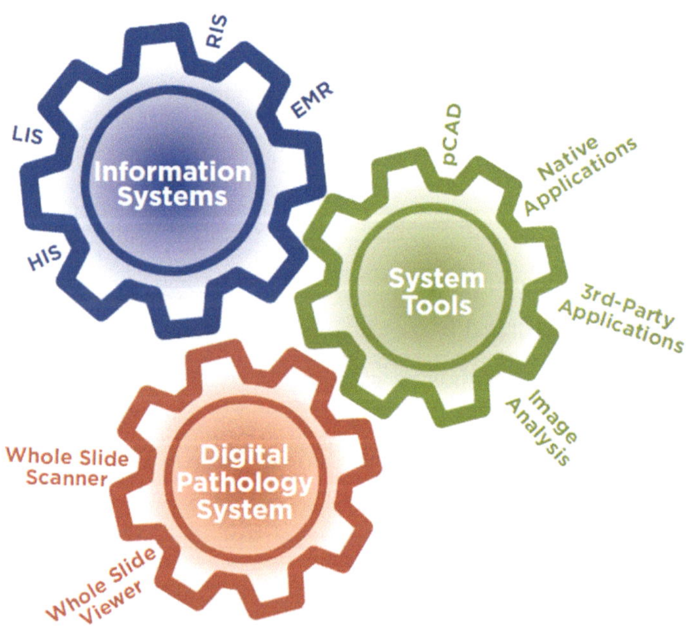

FIGURE 2.1 Digital Pathology Ecosystem.

The digital pathology system integrates into the healthcare information systems (e.g., laboratory information system, electronic medical record, radiology information system, hospital information system), as well as downstream system tools (e.g. pathology computer assisted diagnosis, native applications, third party applications, image analysis software).

In the context of clinical organizations adopting digital pathology, the specific approach can vary due to the complexities of the medical field. Different factors, including risk aversion and specific use cases, influence the adoption strategy. For instance, Memorial Sloan Kettering Cancer Center has adopted digital pathology in phases over the years, including telepathology, intraoperative consultations, and expanded retrospective and prospective scanning for clinical diagnostics (Hanna et al., 2019, 2020; Sirintrapun et al., 2017, 2017b). The comfort level of pathologists in rendering primary diagnoses digitally has significantly increased over time due to regular integration of digital pathology in clinical workflows and growing familiarity with whole slide imaging.

As the technology matures and familiarity increases, pathologist buy-in and comfort levels rise. Studies conducted over time indicate a substantial increase in the comfort of pathologists in rendering primary diagnoses using digital pathology systems. This progress can be attributed to the daily integration of digital pathology into clinical workflows and the growing exposure of trainees to digital pathology during their education. The adoption journey continues to evolve, driven by increasing comfort, technological advancements, and evolving use cases for inclusion of digital workflows in the evolution of pathology.

Sponsorship from leadership

Digital pathology represents a profound technological advancement that touches every facet of pathology practice. The realization of its potential, however, hinges on sponsorship from institutional leadership. In a landscape driven by financial constraints and ever-evolving priorities, leadership support becomes the cornerstone for introducing and maintaining digital pathology initiatives. Importantly, departmental or organizational leadership support bodes well to ensure enterprise wide support, and proper prioritization for digital pathology initiatives (Williams et al., 2017, 2019).

Effective leadership sponsorship is multifaceted. Firstly, leaders who are in decision-making capacity can provide the necessary financial resources to fund the acquisition of cutting-edge equipment and ensuring appropriate personnel are identified. Secondly, leadership communicates a clear vision of the transformative potential of digital pathology to the entire organization, rallying stakeholders around a shared goal. This is essential in cultivating a culture of openness to change. Engaged project sponsors play a pivotal role in nurturing a receptive environment for the implementation process. They align stakeholders, remove barriers, and ensure that necessary resources are available. Advocacy from leaders fosters staff confidence and positive morale in the initiative and communicates its importance. Demonstrating the value of digital pathology through data-driven insights and piloting phased approaches can foster leadership buy-in, reinforcing the significance of this transformation.

Space allocation

As the demand for digital pathology technology such as whole slide scanners continues to rise, many pathology laboratories are faced with the challenge of incorporating these complex devices into their existing laboratory spaces. Laboratories are often constrained by existing setups, and accommodating new equipment like whole slide scanners requires strategic planning. Optimizing this space is paramount to streamline processes and maximize efficiency.

Lean design principles play a crucial role in optimizing the integration of whole slide scanners into laboratory environments that were not originally designed for such technology. Figure 2.2

Assessment of current space utilization

The first step in lean design for repurposing laboratory space is a comprehensive assessment of the current space utilization. This involves identifying inefficient workflows, underutilized areas, and potential bottlenecks. This should include the entire laboratory workflow, from specimen order and receipt to slide distribution, review and reporting, and slide filing. Laboratories need to evaluate their existing layout, considering factors such as the location of existing equipment, specimen processing areas, and the arrangement of workstations. A powerful tool in this endeavor is the Spaghetti diagram, which visually maps staff movements and materials within the laboratory. Process mapping helps identify the sequence of activities and interactions between personnel, equipment, and specimens. This step highlights areas where integrating whole slide scanners could enhance

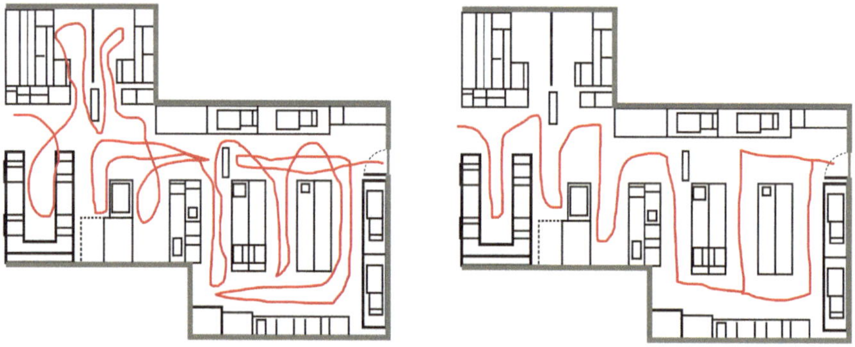

FIGURE 2.2 Lean laboratory space design.

Laboratory blueprint shows a convoluted workflow and inefficient layout of laboratory processes (left). The lean laboratory (right) shows a streamlined workflow and efficient processes.

productivity and streamline processes. Workflow analysis also aids in determining the optimal placement of scanners to minimize unnecessary movement and improve accessibility. This helps pinpoint bottlenecks and inefficiencies in the workflow, guiding decisions on equipment placement. Collaborating with laboratory staff to understand the intricacies of their daily operations is crucial. Such insights enable the creation of ergonomic layouts that facilitate smooth transitions between processing steps. Understanding laboratory logistics involves recognizing peak hours of activity, timing of batch processing, and prioritizing rush cases. The spatial arrangement should ensure that whole slide scanners are located in close proximity to stainers, coverslippers, and drying stations (Baidoshvili et al., 2018; Browning et al., 2020; Hanna et al., 2021). This seamless integration reduces manual intervention and minimizes turnaround times. Moreover, the spatial design must accommodate potential future expansion. As digital pathology's role evolves, the laboratory layout should be flexible enough to incorporate additional equipment. Engaging architects and laboratory designers is instrumental in optimizing spatial utilization. Whole slide scanners require robust technological infrastructure, including high-speed internet connectivity, power supply, and data storage. Laboratories need to collaborate with information technology (IT) experts to ensure seamless integration of these components. Adequate provisions for cable management, network access points, power and internet outlets near scanner locations should be incorporated into the design.

Ergonomics and user experience

The design of the laboratory space should prioritize the ergonomics and user experience of laboratory personnel. Ensuring whole slide scanners are in close proximity to the coverslipping/drying stations safeguards a lean workflow between subsequent stations. Laboratory technicians spend significant hours working at the scanners, making it essential to create an environment that minimizes physical strain, reduces fatigue, and enhances overall productivity. The workstation where technicians operate the whole slide scanners should be designed with attention to user posture and comfort. Adjustable chairs with lumbar support and height-adjustable desks may aid in reducing prolonged discomfort. These features allow users to customize their seating and working positions to prevent discomfort and strain associated with prolonged sitting. Workstations should also provide ample space for positioning monitors, keyboards, and other accessories within easy reach. Tables where whole slide scanners are stationed should have sufficient support to handle the weight-bearing load of the number of scanners placed there, and may further need antivibration measures to mitigate vibration-induced digital artifacts. Review of operations in the floors above and below the intended scanning locations may prove useful to ensure there are no processes that would interfere with the scanning operations.

Digital workflow ergonomics ("Computer workstation ergonomics guidelines," 2000; Emerson et al., 2021; Prabhu et al., 2005): For any user (e.g., pathologist, trainee, or technician) who will be undergoing a digital transformation in their daily activities, there are several ergonomic areas to focus attention. For instance, placement of monitors is crucial to prevent neck, shoulder, and eye strain. Monitors should be positioned at eye level, and the top of the screen should be at eye level to encourage a natural head posture. Adjustable monitor stands or mounts are beneficial to fine-tune monitor height and angle. This allows users to personalize their setup for their specific needs. Additionally, monitors with anti-glare screens and/or blue light filters can minimize eye strain. This reduces the risk of musculoskeletal sequalae and eye fatigue, especially when reviewing many digital slides over long periods of time. Keyboards and input devices (e.g., computer mouse) should be positioned so that users' wrists are in a neutral position, reducing the strain on wrists and forearms. Ergonomic keyboard designs with split layouts and cushioned wrist rests can further enhance comfort and prevent repetitive strain injuries. Ergonomic chairs and footrests can provide support and alleviate pressure on the lower back and legs. Nonslip flooring and antifatigue mats in standing areas can reduce strain when users who use standing desks and need to stand for prolonged periods of time for digital slide review.

Proper lighting is essential to further prevent eye strain and create a comfortable working environment. Task lighting should be adjustable and positioned to illuminate the workspace without causing glare on screens. Glares on screens can also introduce "blind spots" on the monitors, and may result in quality assurance considerations if there is significant glare that prevents the pathologist from reviewing specific fields of view. Ambient lighting may be best suited in a digital pathology workflow, similar to radiology reading rooms (Pollard et al., 2009, 2012; Wade & Brennan, 2004). Proper lighting ensures accurate color perception, minimizes glare and reflections, and enhances detail visibility, while reducing eye strain. Even with optimal ergonomic design, extended periods of static posture can still lead to discomfort. Encouraging and educating users to take regular breaks, stretch, and move around can prevent muscle fatigue and improve circulation. Incorporating standing workstations can provide an alternative to sitting and promote healthier movement patterns. After the integration of whole slide scanners, regularly gather feedback from users about their ergonomic experiences. This feedback can lead to refinements in workstation design and adjustments to optimize comfort and efficiency. By incorporating ergonomic principles into the design of the laboratory space, pathologists and laboratory technicians can work comfortably and efficiently. This not only enhances their well-being but also supports accurate diagnoses and improved patient care. A well-designed ergonomic environment demonstrates the laboratory's commitment to the health and productivity of its staff, creating a positive and conducive atmosphere for embracing digital pathology technology.

Whole slide scanners generate heat and may produce noise during operation. Adequate ventilation and noise isolation measures should be integrated into the design to ensure a comfortable and safe working environment. Furthermore, if an extension of the CLIA certified laboratory, mandatory temperature and humidity measurements are required to be documented to ensure the devices are operating in their intended range.

By assessing current space utilization, optimizing workflows, creating flexible layouts, integrating necessary infrastructure, prioritizing ergonomics, considering environmental factors, and implementing effective training, laboratories can seamlessly integrate digital pathology systems into their existing environments. This approach ensures that the integration process is efficient, user-friendly, and aligned with the laboratory's operational objectives.

Staffing resources

While technology drives the digital pathology revolution, the human element remains the heartbeat of the operation. Effective staffing is a critical factor in the successful implementation and operation of a digital pathology system. The transition to digital pathology introduces new roles, responsibilities, and challenges for laboratory personnel. Adequate staffing allocation, proper training, and ongoing support are essential to ensure a seamless and efficient workflow. Table 2.2

Table 2.2 Stakeholders and roles in digital pathology.

Representative stakeholders and their role in digital pathology	
Stakeholder	Role
Leadership	Support from departmental and institutional leadership is critical to ensure success in implementing digital pathology systems
Pathologists	Super users can champion a digital pathology implementation, as well as drive value of the operation
Trainees	Academic institutions rely on providing foundational training programs for medical students, residents, fellows, observers, rotators, etc. Trainees should be included as a unique perspective to evaluate and provide input in the operation.
Lab staff	Laboratory staff (e.g., histotechnologists, technicians, aids) are vital to the digital pathology operation and can provide tremendous value toward loading and unloading glass slide cartridges from the whole slide scanners
Scan staff	Staff may be dedicated to supporting the scanning operation and include those that can operate scanners, troubleshoot the hardware and software when needed, and provide quality control of the glass and digital slides.
Informatics/IT	Information technology staff dedicated to the department or institution are needed to ensure network connectivity, servers, security, LIS integration, reporting, and enterprise imaging efforts

IT, *Information technology;* LIS, *laboratory information system.*

Empowering staff through training and development ensures seamless workflow execution and harnesses the technology's full potential. Training histotechnologists and laboratory personnel in slide loading, quality assurance, and troubleshooting for whole slide scanners is paramount ("Invisible for a few but essential for many: the role of histotechnologists in the establishment of digital pathology—PubMed," n.d.). Properly trained staff contributes to consistent scanning throughput, timely issue resolution, and effective slide review. Providing learning opportunities will cultivate a culture of adaptability and responsiveness to technological advancements. An optimal staffing model aligns with operational needs. Beyond technical skills, staff should be equipped with problem-solving capabilities and adaptability. Cross-training enables personnel to step in when demand spikes or technical challenges arise. This approach minimizes disruptions and maintains a steady workflow. Additionally, fostering a supportive environment is essential. Open communication channels empower staff to share their insights and challenges. Regular feedback loops enable process refinement and foster a sense of ownership, ensuring that every team member is invested in the digital pathology initiative's success. The introduction of digital pathology may require reallocating existing personnel or hiring new staff with specialized skills. Histotechnologists, laboratory aides, and pathologists' assistants may be trained to handle tasks related to loading glass slide racks into scanners, quality control (QC) of scanned images, and troubleshooting. Pathologists themselves may need training to effectively navigate digital images and use the new technology to its fullest potential.

As whole slide scanners become an integral part of the laboratory, designated staff members must ensure the quality and accuracy of scanned images. Automated software solutions to screen for digital slide artifacts are evolving, however, at this moment are not directly integrated in digital pathology systems. Currently, this is performed via manual review and involves monitoring the scanning process, addressing technical issues, and performing regular QC checks to identify artifacts, focus errors, or other glass slide and digital artifacts (Albuquerque et al., 2022; Ardon et al., 2023; Hosseini et al., 2020; Janowczyk et al., 2019). Pathologists involved in consultation services may require additional support to manage the unique workflow of scanning glass slides received from external institutions. Staff may need to implement methods to handle and label consultation slides appropriately to prevent errors during the scanning process. A team responsible for the maintenance and troubleshooting of digital pathology equipment is crucial to minimize downtime and ensure continuous operation. This team can address technical issues promptly, perform routine maintenance tasks, and collaborate with vendors for timely repairs if needed.

Depending on the laboratory's workload and scanning volume, it might be necessary to consider multiple shifts to maximize scanner utilization and maintain turnaround times. High-volume labs may benefit from round-the-clock scanning to accommodate the demand. Given the digital nature of pathology images, IT support becomes critical for scanning operations. Personnel with expertise in networking, data security, and system integration are necessary to maintain seamless communication between the scanners, the laboratory information system (LIS), and other connected systems. In larger institutions, having dedicated digital pathology specialists can streamline operations. These specialists can assist pathologists in adopting the technology, troubleshoot technical issues, optimize scanning parameters, and provide ongoing training and support. Staff training and education programs should be developed and implemented to ensure all team members are comfortable with the new technology. Training sessions can cover digital slide navigation in the viewer, image quality assessment and metrics to be evaluated, troubleshooting, and adherence to standard operating procedures. Appropriate staffing resources significantly impact the overall workflow and efficiency of the digital pathology system. When staffing levels are inadequate, it can lead to delays in scanning, diminished image quality, and compromised patient care. On the other hand, well-trained and supported staff members contribute to faster turnaround times, accurate diagnoses, and improved patient outcomes. A skilled workforce ensures that technical issues are promptly addressed, reducing system downtime and maintaining laboratory productivity. The staff's expertise in managing digital slides also ensures that images are consistently of high quality, reducing the need for rescanning and maintaining a streamlined workflow.

Scanner considerations

Selecting the appropriate whole slide scanners is a pivotal decision that shapes the efficiency and effectiveness of the digital pathology workflow. The landscape offers a variety of scanner types, each catering to specific laboratory needs, volumes, and budgets (Farahani et al., 2015; Mutter et al., 2022).

High-throughput scanners are tailored for large-volume laboratories, capable of scanning hundreds of slides in a single batch. Continuous loading capabilities streamline processes, enabling a "load and walk away" workflow. These scanners support conventional glass slide formats as well as larger sizes, offering flexibility for diverse scanning needs. Low-throughput scanners are ideal for laboratories with moderate volumes. They take up less space and offer an economical solution for labs taking their initial steps in the digital pathology domain. Some low-throughput scanners excel in specialized tasks, such as immunohistochemistry, fluorescence imaging, or real-time scanning for telepathology. Dynamic-robotic imaging devices offer real-time remote consultation capabilities, enabling pathologists to operate scanners and view slides in real-time. These scanners are well suited for intraoperative consultations and rapid cytologic evaluations. Their versatility in selecting different objectives enhances the viewing experience, enabling efficient z-plane adjustments (e.g., fine focus). Integrated microscopes, while providing lower throughput, offer a cost-effective option for institutions with limited resources. These devices mount digital cameras on traditional microscopes,

Table 2.3 Scanner considerations.

Whole slide scanner considerations	
Imaging mode	Brightfield, fluorescence
Slide throughput	Continuous load, batch processing, slides per hour
Scan speeds	Time per slide scan: prescan time, handling, scan time over glass, file creation, time to view
Slide format (e.g., glass size)	1" × 3," 2" × 3," 6" × 8"
Magnification/resolution	0.5, 0.25, 0.1 μm/pixel
Flexibility in scan area	Single ROI, multiple ROI
Live viewing needs	Collaborative screen sharing, single versus multiuser access
Multilayer support	Z-stack scanning, extended focal imaging, tilted stage
Rescan rates	Percentage of slides needing to be rescanned due to faulty digital image
Whole slide image file format	DICOM, SVS, NDPI, ISYNTAX, MRXS, BIF, TIF
Lab space/countertop weight load	Device countertop space footprint, area surrounding scanner for accessing internal components (e.g., repairs, troubleshooting)
User experience	Touchscreen operation, dedicated workstation with specified software, alerts and notifications, scanner profiles, configuration settings

ROI, *Region of interest.*

facilitating digital imaging without significant infrastructure changes. While operator-dependent, integrated microscopes bridge the gap between conventional and digital microscopy.

Implementing a successful digital pathology workflow requires careful consideration of various factors that influence the choice of whole slide scanners. Each element is impacted by the intended use and in turn impacts the efficiency, accuracy, and user experience of the digital pathology system. Table 2.3

Imaging mode

The choice between brightfield and fluorescence imaging modes depends on the laboratory's specific needs. While brightfield imaging is widely used for routine histopathology, fluorescence imaging is crucial for applications like immunofluorescence staining. Selecting a scanner that supports the required imaging mode ensures compatibility with the laboratory's diverse diagnostic and research requirements, while also assuring the color calibration is within normal limits of the pathologist's standards.

Slide throughput/continuous load

Scanner throughput determines the number of slides that can be digitized within a specific time. Continuous loading capabilities are essential for labs with high slide volumes. High capacity does not essentially reflect high throughput. Whole slide scanners that offer numerous slide bays, however, take a lengthy time to digitize each glass slide may delay clinical turnaround time as slide cartridges are awaiting digitization. High throughput devices will have a low scan time (including slide handling, digitization, and file creation), and offer continuous loading capabilities such that the users can on/offload slide cartridges that have been completed or need scanning, while the scanner is in use. Considerations also include the time required for slide loading, scanning, and unloading, as well as the scanner's ability to handle diverse slide types seamlessly (Zarella & Rivera Alvarez, 2022).

Scan speeds

Scan speed influences the overall turnaround time for digitizing slides. Most whole slide scanner manufacturers use a 15×15 mm scan area standard. Many academic centers may have tissue areas that are significantly larger than that "standard" area, and should adjust those scan times reported by the device manufacturers, as well as expectations. Faster scanning speeds are beneficial to maintain efficiency, however, a balance must be maintained between speed and image quality to ensure accurate diagnostic interpretation.

Slide size/format

Scanners should accommodate various slide sizes and formats commonly used in pathology. This flexibility enables labs to digitize slides of different dimensions without compromising image quality. It is essential to choose a scanner that can handle standard as well as nonstandard slide sizes at scale to assure laboratories with larger glass slide formats can also be digitized. Laboratories with whole mount or other glass slide formats should ensure whole slide scanning devices have features to scan those slides with the same throughout and quality as traditional glass slide formats. Digitization of all glass slides used in the clinical context will ensure that laboratories will be truly "100% digital." Furthermore, to meet clinical demands, institutions may have multiple vendors with different glass slide thicknesses. Attention should be given to ensure glass slides can accommodate the glass slide thicknesses to fit inside the slide cartridges.

Magnification/resolution

The ability to capture high-resolution images is crucial for accurate diagnosis. It is also important to understand the relationship between the magnification of microscopes and the resolution of images generated by camera sensors from whole slide scanners (Sellaro et al., 2013). Scanners differ in the native scan resolution that they offer, depending on the intended use of the device. For example, high throughput scanners typically scan at one magnification (e.g., "40 ×" equivalent magnification), however, hybrid real-time viewing scanners have several objectives for real time viewing of a native magnification. Either system enable pathologists the ability to zoom in and out of specific regions of interest while maintaining image clarity and detail. Pathologists must validate each workflow, including scan magnification for their clinical use case. Digital slide resolution is reported in microns per pixel (μm/pixel). The high the resolution (e.g., better), the lower the μm/pixel. To calculate an estimate of true equivalent scan magnification, divide 10 by the μm/pixel. For instance, a resolution of 40 × is equivalent to 10/0.25 μm/pixel = 40 ×. Equivalent magnifications are not marketed similarly across all scanner manufacturers, it is important for organizations to be aware of this phenomenon and have a technical champion to understand these prior to implementation.

Flexibility in scan area

Some scanners allow for flexible selection of the scan area, where only areas identified to be tissue by the device are digitized versus a single region of interest capturing all nontissue areas (e.g., deadspace) on the slide between pieces of tissue. For scanners that have flexible scan areas, this enables pathologists to focus on specific regions or avoid areas of no interest, however, may increase the probability of missing small pieces of tissue that were not identified by the scanner. This feature may or may not decrease scan times, so it is appropriate to validate the scanner with specimens from the implementing laboratory and review any gained efficiency from scan times. This feature conserves storage space and minimizes file sizes and data transfer times.

Live viewing needs

Pathologists often require live viewing of slides to navigate and assess specific areas in real-time. A scanner's live viewing capabilities are integral for interactive discussions during consultations, consensus meetings, or multidisciplinary conferences. Real-time hybrid whole slide scanners represent a significant advancement in the field of digital pathology, offering dynamic capabilities that extend beyond traditional static scanning. These scanners are designed to enable real-time interaction with the glass slide during such that the pathologist can pan, zoom, and focus one or multiple glass slides in real time. This innovative approach has introduced new possibilities for pathologists and researchers, enhancing their ability to collaborate, make rapid decisions, and improve diagnostic accuracy ("Utilizing novel telepathology system in preclinical studies and peer review—PubMed," n. d.; Zhou et al., 2000).

Multi z-plane support

Many high throughput whole slide scanners opt to digitize glass slides in a single z-plane to optimize quality and efficiency. Certain specimen types may require scanning in multiple z-planes to capture additional focal plane information, similar to fine focus capabilities on the microscope. Newer scanner technologies include extended focal image scanning where the best focal plane for each respective field of view is stitched together across the entire slide, effectively ensuring the slide is scanned with the highest quality for each region of interest (Hildén & Shevchenko, 2023). This may take additional time to digitize, however, saves storage space compared to true multi z-plane scanning. Z-stack scanning captures images at various focal planes, allowing for 3D reconstruction and depth analysis. This is particularly useful for applications such as cytology, hematology (e.g., cellular specimens), or capturing images of thick tissue sections (Eccher & Girolami, 2020; Kim et al., 2022; Lewis et al., 2023; Singh & Ohgami, 2018).

Rescan rates

Rescan rates indicate the efficiency of rescanning slides for QC purposes. Through QC of the digitization process, various stages of QC should detect digital slide defects and promote rescanning of the poorly scanned slide. QC of the glass slide should be completed regardless of scanning efforts as this ensures the highest quality glass slide is first created (e.g., typical laboratory processes). After digitization, slide thumbnail review should be performed to assure that all tissue present on the glass slide is present in the digitized image, and that the barcode was properly decoded. This ensures the digital slide is linked to the patient's case (and more specifically, the stain), in the LIS. In parallel, either manual review or automated QC software should interrogate the pixel review to ensure the slide is scanned in focus, does not show stitching artifacts, banding, or other digital artifacts.

Whole slide image file format

Compatibility with common whole slide image file formats ensures seamless integration with image viewing and analysis software used by pathologists. This is further discussed in the "Standards" section.

Lab space/countertop weight load

The scanner's physical dimensions and weight impact the lab's space planning. Some labs may need to retrofit existing spaces to accommodate whole slide scanners. Ensuring that the lab infrastructure can support the scanner's weight is crucial to prevent structural issues.

User experience

An intuitive user interface and software platform enhance the overall experience for pathologists. User-friendly interfaces simplify digital slide navigation, annotation, and collaboration making it easier for pathologists to analyze high-resolution images. User experience design also plays a role to enhance pathology diagnoses and research. Incorporating user feedback and conducting usability testing are important for continually improving the digital pathology system's design and functionality.

Balancing these factors is vital when selecting a whole slide scanner and image/case management system for digital pathology. The chosen scanner should align with the laboratory's workload, imaging needs, and long-term growth strategy. The digital pathology system should be user friendly and support the organization's intended uses (Retamero et al., 2020). Collaborating with vendors and considering feedback from early adopters can provide valuable insights to make informed decisions that optimize the digital pathology workflow.

Storage

The sheer volume of data generated by whole slide images necessitates robust storage solutions. Institutions implementing digital pathology must balance accessibility with data security, factoring in costs, redundancy requirements, regulatory compliance, and image retention policies. Most organizations will weigh the factors to decide on a digital storage solution to be on-premise versus cloud storage. This decision is based on institutional policies, digital readiness, security, compliance, accessibility, and cost. Cloud-based storage offers facile scalability, enabling remote image viewing and sharing. However, cloud-based digital storage is more complex to consider additional factors such as storage tiers (i.e., affects performance of image viewing), network bandwidth, data retrieval times, and annual costs (e.g., including storage and egress costs) influence the choice of cloud storage providers. Implementing a tiered storage strategy based on image access frequency optimizes storage costs. Cloud-based storage solutions offer convenience to accommodate growing image archives. On-premises storage solutions provide more control over the data and accessibility, but require digital storage management and investment in infrastructure. Either storage solution requires a robust backup and disaster recovery plans safeguard against data loss. Hybrid storage solutions, combining on-premises and cloud storage, offer flexibility and redundancy, ensuring data accessibility even during infrastructure outages, however, may incur more cost. Strategic data storage management is pivotal in ensuring the longevity of a digital pathology workflow. The size and frequency of whole slide image generation dictate storage costs and accessibility. Institutions must consider image retention policies, cloud storage options, data redundancy, and network bandwidth to optimize their digital slide management ("Don't be afraid of storage costs," n.d.; Laurinavicius & Raslavicus, 2012).

Additionally, many institutions are including pathology in the enterprise image management solution discussions. Historically, radiology, dermatology, ophthalmology, and few other specialties were the main focus, however, with the advent of digital pathology and the size of each whole slide image, there is an increasing need for institutions to manage pathology imaging at the enterprise level (Dash et al., 2021; Petersilge, 2020; Roth et al., 2016). Enterprise image management systems provide a robust infrastructure for storing and retrieving digital pathology images. The systems support secure storage, easy accessibility, and organization for efficient review and analysis. Integration with electronic health records (EHR) streamlines data flow, and the technology supports both clinical and de-identification use cases (e.g., research and education). Enterprise image management can be now associated with pathology due to these digital transformations, providing the infrastructure and capabilities necessary to store, manage, secure, and integrate pathology images and data. This relationship enhances the quality of patient care, promotes collaboration among healthcare professionals, and supports advancements in medical research and education. As digital pathology continues to evolve, enterprise image management systems will play an increasingly vital role in optimizing the benefits of this transformative technology.

Software

In the rapidly evolving landscape of digital pathology, software solutions have emerged as pivotal tools in clinical care, research, and education. From sophisticated image viewers to cutting-edge machine learning applications, these software innovations have reshaped the way pathologists interact with digital slides and leverage the vast potential of digital pathology data.

Image viewers: Digital pathology relies on advanced image viewers that enable pathologists to navigate, annotate, and evaluate high-resolution slide images seamlessly. These viewers provide a dynamic platform for pathologists to pan, zoom, and focus in on pathology specimens, annotate regions of interest, and collaborate with colleagues. User-friendly interfaces and intuitive tools have made a plethora of image viewers available for clinical validation, research, or educational use cases.

Quality assurance: Ensuring the accuracy and reliability of digital pathology data is paramount. Quality assurance tools have emerged as a critical component, to check the quality of the generated whole slide image, and support the accuracy of the reported clinical diagnosis. These tools perform automated checks for image consistency, staining quality, and data completeness. They not only bolster diagnostic confidence but also aid in compliance with regulatory standards, such as CAP and CLIA.

Operational dashboards: Managing the operational aspects of a digital pathology laboratory requires real-time insights and efficient workflows. Operational dashboards have become indispensable tools for laboratory administrators. These dashboards provide a holistic view of laboratory processes, including specimen tracking, workload management, and turnaround times related to digital pathology. These dashboards facilitate informed decision-making, optimize resource allocation, and enhance overall laboratory productivity. Figure 2.3

Machine learning applications: Machine learning has made significant inroads into digital pathology, offering powerful tools for image analysis, pattern recognition, and predictive diagnostics (Abels et al., 2019; Harrison et al., 2021; Niazi et al., 2019). Machine learning applications are now capable of identifying subtle anomalies in pathology slides, aiding in cancer detection, grading, and prognostication. Furthermore, these applications continue to evolve, providing pathologists with valuable insights that complement their expertise and drive the field toward more personalized and precise medicine.

From image viewers that empower pathologists with unprecedented visualization capabilities to quality assurance software that upholds data integrity, operational dashboards that optimize workflows, and machine learning applications that redefine diagnostic accuracy, these software solutions are reshaping the future of pathology. As digital pathology continues to advance, the synergy between pathologists and software tools promises to unlock new horizons in healthcare.

Service and maintenance

The journey of digital pathology does not end with implementation—it requires ongoing maintenance to ensure seamless operations. Regular maintenance of glass

FIGURE 2.3 Representative software examples.

(Top left) Example of a digital pathology viewer showing a high-resolution pathology whole slide image and thumbnail viewer. (Top right) Graphical dashboard allowing administrators and laboratory directors at-a-glance data analytics on the scanning operation. (Bottom left) Quality control dashboard solution that detects whole slide images scanned out of focus, with missing tissue, and other scanning artifacts; while passing good quality scans. (Bottom right) Representative heatmap visualization showing red as high probability for tumor and blue as low probability for tumor.

slide label printers, slide scanners, and associated equipment minimizes errors and operational disruptions. Collaborating with vendors to develop software solutions, such as automatically suppressing external barcodes during scanning, enhances workflow efficiency.

A well-structured maintenance schedule encompasses preventive and corrective measures. Preventive maintenance prevents breakdowns through regular check-ups, cleaning, and parts replacement. Regularly scheduled inspections, cleaning, and calibrations help identify and address potential issues before they escalate. These proactive measures extend the lifespan of hardware and minimize unplanned downtime, ultimately contributing to uninterrupted patient care. Corrective maintenance swiftly addresses issues as they arise, minimizing downtime and ensuring continuous workflow. Preventive maintenance is the cornerstone of digital pathology system reliability.

Timely access to technical support is essential for addressing unexpected system issues. Digital pathology vendors often offer comprehensive technical support packages, including helpdesk services, remote diagnostics, and on-site interventions when necessary. Effective technical support ensures that pathologists can rely on their digital pathology systems for accurate and efficient diagnoses. Additionally, it would be beneficial for on-site departmental staffing to be trained on scanner operations and technical troubleshooting in order to minimize downtime.

Close collaboration with vendors is crucial. They provide timely software updates, address emerging challenges, and offer technical support. Regular software updates from vendors are commonplace to keep systems up-to-date and secure. These updates may also introduce new features and improvements that further empower pathologists in their diagnostic and research efforts. User training equips staff with troubleshooting skills, reducing dependency on external support. Vendor-provided training sessions and resources facilitate skill development, ensuring that users harness the full potential of their digital pathology systems.

Compliance and regulatory requirements: Digital pathology systems must adhere to regulatory requirements, such as those set forth by CLIA, CAP, and HIPAA. Clinical validation exercises must be adhered to, and be with proper documentation. Service and maintenance practices must align with these standards to ensure compliance. Regular audits and documentation are essential components of meeting these regulatory obligations. Institutional knowledge and experience play a pivotal role. Documenting troubleshooting procedures, sharing lessons learned, and fostering a collaborative atmosphere among staff amplify the effectiveness of maintenance efforts. The symbiotic relationship between vendors and institutions is a cornerstone of seamless digital pathology operations ("Academic collaborations with industry: lessons for the future—PubMed," n.d.; Pantanowitz et al., 2022).

Service and maintenance play a pivotal role in ensuring the continued functionality, reliability, and security of digital pathology systems. As digital pathology technology advances, robust service and maintenance practices will continue to be integral to its success in revolutionizing pathology diagnostics and patient care.

Honorable mention

Standards

One of the primary functions of standards in digital pathology is to ensure interoperability. Standards are definitely needed to ensure success in implementing a digital pathology system, however, are still evolving and can be later added as an eighth "S." Strategically managing metadata and adopting standardized file formats and/or transmission protocols facilitate efficient image retrieval, sharing, and analysis. Pathology laboratories, healthcare institutions, and technology providers must be able to exchange data seamlessly. Interoperable standards allow for the integration of various digital pathology systems, ensuring that images, metadata, and reports can be shared, accessed, and interpreted consistently. Standards provide a framework for data security, patient privacy, and ethical handling of pathology data. As digital pathology continues to evolve, continued development and adherence to standards ensures that new technologies can seamlessly integrate with existing systems, preventing fragmentation, and enhancing the overall utility of digital pathology solutions. They provide a common foundation upon which developers can build new tools, applications, and algorithms. Metadata tagging aids in organizing and categorizing images, while standardized file formats ensure compatibility across platforms. Both standardization of file formats and comprehensive metadata tagging can streamline image retrieval and sharing. Standardized metadata enriches image categorization, simplifying search and retrieval processes. Furthermore, adopting open standards ensures compatibility with various image viewing platforms, mitigating vendor-specific risks.

Digital imaging and communications in medicine (DICOM) is a widely recognized standard in the healthcare industry. It was originally developed for radiology, however, has since been extended to support pathology images. It defines a standardized format for storing, transmitting, and exchanging medical images and associated data; the standard now includes pathology imaging. DICOM ensures that digital pathology images can be managed within a healthcare enterprise's existing infrastructure, and even more crucial—be interoperable between vendor systems (Herrmann et al., 2018; Kalinski et al., 2012). There are currently three supplements dedicated to digital pathology which are summarized below: Table 2.4

DICOM supplement 122 was developed for the Specimen Module and revised Pathology service-object pair (SOP) classes, and was finalized in June 2008. This supplement introduces a new module within the DICOM standard to address the representation of pathology specimen information in medical imaging. It enhances the standard's capabilities for managing and exchanging data related to

Table 2.4 Key standards related to digital pathology.

IHE PaLM	Image acquisition profiles, ordering and reporting
DICOM 122	Specimen module and revised pathology SOP classes
DICOM 145	Whole slide microscopic image IOD and SOP classes
DICOM 222	Microscopy bulk simple annotations storage SOP class

IHE, *Integrating health exchange;* PaLM, *pathology and laboratory medicine;* DICOM, *digital imaging and communications in medicine;* SOP, *service-object-pair;* IOD, *information object definition.*

specimens, such as specimen type, collection technique, and container information. This is particularly relevant in anatomical pathology, where it allows for the integration of pathology-related information into the DICOM framework, patient metadata into pathology imaging protocols, and interoperability between digital pathology systems.

DICOM supplement 145, finalized in August 2010, focuses on the "Whole Slide Microscopic Image Information Object Definition (IOD) and SOP Classes." This supplement extends the DICOM standard to accommodate whole slide images generated by digital pathology scanners. It defines the structure and encoding for storing, exchanging, and managing these high-resolution digital representations of microscope slides. By standardizing the format and metadata associated with WSI, supplement 145 enables the seamless integration of digital pathology images (e.g., WSI), supporting diagnostic and research applications.

DICOM supplement 222, titled "Microscopy Bulk Simple Annotations Storage SOP Class," was published on July 2021. This supplement focuses on providing standards toward annotations for whole slide images. These annotations typically consist of points, lines, polygons, and simple geometric shapes without complex relationships. They are commonly generated by machine algorithms from high-resolution images of pathology tissue sections. Standardization for these elements can greatly provide interchange of annotations between digital pathology systems, which will be critical for decision support systems and validation of machine learning models to be used for patient care.

Other healthcare messaging standards have also been instrumental in the adoption and implementation of digital pathology. Health Level Seven International (HL7) is a standard development organization focused on healthcare interoperability. While it is known primarily for EHR standards, it has extended its scope to include standards for LIS tasks such as anatomical pathology reporting and whole slide imaging integration. HL7 structured report (SR) templates facilitate the structured reporting of pathology findings, ensuring that pathology reports are structured, consistent, and machine-readable. In addition, DICOM has a SR as a standard for recording clinical imaging observations, however, is used mostly in radiology today.

Integrating the healthcare enterprise (IHE) pathology and laboratory medicine (PaLM) is also a known standard in pathology, which has since extended to include digital pathology. This standard refers to the integration of healthcare enterprise profiles and standards specifically designed to improve interoperability and streamline processes within the realm of pathology and laboratory medicine. IHE PaLM establishes a framework for exchanging information and ensuring seamless communication between various components and systems involved in pathology and laboratory processes. It encompasses several key functionalities, including specimen tracking, laboratory order management, and results reporting. This standard promotes the sharing of critical information, such as specimen identification, test orders, and results, among different systems, reducing errors and enhancing patient care. In relation to digital pathology, there is on-going effort to document whole slide scanner image acquisition profiles and digital pathology ordering and reporting (DPOR) (Dash et al., 2021).

Standards are the foundation upon which digital pathology will further evolve from a promising technology to a transformative force in healthcare. Standards enable interoperability, enhance data quality, support regulatory compliance, and foster innovation. As digital pathology continues to shape the future of pathology diagnostics, education, and research, the development and adoption of standards will remain essential for achieving its full potential.

Conclusion

The digital pathology landscape is multifaceted, demanding a holistic approach to implementation. The seven S's—sponsorship, space, staffing, scanners, storage, software, service—provide a structured framework for successful integration. Additionally, recognizing the significance of service and maintenance, as well as digital storage, sustains the digital pathology ecosystem's resilience. By addressing each facet, institutions can understand how to implement a successful digital pathology implementation. Customizing strategies to fit the unique needs of each organization, while embracing lessons learned by pioneers in the field, transforms the integration process into a manageable endeavor. Pathology transformation propelled by digital pathology extends beyond technology. It rejuvenates pathology practice, enhancing patient care through precision diagnostics and fostering collaboration among healthcare professionals, as well as advocating for medical students to thrive in an inventive medical subspecialty. The path ahead is illuminated by the combined efforts of visionary leadership, dedicated personnel, and cutting-edge technology, and will ultimately set new standards for healthcare excellence (Figs. 2.1–2.3; Tables 2.1–2.4).

References

Abels, E., Pantanowitz, L., Aeffner, F., Zarella, M. D., van der Laak, J., Bui, M. M., Vemuri, V. N. P., Parwani, A. V., Gibbs, J., Agosto-Arroyo, E., Beck, A. H., & Kozlowski, C. (2019). Computational pathology definitions, best practices, and recommendations for regulatory guidance: A white paper from the Digital Pathology Association. *Journal of Pathology*, *249*(3), 286–294. Available from https://doi.org/10.1002/path.5331, http://onlinelibrary.wiley.com/journal/10.1002/(ISSN)1096-9896.

Albuquerque, T., Moreira, A., Barros, B., Montezuma, D., Oliveira, S.P., Neto, P.C., Monteiro, J., Ribeiro, L., Goncalves, S., Monteiro, A., Pinto, I.M., Cardoso, J.S. (2022). Quality control in digital pathology: Automatic fragment detection and counting 2022-. *Proceedings of the annual international conference of the IEEE Engineering in Medicine and Biology Society*. Institute of Electrical and Electronics Engineers Inc., Portugal. doi:10.1109/EMBC48229.2022.9871208

Ardon, O., Labasin, M., Friedlander, M., Manzo, A., Corsale, L., Ntiamoah, P., Wright, J., Elenitoba-Johnson, K., Reuter, V. E., Hameed, M. R., & Hanna, M. G. (2023). Quality management system in clinical digital pathology operations at a tertiary cancer center. *Laboratory Investigation*, *103*(11)100246. Available from https://doi.org/10.1016/j.labinv.2023.100246.

Baidoshvili, A., Bucur, A., van Leeuwen, J., van der Laak, J., Kluin, P., & van Diest, P. J. (2018). Evaluating the benefits of digital pathology implementation: Time savings in laboratory logistics. *Histopathology*, *73*(5), 784–794. Available from https://doi.org/10.1111/his.13691, http://onlinelibrary.wiley.com/journal/10.1111/(ISSN)1365-2559.

Browning, L., Colling, R., Rittscher, J., Winter, L., McEntyre, N., & Verrill, C. (2020). Implementation of digital pathology into diagnostic practice: Perceptions and opinions of histopathology trainees and implications for training. *Journal of Clinical Pathology*, *73*(4), 223–227. Available from https://doi.org/10.1136/jclinpath-2019-206137.

Dash, R. C., Jones, N., Merrick, R., Haroske, G., Harrison, J., Sayers, C., Haarselhorst, N., Wintell, M., Herrmann, M. D., & Macary, F. (2021). Integrating the Health-care Enterprise Pathology and Laboratory Medicine Guideline for Digital Pathology interoperability. *Journal of Pathology Informatics*, *12*(1). Available from https://doi.org/10.4103/jpi.jpi_98_20, https://www.sciencedirect.com/journal/journal-of-pathology-informatics/about/aims-and-scope.

Eccher, A., & Girolami, I. (2020). Current state of whole slide imaging use in cytopathology: Pros and pitfalls. *Cytopathology*, *31*(5), 372–378. Available from https://doi.org/10.1111/cyt.12806.

Emerson, S., Emerson, K., & Fedorczyk, J. (2021). Computer workstation ergonomics: Current evidence for evaluation, corrections, and recommendations for remote evaluation. *Journal of Hand Therapy*, *34*(2), 166–178. Available from https://doi.org/10.1016/j.jht.2021.04.002.

Hanna, M. G., Reuter, V. E., Ardon, O., Kim, D., Sirintrapun, S. J., Schüffler, P. J., Busam, K. J., Sauter, J. L., Brogi, E., Tan, L. K., Xu, B., Bale, T., Agaram, N. P., Tang, L. H., Ellenson, L. H., Philip, J., Corsale, L., Stamelos, E., Friedlander, M. A., … Hameed, M. (2020). Validation of a digital pathology system including remote review during the COVID-19 pandemic. *Modern Pathology*, *33*(11), 2115–2127. Available from https://doi.org/10.1038/s41379-020-0601-5, http://www.nature.com/modpathol/index.html.

Harrison, J. H., Gilbertson, J. R., Hanna, M. G., Olson, N. H., Seheult, J. N., Sorace, J. M., & Stram, M. N. (2021). Introduction to artificial intelligence and machine learning for pathology. *Archives of Pathology and Laboratory Medicine*, *145*(10), 1228–1254. Available from https://doi.org/10.5858/arpa.2020-0541-CP, https://meridian.allenpress.com/aplm/article/145/10/1228/451554/Introduction-to-Artificial-Intelligence-and.

Herrmann, M. D., Clunie, D. A., Fedorov, A., Doyle, S. W., Pieper, S., Klepeis, V., Le, L. P., Mutter, G. L., Milstone, D. S., Schultz, T. J., Kikinis, R., Kotecha, G. K., Hwang, D. H., Andriole, K. P., Iafrate, A. J., Brink, J. A., Boland, G. W., Dreyer, K. J., Michalski, M., … Lennerz, J. K. (2018). Implementing the DICOM standard for digital pathology. *Journal of Pathology Informatics*, *9*(1). Available from https://doi.org/10.4103/jpi.jpi_42_18, http://www.jpathinformatics.org/browse.asp?sabs = n.

Hildén, P., & Shevchenko, A. (2023). Extended depth of field of an imaging system with an annular aperture. *Optics Express*, *31*(7), 11102. Available from https://doi.org/10.1364/oe.484290.

Hosseini, M. S., Brawley-Hayes, J. A. Z., Zhang, Y., Chan, L., Plataniotis, K., & Damaskinos, S. (2020). Focus quality assessment of high-throughput whole slide imaging in digital pathology. *IEEE Transactions on Medical Imaging*, *39*(1), 62–74. Available from https://doi.org/10.1109/TMI.2019.2919722, http://ieeexplore.ieee.org/xpl/RecentIssue.jsp?punumber = 42.

Janowczyk, A., Zuo, R., Gilmore, H., Feldman, M., & Madabhushi, A. (2019). HistoQC: An open-source quality control tool for digital pathology slides. *JCO Clinical Cancer Informatics* (3), 1–7. Available from https://doi.org/10.1200/CCI.18.00157, https://ascopubs.org/doi/pdfdirect/10.1200/CCI.18.00157.

Kalinski, T., Zwönitzer, R., Roßner, M., Hofmann, H., Roessner, A., & Guenther, T. (2012). Digital imaging and communications in medicine (DICOM) as standard in digital pathology. *Histopathology*, *61*(1), 132–134. Available from https://doi.org/10.1111/j.1365-2559.2012.04243.x.

Kim, D., Burkhardt, R., Alperstein, S. A., Gokozan, H. N., Goyal, A., Heymann, J. J., Patel, A., & Siddiqui, M. T. (2022). Evaluating the role of Z-stack to improve the morphologic evaluation of urine cytology whole slide images for high-grade urothelial carcinoma: Results and review of a pilot study. *Cancer Cytopathology*, *130*(8), 630–639. Available from https://doi.org/10.1002/cncy.22595, http://onlinelibrary.wiley.com/journal/10.1002/(ISSN)1934-6638.

Laurinavicius, A., & Raslavicus, P. (2012). Consequences of 'going digital' for pathology professionals-entering the cloud. *Studies in Health Technology and Informatics*, *179*, 62–67. Available from https://doi.org/10.3233/978-1-61499-086-4-62, http://www.iospress.nl/bookserie/studies-in-health-technology-and-informatics/.

Lewis, J. E., Shebelut, C. W., Drumheller, B. R., Zhang, X., Shanmugam, N., Attieh, M., Horwath, M. C., Khanna, A., Smith, G. H., Gutman, D. A., Aljudi, A., Cooper, L. A. D., & Jaye, D. L. (2023). An Automated pipeline for differential cell counts on whole-slide bone marrow aspirate smears. *Modern Pathology*, *36*(2)100003. Available from https://doi.org/10.1016/j.modpat.2022.100003.

Mutter, G. L., Milstone, D. S., Hwang, D. H., Siegmund, S., & Bruce, A. (2022). Measuring digital pathology throughput and tissue dropouts. *Journal of Pathology Informatics*, *13*(1) (), 8. Available from https://doi.org/10.4103/jpi.jpi_5_21, http://www.jpathinformatics.org/browse.asp?sabs = n.

Niazi, M. K. K., Parwani, A. V., & Gurcan, M. N. (2019). Digital pathology and artificial intelligence. *The Lancet Oncology*, *20*(5), e253. Available from https://doi.org/10.1016/S1470-2045(19)30154-8, http://www.journals.elsevier.com/the-lancet-oncology/.

Petersilge, C. A. (2020). The enterprise imaging value proposition. *Journal of Digital Imaging*, *33*(1), 37–48. Available from https://doi.org/10.1007/s10278-019-00293-1, http://www.springer.com/journal/10278.

Pollard, B. J., Samei, E., Chawla, A. S., Beam, C., Heyneman, L. E., Koweek, L. M. H., Martinez-Jimenez, S., Washington, L., Hashimoto, N., & McAdams, H. P. (2012). The effects of ambient lighting in chest radiology reading rooms. *Journal of Digital Imaging*, *25*(4), 520–526. Available from https://doi.org/10.1007/s10278-012-9459-5.

Prabhu, S. P., Gandhi, S., & Goddard, P. R. (2005). Ergonomics of digital imaging. *British Journal of Radiology*, *78*(931), 582–586. Available from https://doi.org/10.1259/bjr/51650129.

Retamero, J. A., Aneiros-Fernandez, J., Aneiros-Fernandez, J., Del Moral, R. G., & Del Moral, R. G. (2020). Microscope? No, thanks: User experience with complete digital pathology for routine diagnosis. *Archives of Pathology and Laboratory Medicine, 144* (6), 672–673. Available from https://doi.org/10.5858/arpa.2019-0355-LE, https://www.archivesofpathology.org/doi/pdf/10.5858/arpa.2019-0355-LE.

Rogers, E. M. (2003). *Diffusion of innovations* (2003). Free Press.

Roth, C. J., Lannum, L. M., Dennison, D. K., & Towbin, A. J. (2016). The current state and path forward for enterprise image viewing: HIMSS-SIIM collaborative white paper. *Journal of Digital Imaging*, *29*(5), 567–573. Available from https://doi.org/10.1007/s10278-016-9887-8.

Sellaro, T. L., Filkins, R., Hoffman, C., Fine, J. L., Ho, J., Parwani, A. V., Pantanowitz, L., & Montalto, M. (2013). Relationship between magnification and resolution in digital pathology systems. *Journal of Pathology Informatics*, *4*(1)21. Available from https://doi.org/10.4103/2153-3539.116866.

Singh, A., & Ohgami, R. S. (2018). Super-resolution digital pathology image processing of bone marrow aspirate and cytology smears and tissue sections. *Journal of Pathology Informatics*, *9*(1). Available from https://doi.org/10.4103/jpi.jpi_56_18, http://www.jpathinformatics.org/browse.asp?sabs = n.

Sirintrapun, S., Rudomina, D., Mazzella, A., Feratovic, R., Alago, W., Siegelbaum, R., & Lin, O. (2017). Robotic telecytology for remote cytologic evaluation without an on-site cytotechnologist or cytopathologist: A tale of implementation and review of constraints. *Journal of Pathology Informatics*, *8*(1). Available from https://doi.org/10.4103/jpi.jpi_26_17, http://www.jpathinformatics.org/browse.asp?sabs = n.

Wade, C., & Brennan, P. C. (2004). Assessment of monitor conditions for the display of radiological diagnostic images and ambient lighting. *The British Journal of Radiology*, *77*(918), 465–471. Available from https://doi.org/10.1259/bjr/18928981.

Williams, B. J., Bottoms, D., & Treanor, D. (2017). Future-proofing pathology: The case for clinical adoption of digital pathology. *Journal of Clinical Pathology*, *70*(12), 1010–1018. Available from https://doi.org/10.1136/jclinpath-2017-204644, http://jcp.bmj.com/.

Zarella, M. D., & Rivera Alvarez, K. (2022). High-throughput whole-slide scanning to enable large-scale data repository building. *Journal of Pathology*, *257*(4), 383–390. Available from https://doi.org/10.1002/path.5923, http://onlinelibrary.wiley.com/journal/10.1002/(ISSN)1096-9896.

Zhou, J., Hogarth, M. A., Walters, R. F., Green, R., & Nesbitt, T. S. (2000). Hybrid system for telepathology. *Human Pathology*, *31*(7), 829–833. Available from https://doi.org/10.1053/hupa.2000.9196, http://www.elsevier.com/inca/publications/store/6/2/3/1/3/9/index.htt.

CHAPTER

Building a business case for digital pathology

3

Orly Ardon

Department of Pathology and Laboratory Medicine, Memorial Sloan Kettering Cancer Center, New York, NY, United States

Introduction

Digital pathology (DP) is proving to be a valuable tool in the diagnosis and management of cancer and other diseases, offering pathology multiple benefits for improved diagnostic accuracy, faster turnaround times, and increased efficiency in pathology workflows (Eloy et al., 2021; Fraggetta et al., 2021; Ho et al., 2014; Jahn et al., 2020; Montezuma et al., 2022; Retamero et al., 2020; Schüffler et al., 2021; Stathonikos et al., 2019; Tizhoosh & Pantanowitz, 2018). The integration of artificial intelligence (AI) and machine learning (ML) algorithms into DP operations is anticipated to increase the adoption and implementation of DP workflows in pathology beyond early adopters in large academic medical centers (Hanna et al., 2020; Marletta et al., 2023; Schüffler et al., 2021).

Despite the potential of DP to modernize patient care, multiple barriers exist in adoption and implementation of DP in clinical use cases. The costs of DP operations vary greatly depending on the size and use cases of an organization, the specific technologies chosen, and the needs and level of automation and integration with all other laboratory systems. Overall, the costs of DP operations can be significant, with the requirement of the set up cost for basic infrastructure—as well as the ongoing operations—viewed as a major barrier for its adoption (Griffin & Treanor, 2017; Hanna et al., 2019; Hanna et al., 2020; Jahn et al., 2020; Lujan et al., 2021; Montezuma et al., 2022; Williams et al., 2017; Williams et al., 2018; Zarella et al., 2018). In 2019 Stathonikos et al. reported that the whole digital operation, including storage, is estimated to cost their department about <3% of their yearly budget, with an expectation that this rate will rise with expansion (Stathonikos et al., 2019). This is especially prohibitive for smaller, nonacademic institutions and for institutions in resource-poor geographic locations. The limited available data on the cost of DP operations complicate the development of business plans that are required for institutional investment in the technology. This task is difficult as many costs are unknown when introducing disruptive technologies, and will differ greatly based on institutional needs, size,

Digital Pathology. DOI: https://doi.org/10.1016/B978-0-443-13809-6.00005-1

and geography (Montezuma et al., 2022; Pinto et al., 2023; Turnquist et al., 2019).

An additional requirement for the development of business plans is the projected revenue or cost avoidance and savings that will offset the investment in DP. In 2010 Walter H. Henricks argued that "revenue from DP activities is not likely to cover the costs, at least in the near term but possibly not ever" (Henricks, 2010). Most later studies describe future savings based on potential cost savings calculations (Baidoshvili et al., 2018; Hanna et al., 2019; Ho et al., 2014; Lujan et al., 2021). Complicating these calculations is the fact that the costs of implementing DP will vary based on institutional size and needs, intended use cases, geographical location, and the pace of DP adoption. In 2014 Ho et al. projected 5-year total cost savings of USD $12.38 million based on anticipated improvements in pathology productivity and histology lab consolidation. Another cost saving of USD $5.35 million was suggested due to improved diagnoses and avoidance of unnecessary treatment costs (Ho et al., 2014). A paper published from Memorial Sloan Kettering Cancer Center (MSK), projected a 5-year USD $1.3 million savings based on comprehensive comparative cost analysis that included off-site storage, slide retrieval costs, vendor labor, scanner hardware and software, whole slide image (WSI) storage, and support staff (Hanna et al., 2019). Williams et al. (2019) suggested that the insourcing of cases has a potential to increase income by up to USD $46 per case (Williams et al., 2019). Vodovnik (2016) demonstrated that DP required a shorter diagnosis time than traditional microscopy and suggested the potential of DP to yield savings in both diagnostic and nondiagnostic tasks. Other publications have highlighted other benefits of DP transformation, yet many of these benefits were not quantifiable and did not point to any cost reduction, additional revenue, or cost avoidance (Vodovnik, 2016).

The stand-alone slide acquisition cost of DP is higher than that of analog microscopy due to the need for an additional digitization step (e.g., scanner, personnel, software) and data storage infrastructure. Each institution will have specific priorities, needs, timelines, and constraints for their adoption and implementation of DP workflows. It is also essential to identify the institutional internal and external scanning needs and workflows and the barriers that they pose for DP adoption. The internal scanning addresses cases produced at the institution either presently, or in previously archived material. The resulting digital images could be used for primary signout, archival needs, education, and research. The external scanning workflows provide solutions for clinical slide consultation, academic research, and industry collaborations. Additional external workflows may be needed for patient and legal requests. Each of these workflows will have different priorities and turnaround time needs, yet they all require trained staff across multiple teams, physical space, and other resources to provide timely quality images to the end user.

The specific infrastructure needs are institution and use case dependent and will dictate the budget required for that specific operation (Hanna et al., 2022).

Understanding the costs that are associated with these is crucial for any team before embarking on any DP operation or expansion effort.

A recent publication from our team offered a real-world retrospective analysis of the fixed infrastructure costs as well as the variable costs that are associated with the MSK DP clinical operations. The breakdown of specific categories can serve as a guide to institutions entertaining DP transformation (Ardon et al., 2023). As departments look to initiate or expand their DP operations, they will also need to structure a business plan detailing their projection investment and expected financial returns. This chapter will explore the costs associated with DP infrastructure and operation, and the financial benefits that it offers. Additional discussion will explore some useful financial terms, the need for a DP cost calculator, and some guidance for the development of a basic business plan for DP adoption and expansion.

Understanding the costs associated with digital pathology

DP infrastructure is required for transforming pathology specimens into digital slides to allow review and use for downstream applications including routine pathology reporting. The needs of institutions such as academic pathology laboratories, commercial labs, and community hospitals will vary, yet for all, DP system infrastructure is required (Hanna et al., 2020; Hanna et al., 2022; Schüffler et al., 2021). The costs required for these DP operations will also vary, and may be calculated in several implementation phases, or as an upfront cost with calculated ongoing operational expense. Recent publications from our team have summarized essential DP resources including hardware, software, data storage, staffing, space, and service agreements (Fig. 3.1) (Ardon et al., 2023; Hanna et al., 2022). We have also identified sponsorship from leadership as a critical component in the six "S" requirements for successful DP operations, assumed to be a crucial factor in providing vision and commitment to resources. A leadership sponsor is likewise a prerequisite for business plan approval and for any continued DP operations. In our opinion, this support must be in place before further investment in technology and related innovations.

Acquisition and ongoing costs

Before embarking on and securing funding for a DP-based investment, each institution needs to assess the existing institutional infrastructure, identify gaps that need to be addressed, and develop an understanding of the costs that are involved with obtaining the required infrastructure. This will include a distinction between their anticipated capital investments costs (i.e., acquisition) and ongoing (i.e., ownership) costs.

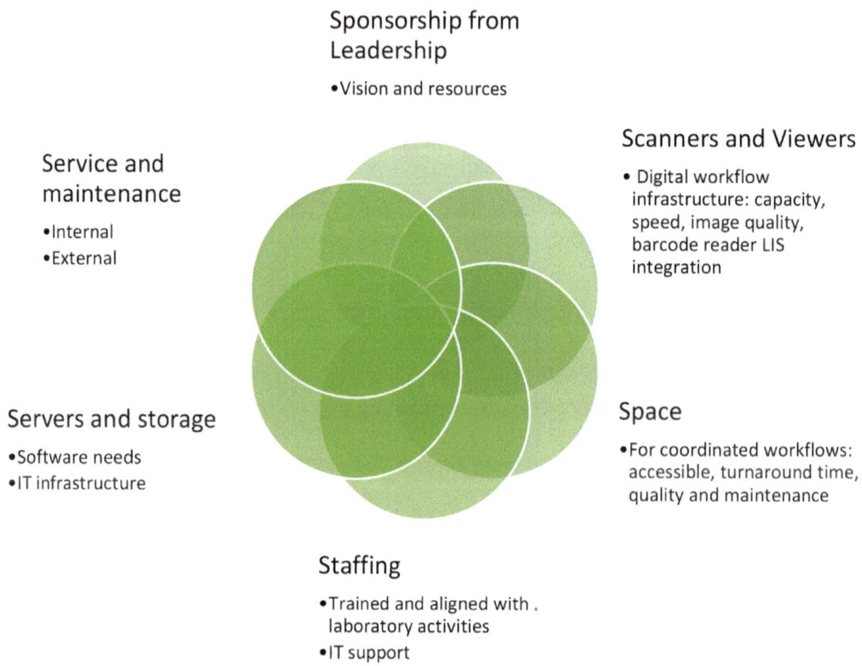

FIGURE 3.1

The six "S" infrastructure requirements for successful whole slide imaging operations (Ardon et al., 2023).

A retrospective study of MSK's DP operation identified costs that were incurred during the 2021 budget year. The authors separated between capital and ongoing costs and offered a template (Fig. 3.2) that could be adapted and used by other institutions (Ardon et al., 2023).

The capital acquisition costs include (1) scanning equipment; (2) servers, workstations, and software development; (3) infrastructure-related employees; (4) space configuration, including furniture; (5) network connectivity; and (6) equipment depreciation. The ownership cost includes the annual operating costs of related staffing, training, space, and utilities as well as service and maintenance costs, both internal and external. Factors such as digital scanner downtime should be included in calculating maintenance costs, as delays in service will have tangible consequences on clinical operations.

The ownership cost can be further divided between fixed costs that are incurred regardless of the volume of slides scanned, and variable costs, which will fluctuate depending on scanner output during the period studied.

A subset of the ongoing costs are also fixed expenses (e.g., scanner depreciation, maintenance, full time employees space) while others are variable costs that include the servers use, data storage, and network bandwidth (Sun et al., 2022).

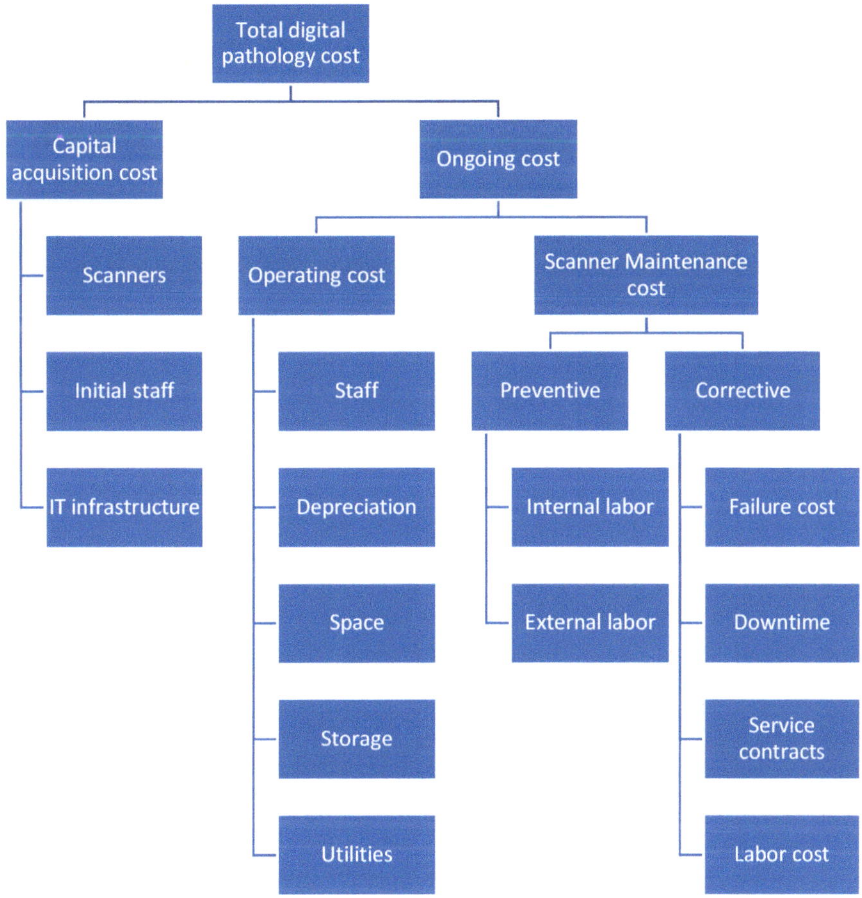

FIGURE 3.2

Identified components of digital pathology operations cost (Ardon et al., 2023).

As with other laboratory operations, careful analysis of capital and operational budgets should be conducted annually to ensure all costs associated with the operation are evaluated and accounted for. These budgets may change as the technology continues to develop (MacMillan, 2003).

Most institutions with established analog pathology operations will employ a hybrid workflow in which both digital and analog pathology are maintained temporarily while developing a phased DP adoption. This phased strategy allows stepwise introduction of various DP workflows, and gradual accrual of experience with DP technology. This results in stepwise accumulation of the benefits of digitization over a number of years (Williams et al., 2019). Smaller, simpler digital workflows, such as the scanning of second opinion cases, can get developed earlier, with more complex, large-scale changes in practice such as use of digital diagnosis for standard primary diagnosis deferred until the laboratory, and diagnosticians, have gained experience in the laboratory workflow, and the digital diagnostic process (Hanna & Ardon, 2023; Williams et al., 2019).

The disadvantage for these parallel, hybrid workflows is that institutions will incur the traditional costs associated with analog microscopy, while migrating to a fully DP operation. Costs for categories such as glass slide shipping, courier expense, and extra-departmental staffing will continue during this transition period and should be estimated and reflected when budgeting pathology operations.

Digital scanners cost

Designated whole slide scanners are necessary for digital workflow infrastructure. Commercial high-throughput scanner models cost multiple USD $100K+ depending on the model, its capacity, and geographic location (Lujan et al., 2021; Patel et al., 2021). For many years, the cost of the scanners seemed stagnant (Isse et al., 2012; Patel et al., 2021), yet starting in 2020, as with other electronic component devices, the global supply challenges resulted in increased cost of scanners (Ardon et al., 2023). Besides capital acquisition, there are other models of financing digital scanners such as instrument leasing or scanning as an onsite service that are now available, and others may emerge as the industry grows. Fraggetta et al. (2021) report that in Italy, it is customary to rent laboratory instruments rather than purchasing them. Their experience shows that an expensive slide scanner can be rented for approximately 5000 €/month (Fraggetta et al., 2021). Institutions may have restrictions on their instrument financing based on their location and source of funding.

Scanners have different slide capacities and throughput and varying technical and quality attributes (Hanna et al., 2020; Hanna et al., 2022; Patel et al., 2021). Each institution should identify the hardware and model that best suits its needs and available resources. Depending on the institutional practice with other laboratory equipment, scanners will assume an annual depreciation cost that should be accounted for in annual budgets. The cost of scanner ownership can be divided by the anticipated number of slides scanned on the instrument to calculate the contribution of the scanner hardware cost per slide. Using this approach, institutions can calculate the anticipated hardware cost contribution to the DP operation and compare between hardware solutions.

$$\text{Annual scanner ownership} = (\text{Depreciated scanner} + \text{Service})/(\text{Scan volumes}(n)) \quad (3.1)$$

As scanners have relatively high fixed costs, maximizing their usage based on slide scanning capacity will result in a reduction in the fixed cost per slide. Scanners can allow daily scanning volumes that exceed their marketed intended capacity after careful planning and increasing scanning operations time to more than one 8-hour scanning shift per day. Continuous load scanners offer the benefit of undisrupted operations with minimal or no supervision, potentially minimizing the need for additional staffing for the operation. Choosing a scanner with the appropriate capacity to a projected volume will allow maximizing its usage and decrease the cost of scanning per slide. Analysis of the hourly and daily anticipated departmental scan needs is a prerequisite for expanding the scanning operating schedule and maximizing the planned scanner capacity (Ardon et al., 2023).

Other scanner-related costs

Scanner service and maintenance

An additional cost that should be budgeted is the digital slide scanner annual maintenance contract. This is a component of the cost of scanner ownership and should be calculated based on vendor-supplied quotes. The cost of the service contracts varies between vendors, level of service, remote and onsite service visit schedule, and parts replacement response times.

Preventative maintenance

Scanners rely on sensitive optical as well as robotic components, and users need to make efforts to prevent failures and downtimes. New WSI users are often surprised by the amount of maintenance that scanners need in order to keep scanners operational, and fail to anticipate the difficulties caused by scanner outages (Mayall et al., 2022). Many of the failures observed in the WSI process are a result of the accumulation of mounting media, glass shards, label fragments inside the scanners, and could be minimized with proper timely cleaning of the instruments. Preventive maintenance that includes the cleaning of the affected instrument parts can dramatically reduce some common failures of the scanners, alleviating the need for unscheduled downtime and more efficient usage. This cost should be added to the digital scanning team staffing needs (Ardon et al., 2023).

Scanner downtime cost

The downtime that is associated with external service visits can be significant. Retrospective calculation of the cost of scanner downtime interruptions should include the actual time the scanner is not operational. The annualized cost of the scanner is used to estimate the monetary cost of the idle scanner. In addition to this cost, the effort of the institutional staff that is involved with the troubleshooting and repair of scanners should also be accounted for. The reliability of scanners should affect the procurement decisions of scanning operations. As the industry matures and newer high throughput scanners enter the market, the technologies and reliability of scanners are likely to improve (Ardon et al., 2023).

Research and development cost

All new laboratory equipment should be tested to ensure suitability for laboratory needs. Scanners should be evaluated, preferably on site prior to commitment to purchase in order to test their performance using laboratory-specific glass slides. These evaluations require shifting available resources, mainly staff time and space, but result in cost avoidance for the purchase of unsuitable equipment. At MSK, we estimated this effort to require an average of 6 weeks of a technician's time to install, train, scan, and review images. Additional time for a senior scientist and a pathologist is needed for each scanner evaluated and these costs should be factored in business plan discussions (Ardon et al., 2023).

Space allocation

Dedicated DP space is essential for coordinated workflows in the lab or in close proximity space to enable turnaround time, quality and ease of troubleshooting and maintenance (Eloy et al., 2021; Hanna et al., 2022). An overlooked laboratory space need that should be accounted for when transitioning to DP operations is for pre- and postscan activities. WSI operations require slide prep areas, workstations for image reviews, and multiple racks for staging folders of cases for different scanning needs. These areas should ideally be in close proximity to the scanning activities to maximize efficiency. Additional research scanning dedicated space is needed for separating between clinical and research/development activities. Most pathology operations are under pressure to minimize their physical space footprint, and the cost of pathology space should be accounted for when making budgetary projections for DP operations. Finally, glass slide storage expense can be minimized once departments adopt DP (Hanna et al., 2019; Lujan et al., 2021). This cost saving benefit of DP can be calculated and added to the financial projections when making future projections and developing a DP business plan.

Staffing

The digital scanning process involves a handoff between multiple teams, from accessioning to the histology lab, and slide file room (Ardon et al., 2021). Digital scanning staff is responsible for ensuring successful quality scanning as well as postscan review, and needs to be properly trained for specific workflows. Scanning team staff may oversee prospective scanning in the histology lab and/or handle educational and research requests from trainees, pathologists, and other stakeholders. In other institutions, dedicated staff may be tasked with prepping slides for retrospective scanning, including slides submitted from outside institutions that need manual relabeling and cleaning. An understanding of the analog pathology workflows and organizational structure is therefore a prerequisite to assess the gaps in staffing that need to be addressed once digital workflows are adopted. The daily tasks will vary depending on the use case and the technologies in use at the institution, and each center will need to estimate labor needs based on specific workflows and use cases, size of operations, hours of operations, etc.

In general, digital scan staff tasks include technical as well administrative, training, and regulatory tasks. Scanning staff may work in the histology lab area or in separate designated dry labs. Research and education scanning may involve additional de-identification steps that the clinical workflows do not require. Fig. 3.3 summarizes the digital scanning team time allocation for the various tasks of an average workday at MSK as an example of variety of tasks expected in a large digital scanning operation (Ardon et al., 2023). The manual preparation of slides for scanning is the single most time-consuming task for the department, accounting for 28% of staff time. The postscan quality control (QC) review is the second most time-consuming and we estimated a need of 1 full time equivalent (FTE) to review images from three to four whole slide scanner devices (Ardon et al., 2023). Most other tasks are manual and could benefit from future automation. Interestingly, the actual loading and unloading of the scanners takes less than 5% in an average day for the team. This time allocation distribution will vary with changes in pathology workflow automation, size of scanning operations, daily shift coverage, administrative support, as well as the type of scanners used, and departmental specific QC needs.

For example, the ratio of scanning team members (or scan techs) to scanners varies across different institutions and technologies. Lujan et al. estimated this ratio range to be between 0.3 to 1 FTE per scanner (Lujan et al., 2021).

Additional staff from other teams that support some aspects of the digital scanning operations include slide file room attendants and administrative staff. These tasks include providing slides for scanning, archival retrieval, and the return of slides after scanning.

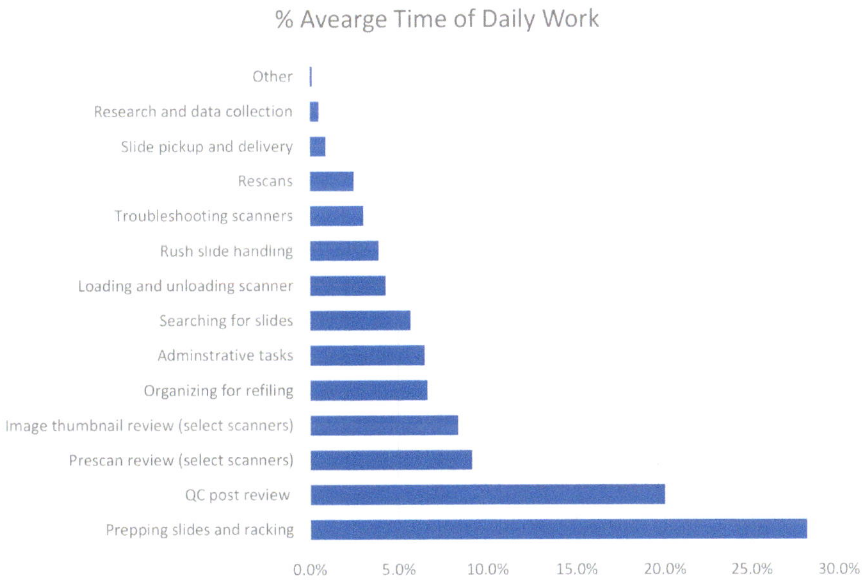

FIGURE 3.3

The percentage of digital scan staff time at Memorial Sloan Kettering Cancer Center allocated to different tasks on an average operation day.

Most laboratories calculate labor costs such as salaries and benefits as 60%–70% of total expense budget (MacMillan, 2003). These costs will vary monthly depending on the total scan volume and will fluctuate based on actual patient hospital visits, number of workdays in a month, holiday schedule, etc. Many of the WSI daily tasks identified above require manual handling of slides and equipment. These include physical glass slide searches, cleaning slides and slide transfer, scanner maintenance and troubleshooting. Other tasks, mainly pre- and postscan image review, are currently manual steps, but are attractive candidates for future automation development. Automation will increase the quality of the WSI workflows and offer cost savings by removing mundane tedious tasks (Boyd & Hawker, 2012).

Staffing cost may also include the need for 24-hour coverage for some operational needs, maximizing the use of the hardware and space by extending the operations and alleviating the need to obtain additional digital slide scanners and laboratory space. An indirect advantage for this continuous operation is the intangible job satisfaction for the staff who are offered a flexible schedule that fits their personal circumstances.

DP requires adequate staff to support the pathologist clinical digital workflow including slide viewing solutions, workstation support, laboratory information systems (LIS) integration support, networking, and new instrument onboarding processes. The staff includes individuals from pathology inofrmation technologies (IT) who directly support the scanning operations (Schüffler et al., 2021). These support staff members should be accounted for when crafting financial plans. Likewise, regulatory compliance staffing should also be accounted for when institutions transition DP to clinical systems. The establishment and validation of digital workflows to ensure quality clinical operations should be accounted for, as with any other new test and laboratory equipment. Additional costs may be associated with scanner validation and any related research and development work before initiating clinical scanning. The clinical digital transformation requires planning and verification/validation studies including writeup and submission for regulatory agencies per College of American Pathologists (CAP) guidelines (Hanna et al., 2020; Pantanowitz et al., 2013; Evans et al., 2021). This effort involves input from multiple pathologists, technicians, and regulatory personnel. The regulatory cost includes periodic testing of the scanners and establishing preventative maintenance schedules and training modules that require dedicated staff time. Additional time is required for pathologist glass and digital slide review as well as discordance analysis and documentation. The estimated pathologist and scanning team time will be dependent on the scale of the validation.

Servers and data storage

One of the largest costs associated with DP is the IT infrastructure that is necessary to move and store the images (Lujan et al., 2021). This infrastructure includes data transfer networks, data storage solutions, workstations, and peripheral equipment such as QR code scanners and label printers.

DP may require the preservation of a large number of digital images for an extended time depending on retention guidelines and hospital operations decisions (CAP Accreditation Program. Anatomic Pathology Checklist. College of American Pathologists, 2022). The estimated size of a 1 × 3″ DP image scanned in one z-plane is 2 GB. Sun et al. (2022) estimated that storage requirements in China, where there are as many as 50 million new pathological examinations every year, will require an annual 50–250 PB storage capacity. This storage cost will vary among institutions, as data storage considerations depend on institutional policy regarding the longevity of patient data. At MSK, where more than 6.3 MM slides have been scanned to date (July 2023), WSI are hosted on a centralized data share in the hospital's data center. The institution allocated a total of 9 PB for slide storage including similar redundant storage at an additional site that could be used for data recovery in case of a disaster (Schüffler et al., 2021). The cost of this storage infrastructure amounts to USD $1M per PB (Ardon et al., 2023). With the arrival of commercial cloud service providers, outsourcing hardware infrastructure and software operation to third parties has become a viable and attractive alternative (Juhasz, 2021). As digital slide data volume increases steadily, on-premises server storage could become difficult to sustain, while cloud storage offers on-demand storage solutions. The question of whether using the cloud is indeed a cost-effective solution is still debated, but pathology departments will need to understand the estimated monthly data usage to determine the cost of storing data locally versus in the cloud.

Digital pathology specific software solutions infrastructure

DP operations at scale require multiple software solutions to enable efficiency and maintain quality operations. Some digital slide scanners are marketed and distributed with minimal software, while other systems are bundled with elaborate software platforms to host and organize large image sets as well as manage many requirements of a digital workflow (Lujan et al., 2021; Patel et al., 2021). Specific software is used to manage, view, manipulate, and QC the digital images. As with other components of the DP system, the needs for software will be determined by the intended use cases in the institution.

Early adopters of DP had to identify and develop solutions to enable DP workflows as commercial tools were not widely available. At MSK, such software included a single vendor-agnostic digital slide viewer and a service linking clinical pathology data with research to enable the compilation of large-scale datasets for computational pathology (Schüffler et al., 2021). Other tools developed included a pathologist worklist, a scanning dashboard, and image QC review dashboard (Ardon et al., 2023).

These developments required investment in capital expenses that are part of the systems hardware and software infrastructure, as well as the ongoing needs for IT solutions to support new scanner technologies and operational needs, periodic scanner and network updates, and updated digital storage migration.

There are now a range of commercial software solutions that are offered by multiple vendors to manage, view, and manipulate digital images. Institutions should understand their preferred software needs and evaluate different commercial options to ensure integration of image management systems with their electronic medical record (EMR) or LIS. Understanding the capabilities and costs of the different solutions will lead to including these software costs in any DP investment discussion.

Overhead expenses

Additional expense that may be incurred by DP operations include space-associated expenses such as utilities and administrative salary allocation. These costs will vary among institutions and should be included in the financial analysis of the DP business case.

Total cost of infrastructure

Institutions that have existing DP operations can analyze the actual incurred costs of their DP operations and use that analysis to estimate their future costs. Such aggregated cost data from MSK was published with the estimated contribution of each cost category to the total cost of DP operations in the 2021 budget year (Ardon et al., 2021). In the study, annualized costs were grouped into six categories: scanner acquisition, related IT infrastructure, vendor scanner service, initial staffing for the infrastructure stages, IT staff, and digital scanning team (Fig. 3.4).

A review of the aggregated data of the 2021 operations costs demonstrates that the highest costs associated with our DP operations stem from infrastructure costs, those of IT hardware and software (29%), and the staff involved in setting the infrastructure (21%), while scanner acquisition contributed only 18% to the annualized costs of the 2021 operations. Additional ongoing costs include those of the digital scanning team staff (15%), IT staff (8%), and cost of scanner service agreements (9%).

Pinto et al. (2023) surveyed 127 anatomic pathology laboratories in Asia and Europe, regarding their clinical DP adoption and reported that a substantial number of respondents chose not to disclose their institution's expenses for DP implementation. Of those that did share, most (63%) spent USD $100,000–1,000,000, 22% institutions spent between USD $1,000,000 and $5,000,000, and 2% spent more than USD $5,000,000. Interestingly, most survey respondents expected a return on their investment either in the medium or long term (36%), while approximately one-fifth admitted they did not expect any returns (21%) (Pinto et al., 2023).

It is important to keep in mind that historic costs should only serve as a guide, as costs fluctuate in time, new technologies, market forces, and geographical locations. As with other specifics in this chapter, the total cost may vary greatly between two similar institutions with similar use cases and geographical locations.

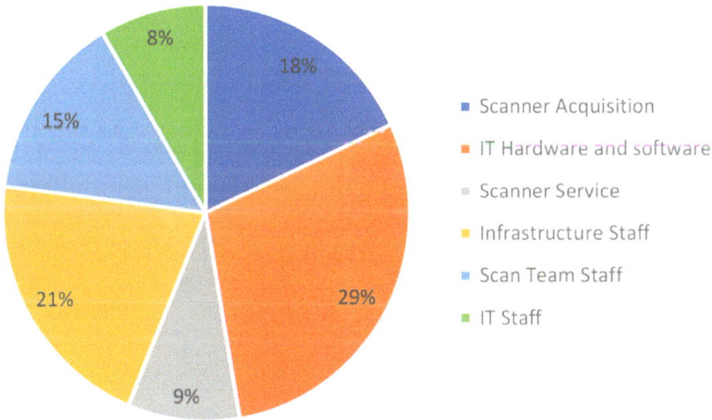

FIGURE 3.4

Pie chart representing percentage distribution of DP cost categories annualized for the year 2021 (Ardon et al., 2023).

Financial benefits of digital pathology

There are many direct quality improvements to patient care that are recognized by developing DP capabilities in organizations. These include the integration of pathology in patient digital records, adding to pathologist diagnostic accuracy by integrating computational pathology, improving turn-around time (TAT) for case signout, making world class pathology available to underserved populations, gaining ability to consult for rare pathologies, and enabling operations during emergency times.

The calculation of DP benefits should include the direct benefits and the indirect benefits that are associated with the migration to DP. These benefits will have different impact at the staff, pathologist, departmental, and institutional level. Fig. 3.5 summarizes identified direct benefits of DP, including additional revenue, cost savings, and improved patient care, as well as the indirect benefits, which include increased efficiency, emergency preparedness, recruitment, collaborations, and leadership in innovation. These are further divided between tangible and intangible benefits.

The indirect benefits include some unknown fiscal benefits. Tangible benefits are easier to account for than those of intangible benefits, which are nonphysical in nature. These are based on knowledge assets that are acquired through research and development, human capital, supply chain volume and product distribution systems, brands, software investments, and the organization of a business. These assets are increasing components of the total assets of modern firms (Barker et al., 2022). Despite the growth in importance, few or none of most organizations' intangibles appear as assets on balance sheets even though they affect an organization's value (Barker et al., 2022).

It is often hard to measure and quantify the value of intangible benefits of disruptive technologies, including those of DP (Ardon et al., 2023). These indirect benefits to institutions are sometimes overlooked yet should be included in the financial discussions of DP adoption and expansion. Examples of intangible benefits include items such as attracting innovative faculty and staff, adding job satisfaction, and fighting burnout among pathologists. The cost of advertising and interviewing faculty candidates, for example, is estimated at thousands of dollars per candidate and can get as high as three to four times the position's salary when including advertisement and search committee time commitment and interviews (Navarra, 2022). Offering work-life balance with DP-enabled remote work can help attract and retain talent, avoiding turnover among pathologists. In addition, there is the promise of potential revenue to the institution from innovative digital tools, increasing global presence, and patient outreach in underserved communities.

Academic institutions can realize the added intangible benefit of using WSI for education and research opportunities. The ability to provide state-of-the-art patient care cannot be easily quantified but is in line with the need for healthcare to innovate and provide quality care. This can also lead to quality faculty and trainee applicants who can further increase their department's value through leadership and productivity. More importantly, providing superior patient care may be challenging in a time when anticipated population aging, incidence of cancers,

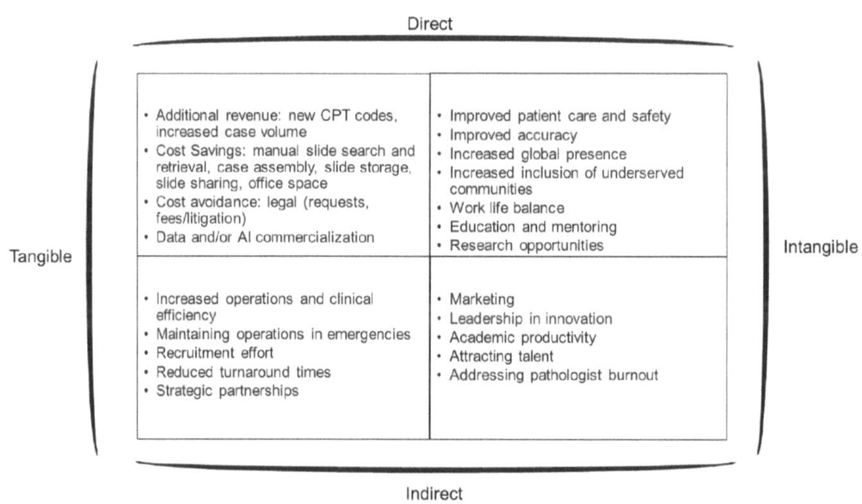

FIGURE 3.5

Quadrant graph displaying potential DP benefits. Direct benefits versus indirect ones, tangible benefits versus intangible ones.

and histopathology specimen volume are growing. There are numerous forecasts of pathologist shortages, and DP is one method to ameliorate this situation (Hanna et al., 2019; Jahn et al., 2020; Lujan et al., 2021). Departments that do not integrate digital solutions into their practice may find it difficult to provide state of the art services to their patients (i.e., telepathology, image analysis, virtual tumor boards).

Each institution must evaluate its list of DP benefits and prioritize those based on the institution type and mission before starting to analyze the associated costs.

Additional revenue

As with other new technologies, there is limited data regarding the projected revenues that are associated with DP implementation for pathology departments and healthcare systems.

It is anticipated that DP implementation will allow institutions to increase their pathology case volume.

DP offers the potential for increased consultation case revenue as digital images could be shared easily through expanded geographies and allow expert second opinions. International pathology consultations often result in patients traveling to the diagnosing institution to seek therapy, resulting in additional revenue for the healthcare system. Lujan at al. listed potential revenue from DP images including WSI-based image analysis, computational pathology, precision medicine, drug development, and clinical trials as well as big data (Lujan et al., 2021). Additional revenue can be derived from educational and training initiatives in international markets (Aeffner et al., 2018). Yet, despite an expected revenue, to date, there is limited data on real world revenue numbers that can be attributed to the adoption of DP.

A recent development that will contribute to additional revenue to healthcare systems adopting DP workflows in the United States is the new current procedural terminology (CPT) codes for DP, which allow billing for the additional time and effort required to digitize glass slides and provide remote consultations. Currently there is one (88,361) CPT code that is approved for computer-assisted immunohistochemistry (IHC) quantification (Billing, 2023). In the future, additional approvals yielding direct tangible benefits will include revenue generated from CPT codes now being added by the American Medical Association as CPT Category III code set, temporary codes for emerging technologies, services, procedures, and service paradigms (Category III Codes, 2023). These codes are effective since January 1, 2024. Their impact on departments that have implemented DP is currently unknown but should help in guiding future financial cost benefit discussions.

The exact impact and amount of revenue that a pathology department may expect to be reimbursed will vary based on the number of cases that are digitized and used for clinical diagnosis and the reimbursement rates set by payers. This future potential reimbursement offers an incentive for departments that have not yet begun their DP journey to invest in digital scanning technologies.

Cost avoidance and savings

WSI offers cost avoidance and cost savings that are derived from using WSI instead of glass slides (Griffin & Treanor, 2017; Hanna et al., 2019; Lujan et al., 2021; Mayall et al., 2022). These costs could be subtracted from the actual incurred DP costs when building a business case for DP. Most cost-efficiency analyses are estimated and derived from academic single-center experiences that cannot easily be generalized (Jahn et al., 2020).

Slide handling and archiving efficiencies can be achieved with a DP workflow (Lujan et al., 2021). Pathology departments store local glass slide in an onsite archive for recent cases, with older cases stored in an off-site archive for long-term storage (Hanna et al., 2019). In the United States, these storage requirements to "retain histopathology slides for at least 10 years from the date of examination" are a federal regulation (CFR 493.1105 – Standard: Retention requirements, 2022). Several steps in the analog pathology signout workflow are manual. These include sorting and matching glass slides to corresponding accessioned cases in the LIS, followed by manual delivery to pathologists.

With a digital workflow, glass slides can be digitized soon after slides are cover-slipped and dry. Compatible slide racks allow direct loading from the drying chamber to a digital slide scanner, avoiding the tedious step of slide collation and case assembly. Glass slides can get filed directly after scanning, avoiding the need for manual sorting and collating slides after pathologists' review. This task will be minimized when the digital transformation is complete. At MSK, laboratory aides were performing this task for all cases, with an estimate of 18 hours/day, resulting in an annual expense of USD $180K in 2021 (Ardon et al., 2023). That year, the slide requests from the slide file room were reduced to ~540/month with an estimated cost saving of ~ USD $142K per year.

Hanna et al. (2019) calculated that glass slide requests from the MSK slide archive and an off-site surgery center showed a 93% and 97% decrease. With this decrease in slide requests from the department slide library, three full-time employees from the slide file room were permanently re-trained and upskilled to join the DP operations workflow. That study projected a 5-year USD $1.3 million savings based on a cost analysis of a fully digital workflow that included off-site storage, slide retrieval costs, vendor labor, scanner hardware and software, WSI storage, and support staff (Hanna et al., 2019). Other cost savings identified are opportunities for space savings for multiheaded scope rooms and pathologists' offices situated close to the laboratory, courier costs, and legal case handling costs.

Retamero et al. reported that following full DP implementation at their Granada multicenter operations, redundant analog workflows were eliminated and resulted in histotechnician time savings, from 3 FTEs to less than 1 (Retamero et al., 2020).

In 2014 Ho et al. projected 5-year total cost savings of USD $12.38 million based on anticipated improvements in pathology productivity and histology lab consolidation (Ho et al., 2014). Another cost saving of $5.35 million was suggested due to improved diagnoses and avoidance of unnecessary treatment costs. More recent studies describe future savings based on cost-saving calculations (Baidoshvili et al., 2018; Hanna et al., 2019; Ho et al., 2014; Lujan et al., 2021). Examples include >19 working hours saved per day with digital slides in the Netherlands (Baidoshvili et al., 2018), annual savings of $CA 26,000 in courier costs, $CA 60,000 travel and $CA 45,000 in accommodation, meals, and car rental expense in Canada (Evans et al., 2022), and USD $24/courier trip between UCLA's pathology sites (Chong et al., 2019). Vodovnik reported that DP may lead to shorter diagnosis time than traditional microscopy and suggested the potential of DP to yield savings in both diagnostic and nondiagnostic tasks (Vodovnik, 2016). Other publications have highlighted additional benefits of DP transformation, yet some were not quantifiable and did not lead to any cost reduction, additional revenue, or cost avoidance (Ardon et al., 2021; Lujan et al., 2021; Yousif et al., 2022).

DP also allows multisite institutions to centralize their core laboratory that will receive specimens, process, and scan slides, and manage the distribution and storage of digital images. Centralizing the lab offers cost reductions in laboratory instrumentation and infrastructure, slide storage facilities, and elimination of redundant technician positions (Hamilton et al., 2014; Ho et al., 2014).

Hybrid analog and digital pathology workflows

A hybrid DP workflow is adopted in many institutions that are transitioning to digital workflows (Griffin & Treanor, 2017; Hanna et al., 2022). Our institution implemented prospective biopsy slide scanning in 2020 (Ardon et al., 2021; Hanna et al., 2020) and added additional prospective slide scanning as the scanning infrastructure capacity increased while maintaining all analog glass workflows. Other institutions reported transitioning to a fully digital workflow in a short period of time (Baidoshvili et al., 2023; Haghighi et al., 2021; Lujan et al., 2021; Retamero et al., 2020; Stathonikos et al., 2019). Pinto et al. conducted a survey of the status of DP implementation in Europe and Asia. Their findings show that for 51% of the institutions start by digitizing biopsies and/or cases from select subspecialities, and 60% of the institutions digitized cases for educational purposes and/or for discussion at tumor boards. Only 17% of the institutions immediately started digitizing all their cases after installation of the WSI. A total of 19% of the laboratories currently digitize all their histology routine, including IHC, and another 21% of the laboratories surveyed expected to reach this point within 2 years (Pinto et al., 2023).

Institutions that are transitioning to DP are incurring additional costs while using hybrid manual-DP workflows. In a phased DP adoption approach, there is an initial need for additional resources to support the digital workflow while maintaining some of the analog microscopy slide handling by multiple pathology teams. For example, tasks such as organizing, refiling, searching for glass slides, rush slide handling, slide pickup, and delivery amounted to a total of 17.2% of MSK digital scanning team time allocation. Other slide handling costs that could be avoided were identified in the slide file room, pathologist, office coordinators, and additional support staff (Ardon et al., 2023).

This type of hybrid operation will complicate the calculation of the DP costs for the department. Some of the benefits that result in increased costs during the hybrid period stem from the added digital workflow staff required for maintaining pathologist workflows and meeting turnaround times while adding digital scanning steps to the laboratory workflows. Slide transport costs, manual slide handling, and current hospital spaces that are used for manual workflows—such as multiheaded scope rooms—are still required in a hybrid workflow. Institutions should be prepared and attempt to minimize a lengthy transition from analog to DP workflows if possible. Table 3.1 outlines some of the tangible and intangible benefits of DP with their potential impact on pathologists, staff, departments, and the organization, as well as potential increases in costs when the institution adopts a phased hybrid workflow. Due to the offsetting benefits and costs, immediate cost savings may not be realized when constructing a financial analysis for a business plan. These direct labor costs should be factored in and reduced annually as institutions migrate to fully digital workflows. Certainly, this may vary based on the scope and size of DP operation, but working through this type of analysis will help an organization with financial planning for their proposed operation.

Creating a business plan

A business plan is a document that outlines the goals, strategies, and financial projections of a business. It is essential for clarifying the DP operations vision and objectives, and for outlining a blueprint on how to achieve the digital transformation. The DP business plan will be used to help guide internal decisions, plan budgets, measure progress and success, and obtain leadership support. The plan may also be used for attracting external funding from investors, for developing marketing and sales strategies, and for setting short- and long-term goals and objectives. Well-researched and written business plans can help departments avoid costly mistakes and increase their chances of successful DP implementation.

Writing any scientific business plan requires a combination of scientific knowledge and business acumen. For the DP business plan, it will be best done with teams of pathologists, laboratory staff, and finance team members. The business plan should include the identification of any potential funding sources and any capital expenditures that will be required for DP onboarding and implementation. Additionally, the business plan requires an operational plan outlining the required resources and personnel, including staffing, facilities, and equipment. The DP business plan should be written in a clear and concise manner, using scientific language where appropriate, and include relevant charts, graphs, and other aids to support the data presented.

Williams et al. outlined a DP business case structure that could be used as an example for other institutions (Williams et al., 2019). Institutions may have their own template requirements, but in general, a template will include:

1. An executive summary with an overview and key points.
2. Departmental/laboratory introduction describing current analog workflows and proposed DP alternative workflows and strategy. This should include benefits and risk of not investing in the technologies.
3. A market analysis describing DP adoption and market trends, with a description of the competition if available, including all tangible and intangible benefits of the plan.
4. Detail of the scope of institutional investment required, and the principal DP components of this investment. This analysis can include different DP use cases and configurations and an optional-phased approach analysis.
5. Annual financial projections following the DP investment and change in operations.
6. Summary that will include concise expected departmental and institutional outcomes and patient care improvements.

While writing the business plan, the team should identify potential risks and challenges that may impact the DP operation. This is also an opportunity for pathology departments to develop contingency plans to address and mitigate any risks. As in the case of other disruptive healthcare innovations, DP technology implementation could benefit from a phased approach that can help reduce the cost of implementation by breaking down the project into smaller phases (Ardon et al., 2023; Hanna et al., 2022; Williams et al., 2019). This phased, hybrid approach that was described earlier, can allow small-scale testing of specific DP use cases using specific scanning technology in a specific location, while being able to manage staff training and pilot new workflows. The phased approach can help mitigate risks associated with implementing new, expensive technologies in a regulated healthcare setting with limited funding resources. It can enable incremental progress and allow identification and addressing of potential problems associated with digital workflows. This may help achieve more favorable outcomes and gain support from pathology department stakeholders. A phased approach also allows greater flexibility in adapting to periodic changes in scanning technology, laboratory operations, or pathologist workflows, and aids in developing and customizing change management initiatives and addressing resistance to change. Phased assessment of interoperability between scanners, image

Table 3.1 Digital pathology tangible and intangible benefits to institutions, departments, pathologists, and staff.

Digital pathology benefit impact	Tangible benefits				Intangible benefits		Intermediate increase in cost with hybrid workflows
	Revenue	Cost saving	Cost avoidance	Increased productivity	Patient care improvement	Other intangible	
Staff							
Lost slide search avoidance			x	x	x		
Decreased slide requests from archive			x	x	x		
Decrease in manual case assembly			x	x			x
Decrease in administrative slide handling			x	x			x
Decrease in slide sorting			x	x			x
Decrease in legal requests handling		x	x		x		x
Pathologist							
Added automation/ decision support tools				x	x		x
Improved accuracy		x	x	x	x	x	
Improvement in diagnostic turnaround time				x	x		x
Potential for insourcing of external consultations	x				x	x	
Option for slide sharing	x	x		x	x		
Option for work sharing		x		x			

Better work Life balance			x	x		x	
Better work satisfaction					x	x	x
Department							
Additional revenue: new CPT codes	x				x		
Avoidance of inclement weather/emergencies delays				x	x		
Decreased recruitment cost		x	x				
Potential for attracting pathologists from additional areas		x				x	
Decrease in slide transport cost		x		x			x
Increase in academic publications					x	x	
Ability to attract trainees		x				x	
Ability to attract tech savvy workforce		x				x	
Decrease in burnout (pathologists and staff)			x			x	
Reduced multiheaded scope room need/cost		x					x
Institution							
Legal (settlements/fees/ litigation)			x			x	x
Office space savings		x					x

(Continued)

Table 3.1 Digital pathology tangible and intangible benefits to institutions, departments, pathologists, and staff. *Continued*

Digital pathology benefit impact	Tangible benefits				Intangible benefits		Intermediate increase in cost with hybrid workflows
	Revenue	Cost saving	Cost avoidance	Increased productivity	Patient care improvement	Other intangible	
Potential for commercializing digital tools	x						
Potential for AI commercialization	x						
Patient/ public facing marketing						x	
Leadership in innovation	x				x	x	
Strategic partnerships	x						
Increased global presence	x				x	x	
Increased inclusion of underserved communities					x	x	

Intermediate cost increase is anticipated for some of the benefits during the hybrid analog to fully digital workflows transition period.

viewers, LIS and other components of the DP tools may also be desirable. Phased adoption of DP offers many benefits to institutions, though as discussed earlier, a lengthy adoption period may not afford the same cost savings benefits of DP workflows (Ardon et al., 2021; Williams et al., 2019). The business plan can include proposals for several phased digital transformations of specific use cases, outlining the advantages and disadvantages of each plan, with the financial data to support each of these scenarios.

Digital pathology cost calculators

As detailed in the previous sections, the disruptive nature of DP workflows requires a thorough evaluation of associated costs to estimate resource allocation and facilitate informed decision-making. Unfortunately, not all costs can be easily identified or are available for institutions as they begin the DP exploration stage.

Given the lack of funding and cost data, institutions may use available departmental funding to initiate digital scanning initiatives before developing a more elaborate business plan for their growing DP operation. Stathonikos et al. used an available funding approach where they calculated a 5-year total budget as their hospital did not provide an extra budget for technological innovations. They then used the budget and relied on vendors' interest in supplying those given the list market prices for scanners and workflow software as part of a European tender process (Stathonikos et al., 2019).

DP cost calculators are valuable tools for pathology departments considering adopting DP. There are several DP cost calculators available on DP hardware and vendor websites. These calculators typically require users to input information about their scanning needs, the type of DP technologies under consideration, and the volume and revenue that they expect to receive. The calculator will then generate an estimate of the total cost of DP for the department.

Users should be aware that the calculators may not have the most updated information about the costs of DP and may be biased by the vendor hosting the website. It is recommended that the calculator be populated to account for the specific use cases and needs of the pathology department with realistic projections of slide volumes and potential revenues that the department may realize. By carefully considering all costs such as equipment, IT infrastructure, personnel, maintenance, and potential cost savings, in addition to all potential financial benefits, pathology departments will be to get realistic estimates for the cost of their proposed DP initiative. Recently, the Digital Pathology Association develped a calculator that is free of commercial bias (Ardon et al., 2024).

Understanding financial indicators

One of the essential components of a business plan is a financial indicator or metric that will allow institutions to make financial investment decisions in a technology or operations such as DP.

There are multiple financial metrics used to judge efficiency of an investment that are used for business plan discussions, and the choice of which financial term to use depends on the specific investment being considered and the institutional preference. The three most common terms are return on investment (ROI), net present value (NPV), and internal rate of return (IRR). Other terms used for investments include terms such as benefit cost ratio (BCR) and payback period (PBP) depending on institutional preference.

ROI is a widely used term which is essentially a measure of the profitability of an investment. It is calculated as a ratio of the net profit from an investment divided by the cost of the investment. For example, if an investment in digital scanner infrastructure costs USD $500,000 and generates USD $100,000 in profit, the ROI for the institution would be 20%. This ratio is sufficient if an investment is relatively short term.

NPV is the difference between the present value of cash inflows and the present value of cash outflows over a period of time. The present value is calculated by discounting future cash flows to their current value. For example, if a new scanner investment is expected to generate USD $500,000 in cash flows over 7 years of the scanner's use, the NPV would be the sum of the discounted cash flows for those years. NPV is a more appropriate measure of profitability for a long-term investment.

BCR is the ratio between the cost and benefit of the project for analysis as it is shown by the present value of benefit expected divided by present value of cost. This is a more comprehensive, long-term analysis of tangible and intangible benefits than the ROI calculation.

PBP is defined as the number of years required to recover the original cash investment in a project. It is the period of time at the end of which the investment has produced sufficient net revenue to recover its investment costs.

Internal rate of return (IRR) is another metric that identifies the discount rate that makes the net present value of an investment equal to zero. It is essentially the rate of return that an investment must earn in order to break even. IRR is a useful metric of profitability when the discount rate is uncertain.

ROI is the most widely used term in DP discussions (Lujan et al., 2021; Toolkit for Using the AHRQ Quality Indicators, 2017; Williams et al., 2019). It is traditionally used as a performance management tool for strategic goal setting and for meeting product quality specifications. In healthcare, ROI has been widely used to evaluate financial value of a program postimplementation (Thusini et al., 2022). These traditional definitions and uses of ROI are not disputed. However, as ROI moved from commerce to healthcare frontlines, it became more than a metric, and is considered a concept of returns or gains from an investment (Thusini et al., 2022). Price et al. argues that the focus on the ROI approach fails to recognize the impact of a diagnostic test result on the other stakeholders involved in delivering care to the patient across the care pathway. Any assessment of the return on investment from a diagnostic service should therefore be undertaken together with a value proposition established for the service (Price et al., 2020).

The DP operation requires investments in infrastructure, and there is an annual cost that will be associated with the operations. In the case of healthcare infrastructure building, when disrupting pathology practice—as in the case of DP—ROI profitability metric may be an insufficient measure of the benefits to patient care. The Agency for Health Research and Quality (2016) has published ROI guidance for use "when a hospital invests in a new program, quality improvement intervention, or technology." They suggest that it can be used as a planning tool or as an evaluation tool for service improvement projects, and thus it can be seen as an evaluation tool that can be applied more broadly to any aspect of a service, where investment in any resource can be evaluated against any outcome (Price et al., 2020; Toolkit for Using the AHRQ Quality Indicators, 2017).

The financial analysis for DP investment should evaluate both the costs and all the qualitative and quantitative benefits to patient care. As DP has many benefits to institutions, pathology departments, pathologists, and patients, some of these benefits are not tangible and therefore cannot be easily used to calculate an ROI (Fig. 3.5). A more accurate approach should be the calculation of a benefit-cost ratio, which takes into account all the tangible as well as intangible costs and benefits of the investment in the DP-specific program.

Discussion

Pathology has always been a medical specialty at the front lines of automation (e.g., electronic reporting, speech recognition, image analysis, structured reporting). Due to lack of affordable and fast scanners and infrastructure to handle large image files, pathology lagged several decades behind radiology in adopting a digital workflow (Flach et al., 2022). In the past few years, the field has seen greater DP adoption, along with studies describing pandemic-induced adoption, highlighting the benefits of DP for continuous pathology operations in public health emergencies (Ardon et al., 2021; Lujan et al., 2021; Williams et al., 2020; Yousif et al., 2022).

The high cost of digital infrastructure and resources is still cited as a barrier for entry into DP implementation; yet, there have only been limited studies that assessed the cost of DP at scale (Ardon et al., 2023; Baidoshvili et al., 2018; Hanna et al., 2020, 2019; Ho et al., 2014; Retamero et al., 2020). Most cost-efficiency analyses are estimated and derived from academic single-center experiences that cannot easily be generalized (Jahn et al., 2020). A recent publication from our group used 2021 historical costs and metrics of the MSK DP operations to allow perspective of the magnitude of the expense, though they may not accurately translate to other institutions' use cases (Ardon et al., 2023). The study identified the essential investment in IT hardware and software as the largest cost category, accounting for 33% of the total DP annual budget, followed by the cost of scanner acquisitions and service (31%). Regarding the benefits of DP, there have been several publications that looked at the value-added benefits of DP other than those of cost savings, including in the areas of patient care, education, institution reputation, and research opportunities. While there is a growing acceptance that WSI has advantages to glass slides, nevertheless, the cost savings are hard to demonstrate.

To ensure successful integration of DP and avoid costly failures or unnecessary investments, much planning and consideration should be given to institutional DP needs and added-value assessment for the specific institution. Additional consideration should be given to the departmental DP readiness, including stakeholder acceptance and engagement. Planning a phased DP implementation would be preferred for some institutions that are still not fully ready for DP disruption. This should include additional IT resource assessment and any involvement from teams outside of pathology. The business plan should include—as with adoption of any disruptive digital technology—a detailed cost-benefit analysis of DP migration from analog workflows. An attempt should be made to start with small scanning operations while planning expansions as needed and adding more technologies and staff when appropriate. This slow adoption and investment in digital technologies will avoid the large investment in unused instruments that take up laboratory space and require expensive service agreements. Slower adoption also allows the institution to ramp up scanning operations and explore the best newly available technologies for its specific scan needs and obtaining synergies between scanning priorities to the pathology most urgent use cases.

Most hospital management focuses primarily on short-term expenses and on the "profitable market," rather than investing in disruptive innovation and technology, which almost always is more costly in the beginning compared to incremental evolutionary technology (Cadamuro, 2023). An example of such healthcare innovation is the adoption of EMRs that allows end-users centralized access to patient data for decision making. The investment in EMRs and related technologies has primarily been driven by their potential to improve patient safety and process efficiency (Black et al., 2011). However, the initial cost of the EMRs was cited as the most common barrier for adoption (Kruse et al., 2016; White et al., 2023). The promise of the availability of patient WSI data offers a similar healthcare efficiency and patient safety improvement, and along all the other benefits of DP should help guide stakeholders to future DP investments.

References

Aeffner, F., Blanchard, T. W., Keel, M. K., & Williams, B. H. (2018). Whole-slide imaging: The future is here. *Veterinary Pathology*, *55*(4), 488–489. Available from https://doi.org/10.1177/0300985818771146, http://vet.sagepub.com/content/by/year.

Ardon, O., Asa, S. L., Lloyd, M. C., Lujan, G., Parwani, A., Santa-Rosario, J. C., Van Meter, B., Samboy, J., Pirain, D., Blakely, S., & Hanna, M. G. (2024). Understanding the financial aspects of digital pathology: A dynamic customizable return on investment calculator for informed decision-making. *Journal of pathology informatics*, *15*, 100376. Available from https://doi.org/10.1016/j.jpi.2024.100376.

Ardon, O., Klein, E., Manzo, A., Corsale, L., England, C., Mazzella, A., Geneslaw, L., Philip, J., Ntiamoah, P., Wright, J., Sirintrapun, S. J., Lin, O., Elenitoba-Johnson, K., Reuter, V. E., Hameed, M. R., & Hanna, M. G. (2023). Digital pathology operations at a tertiary cancer center: Infrastructure requirements and operational cost. *Journal of Pathology Informatics*, *14*, 100318. Available from https://doi.org/10.1016/j jpi.2023.100318.

Ardon, O., Reuter, V. E., Hameed, M., Corsale, L., Manzo, A., Sirintrapun, S. J., Ntiamoah, P., Stamelos, E., Schueffler, P. J., England, C., Klimstra, D. S., & Hanna, M. G. (2021). Digital pathology operations at an NYC tertiary cancer center during the first 4 months of COVID-19 pandemic response. *Academic Pathology*, *8*. Available from https://doi.org/10.1177/23742895211010276.

Baidoshvili, A., Khacheishvili, M., van der Laak, J. A. W. M., & van Diest, P. J. (2023). A whole-slide imaging based workflow reduces the reading time of pathologists. *Pathology International*, *73*(3), 127–134. Available from https://doi.org/10.1111/pin.13309, http://onlinelibrary.wiley.com/journal/10.1111/(ISSN)1440-1827.

Baidoshvili, A., Bucur, A., van Leeuwen, J., van der Laak, J., Kluin, P., & van Diest, P. J. (2018). Evaluating the benefits of digital pathology implementation: Time savings in laboratory logistics. *Histopathology*, *73*(5), 784–794. Available from https://doi.org/10.1111/his.13691, http://onlinelibrary.wiley.com/journal/10.1111/(ISSN)1365-2559.

Barker, R., Lennard, A., Penman, S., & Teixeira, A. (2022). Accounting for intangible assets: Suggested solutions. *Accounting and Business Research*, *52*(6), 601–630. Available from https://doi.org/10.1080/00014788.2021.1938963.

Billing Coding, MolDX: Immunohistochemistry (IHC) Indications for Breast Pathology (A54271). Accessed 2023.

Black, A. D., Car, J., Pagliari, C., Anandan, C., Cresswell, K., Bokun, T., McKinstry, B., Procter, R., Majeed, A., Sheikh, A., & Djulbegovic, B. (2011). The impact of eHealth on the quality and safety of health care: A systematic overview. *PLoS Medicine*, *8*(1), e1000387. Available from https://doi.org/10.1371/journal.pmed.1000387.

Boyd, J. C., & Hawker, C. D. (2012). *Automation in the clinical laboratory* (pp. 469–485). Elsevier BV. Available from http://doi.org/10.1016/b978-1-4160-6164-9.00019-6.

Cadamuro, J. (2023). Disruption vs. evolution in laboratory medicine. Current challenges and possible strategies, making laboratories and the laboratory specialist profession fit for the future. *Clinical Chemistry and Laboratory Medicine*, *61*(4), 558–566. Available from https://doi.org/10.1515/cclm-2022-0620.

CAP Accreditation Program. Anatomic Pathology Checklist. College of American Pathologists. (2022). Unpublished content.

Category III Codes. (2023). American Medical Association. Accessed 7/5/2023. https://www.ama-assn.org/practice-management/cpt/category-iii-codes.

CFR 493.1105 – Standard: Retention requirements. (2022). https://www.ecfr.gov/current/title-42/chapter-IV/subchapter-G/part-493/subpart-J/section-493.1105.

Chong, T., Palma-Diaz, M. F., Fisher, C., Gui, D., Ostrzega, N. L., Sempa, G., Sisk, A. E., Valasek, M., Wang, B. Y., Zuckerman, J., Khacherian, C., Binder, S., & Wallace, W. D. (2019). The California telepathology service: 'UCLA's experience in deploying a regional digital pathology subspecialty consultation network. *Journal of Pathology Informatics*, *10*(1), 31. Available from https://doi.org/10.4103/jpi.jpi_22_19.

Eloy, C., Vale, J., Curado, M., Polónia, A., Campelos, S., Caramelo, A., Sousa, R., & Sobrinho-Simões, M. (2021). Digital pathology workflow implementation at IPATIMUP. *Diagnostics*, *11*(11), 2111. Available from https://doi.org/10.3390/diagnostics11112111.

Evans, A. J., Brown, R. W., Bui, M. M., Chlipala, E. A., Lacchetti, C., Milner, D. A., Pantanowitz, L., Parwani, A. V., Reid, K., Riben, M. W., Reuter, V. E., Stephens, L., Stewart, R. L., & Thomas, N. E. (2022). Validating whole slide imaging systems for diagnostic purposes in pathology guideline update from the College of American Pathologists in collaboration with the American Society for Clinical Pathology and the Association for Pathology Informatics. *Archives of Pathology and Laboratory Medicine*, *146*(3), 440–450. Available from https://doi.org/10.5858/arpa.2020-0723-CP.

Flach, R. N., Fransen, N. L., Sonnen, A. F. P., Nguyen, T. Q., Breimer, G. E., Veta, M., Stathonikos, N., van Dooijeweert, C., & van Diest, P. J. (2022). Implementation of Artificial Intelligence in Diagnostic Practice as a Next Step after Going Digital: The UMC Utrecht Perspective. *Diagnostics*, *12*, 1042. Available from https://doi.org/10.3390/diagnostics12051042.

Fraggetta, F., Caputo, A., Guglielmino, R., Pellegrino, M. G., Runza, G., & L'imperio, V. (2021). A survival guide for the rapid transition to a fully digital workflow: The "caltagirone example". *MDPI, Italy Diagnostics*, *11*(10). Available from https://doi.org/10.3390/diagnostics11101916, https://www.mdpi.com/2075-4418/11/10/1916/pdf.

Griffin, J., & Treanor, D. (2017). Digital pathology in clinical use: Where are we now and what is holding us back? *Histopathology*, *70*(1), 134–145. Available from https://doi.org/10.1111/his.12993.

Haghighi, M., Tolley, J., Schito, A. N., Kwan, R., Garcia, C., Prince, S., Harpaz, N., Thung, S. N., Craven, C. K., Cordon-Cardo, C., & Westra, W. H. (2021). Whole slide imaging for teleconsultation: The Mount Sinai Hospital, Labcorp Dianon, and Philips collaborative experience. *Journal of Pathology Informatics*, *12*(1), 53. Available from https://doi.org/10.4103/jpi.jpi_74_21.

Hamilton, P. W., Bankhead, P., Wang, Y., Hutchinson, R., Kieran, D., McArt, D. G., James, J., & Salto-Tellez, M. (2014). Digital pathology and image analysis in tissue biomarker research. *Methods*, *70*(1), 59–73. Available from https://doi.org/10.1016/j.ymeth.2014.06.015, http://www.elsevier.com/inca/publications/store/6/2/2/9/1/4/index.htt.

Hanna, M. G., & Ardon, O. (2023). Digital pathology systems enabling quality patient care. *Genes, Chromosomes and Cancer*. Available from https://doi.org/10.1002/gcc.23192.

Hanna, M. G., Reuter, V. E., Samboy, J., England, C., Corsale, L., Fine, S. W., Agaram, N. P., Stamelos, E., Yagi, Y., Hameed, M., Klimstra, D. S., & Joseph Sirintrapun, S. (2019). Implementation of digital pathology offers clinical and operational increase in efficiency and cost savings. *Archives of Pathology and Laboratory Medicine*, *143*(12), 1545–1555. Available from https://doi.org/10.5858/arpa.2018-0514-OA, https://www.archivesofpathology.org/doi/pdf/10.5858/arpa.2018-0514-OA.

Hanna, M. G., Ardon, O., Reuter, V. E., Sirintrapun, S. J., England, C., Klimstra, D. S., & Hameed, M. R. (2022). Integrating digital pathology into clinical practice. *Modern Pathology*, *35*(2), 152–164. Available from https://doi.org/10.1038/s41379-021-00929-0, http://www.nature.com/modpathol/index.html.

Hanna, M. G., Reuter, V. E., Ardon, O., Kim, D., Sirintrapun, S. J., Schüffler, P. J., Busam, K. J., Sauter, J. L., Brogi, E., Tan, L. K., Xu, B., Bale, T., Agaram, N. P., Tang, L. H., Ellenson, L. H., Philip, J., Corsale, L., Stamelos, E., Friedlander, M. A., ... Hameed, M. (2020). Validation of a digital pathology system including remote review during the COVID-19 pandemic. *Modern Pathology*, *33*(11), 2115–2127. Available from https://doi.org/10.1038/s41379-020-0601-5, http://www.nature.com/modpathol/index.html.

Hanna, M. G., Parwani, A., & Sirintrapun, S. J. (2020). Whole slide imaging: Technology and applications. *Advances in Anatomic Pathology*, *27*(4), 251–259. Available from https://doi.org/10.1097/PAP.0000000000000273, http://www.anatomicpathology.com.

Henricks, W.H. (2010). Not so fast? Concrete considerations for digital move. http://www.captodayonline.com/Archives/0110/0110ha_not_so_fast.html

Ho, J., Ahlers, S. M., Stratman, C., Aridor, O., Pantanowitz, L., Fine, J. L., Kuzmishin, J. A., Montalto, M. C., & Parwani, A. V. (2014). Can digital pathology result in cost savings? A financial projection for digital pathology implementation at a large integrated health care organization. *Journal of Pathology Informatics*, *5*(1), 33. Available from https://doi.org/10.4103/2153-3539.139714.

Isse, K., Lesniak, A., Grama, K., Roysam, B., Minervini, M. I., & Demetris, A. J. (2012). Digital transplantation pathology: Combining whole slide imaging, multiplex staining and automated image analysis. *American Journal of Transplantation*, *12*(1), 27–37. Available from https://doi.org/10.1111/j.1600-6143.2011.03797.x.

Jahn, S. W., Plass, M., & Moinfar, F. (2020). Digital pathology: Advantages, limitations and emerging perspectives. *Journal of Clinical Medicine*, *9*(11), 3697. Available from https://doi.org/10.3390/jcm9113697.

Juhasz, Z. (2021). Quantitative cost comparison of on-premise and cloud infrastructure based EEG data processing. *Cluster Computing*, *24*(2), 625–641. Available from https://doi.org/10.1007/s10586-020-03141-y.

Kruse, C. S., Kristof, C., Jones, B., Mitchell, E., & Martinez, A. (2016). Barriers to electronic health record adoption: A systematic literature review. *Journal of Medical Systems*, *40*(12). Available from https://doi.org/10.1007/s10916-016-0628-9, http://www.wkap.nl/journalhome.htm/0148-5598.

Lujan, G. M., Savage, J., Shana'ah, A., Yearsley, M., Thomas, D., Allenby, P., Otero, J., Limbach, A. L., Cui, X., Scarl, R. T., Hardy, T., Sheldon, J., Plaza, J. A., Whitaker, B., Frankel, W., Parwani, A. V., & Li, Z. (2021). Digital pathology initiatives and experience of a large academic institution during the coronavirus disease 2019 (COVID-19) pandemic. *Archives of Pathology and Laboratory Medicine*, *145*(9), 1051–1061. Available from https://doi.org/10.5858/arpa.2020-0715-SA, https://allen.silverchair-cdn.com.

Lujan, G., Quigley, J. C., & Hartman. (2021). Dissecting the business case for adoption and implementation of digital pathology: A white paper from the digital pathology association. *Journal of Pathology Informatics*, *12*17. Available from https://doi.org/10.4103/jpi.jpi_67_20.

MacMillan, D. H. (2003). Elements of a typical laboratory budget. *Laboratory Medicine*, *34*(7), 515–519. Available from https://doi.org/10.1309/99F7NNJ09M3TGQHW, http://labmed.oxfordjournals.org/content/by/year.

Marletta, S., Pantanowitz, L., Santonicco, N., Caputo, A., Bragantini, E., Brunelli, M., Girolami, I., & Eccher, A. (2023). Application of digital imaging and artificial intelligence to pathology of the placenta. *Pediatric and Developmental Pathology*, *26* (1), 5–12. Available from https://doi.org/10.1177/10935266221137953.

Mayall, F., Smethurst, H. B., Semkin, L., Mandalia, T., Sohail, M., Hadden, R., & Biddlestone, L. (2022). A feasibility study of multisite networked digital pathology reporting in England. *Journal of Pathology Informatics*, *13*(1), 4. Available from https://doi.org/10.4103/jpi.jpi_61_21, http://www.jpathinformatics.org/browse.asp?sabs = n.

Montezuma, D., Monteiro, A., Fraga, J., Ribeiro, L., Gonçalves, S., Tavares, A., Monteiro, J., & Macedo-Pinto, I. (2022). Digital pathology implementation in private practice: Specific challenges and opportunities. *Diagnostics*, *12*(2), 529. Available from https://doi.org/10.3390/diagnostics12020529.

Navarra, K. (2022). The real costs of recruitment. SHRM. https://www.shrm.org/resourcesandtools/hr-topics/talent-acquisition/pages/the-real-costs-of-recruitment.aspx.

Pantanowitz, L., Sinard, J. H., Henricks, W. H., Fatheree, L. A., Carter, A. B., Contis, L., Beckwith, B. A., Evans, A. J., Otis, C. N., Lal, A., & Parwani, A. V. (2013). Validating whole slide imaging for diagnostic purposes in pathology: Guideline from the College of American Pathologists Pathology and Laboratory Quality Center. *Archives of Pathology and Laboratory Medicine*, *137*(12), 1710–1722. Available from https://doi.org/10.5858/arpa.2013-0093-CPUnitedStates, http://www.archivesofpathology.org/doi/pdf/10.5858/arpa.2013-0093-CP.

Patel, A., Balis, U. G. J., Cheng, J., Li, Z., Lujan, G., McClintock, D. S., Pantanowitz, L., & Parwani, A. (2021). Contemporary whole slide imaging devices and their applications within the modern pathology department: A selected hardware review. *Journal of Pathology Informatics*, *12*(1), 50. Available from https://doi.org/10.4103/jpi.jpi_66_21.

Pinto, D. G., Bychkov, A., Tsuyama, N., Fukuoka, J., & Eloy, C. (2023). Exploring the adoption of digital pathology in clinical settings - insights from a cross-continent study. *medRxiv*. Available from https://doi.org/10.1101/2023.04.03.23288066, https://www.medrxiv.org/.

Price, C. P., McGinley, P., & John, A. S. (2020). What is the return on investment for laboratory medicine? the antidote to silo budgeting in diagnostics. *British Journal of Health Care Management*, *26*(6). Available from https://doi.org/10.12968/bjhc.2019.0075. Available from: http://www.magonlinelibrary.com/loi/bjhc.

Retamero, J. A., Aneiros-Fernandez, J., & del Moral, R. G. (2020). Complete digital pathology for routine histopathology diagnosis in a multicenter hospital network. *Archives of Pathology and Laboratory Medicine*, *144*(2), 221–228. Available from https://doi.org/10.5858/arpa.2018-0541-OA, https://www.archivesofpathology.org/doi/pdf/10.5858/arpa.2018-0541-OA.

Schüffler, P. J., Geneslaw, L., Yarlagadda, D. V. K., Hanna, M. G., Samboy, J., Stamelos, E., Vanderbilt, C., Philip, J., Jean, M. H., Corsale, L., Manzo, A., Paramasivam, N. H. G., Ziegler, J. S., Gao, J., Perin, J. C., Kim, Y. S., Bhanot, U. K., Roehrl, M. H. A., Ardon, O., … Fuchs, T. J. (2021). Integrated digital pathology at scale: A solution for clinical diagnostics and cancer research at a large academic medical center. *Journal of the American Medical Informatics Association*, *28*(9), 1874–1884. Available from https://doi.org/10.1093/jamia/ocab085, http://jamia.oxfordjournals.org/content/22/e1.

Stathonikos, N., Nguyen, T. Q., Spoto, C. P., Verdaasdonk, M. A. M., & van Diest, P. J. (2019). Being fully digital: perspective of a Dutch academic pathology laboratory.

Histopathology, *75*(5), 621–635. Available from https://doi.org/10.1111/his.13953. Available from: http://onlinelibrary.wiley.com/journal/10.1111/(ISSN)1365-2559.

Sun, K., Gao, Y., Xie, T., Wang, X., Yang, Q., Chen, L., Wang, K., & Yu, G. (2022). A low-cost pathological image digitalization method based on 5 times magnification scanning. *Quantitative Imaging in Medicine and Surgery*, *12*(5), 2813–2829. Available from https://doi.org/10.21037/qims-21-749.

Thusini, S., Milenova, M., Nahabedian, N., Grey, B., Soukup, T., Chua, K.-C., & Henderson, C. (2022). The development of the concept of return-on-investment from large-scale quality improvement programmes in healthcare: an integrative systematic literature review. *BMC Health Services Research*, *22*(1). Available from https://doi.org/10.1186/s12913-022-08832-3.

Tizhoosh, H. R., & Pantanowitz, L. (2018). Artificial intelligence and digital pathology: Challenges and opportunities. *Journal of Pathology Informatics*, *9*(1), 38. Available from https://doi.org/10.4103/jpi.jpi_53_18.

Toolkit for Using the AHRQ Quality Indicators. (2017). https://www.ahrq.gov/patient-safety/settings/hospital/resource/qitool/index.html

Turnquist, C., Roberts-Gant, S., Hemsworth, H., White, K., Browning, L., Rees, G., Roskell, D., & Verrill, C. (2019). On the edge of a digital pathology transformation: Views from a cellular pathology laboratory focus group. *Journal of Pathology Informatics*, *10*(1), 37. Available from https://doi.org/10.4103/jpi.jpi_38_19.

Vodovnik, A. (2016). Diagnostic time in digital pathology: A comparative study on 400 cases. *Journal of Pathology Informatics*, *7*(1), 4. Available from https://doi.org/10.4103/2153-3539.175377.

White, N. M., Carter, H. E., Kularatna, S., Borg, D. N., Brain, D. C., Tariq, A., Abell, B., Blythe, R., & McPhail, S. M. (2023). Evaluating the costs and consequences of computerized clinical decision support systems in hospitals: A scoping review and recommendations for future practice. *Journal of the American Medical Informatics Association*, *30*(6), 1205–1218. Available from https://doi.org/10.1093/jamia/ocad040.

Williams, B. J., Lee, J., Oien, K. A., & Treanor, D. (2018). Digital pathology access and usage in the UK: Results from a national survey on behalf of the National Cancer Research 'Institute's CM-Path initiative. *Journal of Clinical Pathology*, *71*(5), 463–466. Available from https://doi.org/10.1136/jclinpath-2017-204808, http://jcp.bmj.com/.

Williams, B. J., Bottoms, D., Clark, D., & Treanor, D. (2019). Future-proofing pathology part 2: Building a business case for digital pathology. *Journal of Clinical Pathology*, *72*(3), 198–205. Available from https://doi.org/10.1136/jclinpath-2017-204926, http://jcp.bmj.com/.

Williams, B. J., Williams, B. J., Treanor, D., & Treanor, D. (2020). Practical guide to training and validation for primary diagnosis with digital pathology. *Journal of Clinical Pathology*, *73*(7), 418–422. Available from https://doi.org/10.1136/jclinpath-2019-206319, http://jcp.bmj.com/.

Williams, B. J., Bottoms, D., & Treanor, D. (2017). Future-proofing pathology: The case for clinical adoption of digital pathology. *Journal of Clinical Pathology*, *70*(12), 1010–1018. Available from https://doi.org/10.1136/jclinpath-2017-204644, http://jcp.bmj.com/.

Yousif, M., Pantanowitz, L., & Hassell, L. (2022). Impact of COVID-19 on the adoption of digital pathology. *Digital innovation for healthcare in COVID-19 pandemic: Strategies and solutions* (pp. 95–107). United States: Elsevier. Available from https://www.sciencedirect.com/book/9780128213186.

Zarella, M. D., Bowman;, D., Aeffner, F., Farahani, N., Xthona;, A., Absar, S. F., Parwani, A., Bui, M., & Hartman, D. J. (2018). A practical guide to whole slide imaging: A white paper from the Digital Pathology Association. *Archives of Pathology and Laboratory Medicine*, *143*(2), 222–234. Available from https://doi.org/10.5858/arpa.2018-0343-ra.

The Agency for Health Research and Quality (2016) Toolkit for Using the AHRQ Quality Indicators. https://www.ahrq.gov/patient-safety/settings/hospital/resource/qitool/index.html Accessed 6/24/2023

Evans, A.J., Brown, R.W., & Bui, M.M., et al. (2021). Validating whole slide imaging systems for diagnostic purposes in pathology: Guideline update from the College of American Pathologists in collaboration with the American Society for Clinical Pathology and the Association for Pathology Informatics. *Archives of Pathology and Laboratory Medicine*. https://doi.org/10.5858/arpa.2020-0723-CP Published online May 18.

CHAPTER

Preanalytic factors in digital pathology implementation

4

Peter Ntiamoah, George Ayob, David Beckles, Rene Serrette and Grecia Castro
Department of Pathology and Laboratory Medicine, Memorial Sloan Kettering Cancer Center, New York, NY, United States

Introduction

Histology provides the fundamental basis for digital pathology so in this chapter, we start with a basic introduction to histology, looking at specimen receipts, tissue handling, and processing, and the potential impact of histological preparation on whole slide imaging. The National Society of Histotechnology (NSH) defines histotechnology, which consists of histology and histopathology, as a science that centers on the microscopic detection of tissue abnormalities for disease diagnosis and the treatment of diseases (NSH, 2001). Histology is the microscopic study of cellular morphology, anatomy, and architecture of normal tissues. Histopathology, on the other hand, is a branch of histology that studies human and animal tissues that are affected by disease. Histotechnology utilizes specific reagents, dyes, and chemical reactions to prepare surgical specimens for microscopic review.

Some of the specimens such as fine needle aspirates, smears, or body fluids that are received in histology laboratories are meant for cytopathology studies. Other specimens that include biopsies, surgical resections, and autopsy tissues are sent for histopathologic examination and analysis (Cotter & Loda, 2017). In addition to human tissues, histology laboratories may also receive medical implants for gross examination or evaluate the immunological process around the implanted device. Such implants or hardware may not require any further processing, after their initial gross examination. Histotechnology involves five basic steps—specimen examination and fixation, tissue processing, embedding, tissue sectioning, and staining to prepare tissues to be viewed under a microscope (NSH, 2001). Histotechnologists combine their training and acquire skills with their knowledge of biology, anatomy, basic science, and chemical, immunological, or molecular biology to produce stained slides for pathological review.

Digital Pathology. DOI: https://doi.org/10.1016/B978-0-443-13809-6.00006-3

Basic histology workflow

Accession

Accession is a crucial aspect of the histological workflow as each specimen received in the histology laboratory is given a unique identifier (accession number). For laboratories with well-established laboratory information systems (LIS), the accession number is often generated once the patient's demographic information is entered or imported into the LIS and linked to the received specimen. Despite the downstream checkpoints that could detect accession errors, medical errors in specimen defects, inaccurate specimen labeling, wrong demographic information, or missing patient identifiers that are not caught at the point of accession may perpetuate throughout the subsequent laboratory processes, and would not be identified (Layfield & Anderson, 2010). A formal error detection plan is highly recommended at the point of specimen receipt and accession to help minimize surgical pathology specimen identification errors and missing crucial clinical history that could affect the completeness and accuracy of pathology reports (Nakhleh, 2006, 2008; Nakhleh & Zarbo, 1996). Most accrediting agencies such as the College of American Pathologists (CAP) and The Joint Commission place significant emphasis on patient safety and the implementation of the applicable cardinal standards starts at accession.

Instead of automatically assigning the accession number in the LIS on demand, some laboratories apply preprinted or prepurchased accession numbers to the requisition, upon receipt, and link that preprinted number to the case. Laboratories without any LIS use logbooks and manually assign the accession numbers which is prone to preanalytic and diagnostic errors (Alenazi & Bugis, 2022). Both preprinted or prepurchased accession labels and on-demand accession labels may be bar-coded to facilitate downstream laboratory processes such as specimen tracking, standardization, quality management, and laboratory automation (Hanna & Pantanowitz, 2015). Thus the introduction of barcoded labels during accessioning does not only prevent manual labeling, it reduces patient specimen and laboratory testing identification errors, and improves efficiency, it also allows easy adoption of future technologies in histopathology laboratories (Hanna et al., 2022; Snyder et al., 2012; Yu et al., 2019). While the accession numbers serve as the unique identifier for each received specimen, the implementation of barcoded labels on the specimen containers and patient requisition forms at the time of accessioning reduces the error rates and provides efficient workflow in the analytic and postanalytic phases (Snyder et al., 2012; Yu et al., 2019).

Specimen grossing and fixation

Histopathological specimens—biopsies, routine surgical specimens, and autopsies received for clinical, or research evaluations are measured, weighed (where applicable), described, and adequately sampled during grossing by well-qualified personnel, under the supervision of a qualified pathologist. Qualified pathologists or pathologist trainees (residents and fellows) could also do the grossing of most surgical specimens. Some specimens are received fresh and others are received in one of the four chemical fixatives—precipitating agents (alcohols and ketones), cross-linking agents (aldehydes), oxidizing agents, or metallic groups (Jalali et al., 2023). Fresh specimens sent to pathology may have multiple test requests on the sample and as such a thorough review of the accompanied requisition form before the specimen is exposed to the desired fixative. For instance, part of the received specimen may be needed for tumor banking, flow cytometry, cytogenetics, electron microscopy, or immunofluorescence analysis—specimens meant for these testing should not be exposed to the routine histology fixatives such as 10% neutral buffered formalin (NBF). Specimens meant for electron microscopy may be fixed in glutaraldehyde (Wisse et al., 2010) and those for direct immunofluorescence specimens should be placed in Michel's solution (Woollons et al., 1999).

Almost all biopsies are received in fixatives and require little to no dissections. Biopsy specimens are usually measured, described, wrapped in biopsy paper, or placed between two biopsy sponges, and then placed into standard tissue cassettes for processing. Multiple biopsy cores in the same specimen container may be separated into multiple cassettes (not more than two cores per cassette) to allow judicious use of tissue for downstream testing. Although the separation of cores into multiple cassettes increases the workload for the histopathology laboratory and pathologists who review the slides, such practice is good for quality patient care. For instance, biopsy cores split into multiple cassettes could preserve materials for multiple tests—one block may be used for immunohistochemistry or histochemistry testing while the remaining block(s) from the same biopsy procedure could be used for molecular testing. Failure to separate multiple biopsy cores into separate cassettes at the time of grossing may lead to inadequate materials for ancillary testing and ultimately require a second biopsy for additional materials. The number of biopsy cores per cassette must be defined by the laboratory and the decision must be based on tissue type, number of routine cores, and anticipated downstream testing. Procedures for thoracic biopsies may yield only one to two thin cores and these scanty materials may be all that is available for routine hematoxylin and eosin (H&E), multiple immunohistochemistry/histochemical stains, and most often, molecular testing. Similarly, prostate biopsies may yield multiple cores from the base, mid, and apex areas of the prostate, and placing each core into a separate cassette helps with proper embedding and tissue preservation. Although pooling of multiple prostate biopsies into one paraffin block may not reduce the core lengths of the core (Tolonen et al., 2015), such practice does not usually preserve the tissues for additional testing. Separating the cores into different cassettes makes it relatively easier to perform all the necessary tests, without exhausting the materials. Gastrointestinal biopsies from the stomach or colon areas, in contrast, may have multiple pieces in one container so placing one piece of tissue into each cassette will lead to an unnecessary increase in workload for both the laboratory and the reviewing pathologist, without any value to quality patient care. In such cases, multiple pieces of tissues could be placed into one cassette. Another area of interest is breast biopsy—cores suspicious of containing breast calcifications should be placed into a dedicated cassette for easy identification of the calcifications during slide review. The biopsy materials free of calcification should be split (if needed) and placed into separate cassettes. Such a practice of knowing which block contains calcification can save the pathologist some time. Portions of biopsies received fresh may be submitted for flow cytometry or cytogenetics studies, depending on the clinical question.

Most surgical specimens and tissues from autopsies, on the contrary, are received fresh at the histopathology laboratories and may require several grossing manipulation techniques such as specimen opening and sectioning, inking, weighing, seeking for radioactive seed localization (where applicable), bread loafing (where needed), and prepping the tissue for fixation (Westra et al., 2003). For biopsies, tissue fixation may begin at the point of tissue collection as the acquired biopsy tissue is directly placed into the desired fixative. The fixation time may begin at the point of grossing or tissue manipulation for surgical and autopsy specimens so the documentation of the cold ischemic time (the difference between the specimen collection time and the start of fixation time) may become important for surgical specimens, especially for hormone receptor testing on breast tissues (Khoury, 2018; Li et al., 2013; Yildiz-Aktas et al., 2012).

The purpose of tissue fixation is to stop endogenous enzymic activities, prevent autolysis and tissue putrefaction, convert the consistency of the normal cells from semifluid (sol) to an irreversible semisolid (gel), and preserve proteins in their in vivo relationship to each other. Most histopathology laboratories fix tissues with 10% NBF which converts hemoglobin to hematin and stabilizes protein through crosslinking via methylene bridges (O'Rourke & Padula, 2016). Tissue fixation rate and its effectiveness would depend on the temperature of the fixative, the penetration rate of the fixative, the size of the specimens, the pH of the solution, and the duration of the tissue in the fixative (Lott et al., 2015). As a general recommendation, all clinical specimens must be fixed for 6–72 hours, when using 10% NBF. Although the CAP temporarily removed this recommendation (ANP. 10039) from its 2021 revised accreditation standard due to insufficient supporting evidence at the time, it is still the best practice to adopt in fixing samples for 6–72 hours. Additionally, fatty tissues such as the breast, spleen, and colon; and dense tissues like the cervix and uterus may require extended fixation times. Pronounced larger or heavier specimens, dense tissues, and those with overlying skin may require longer fixation time, as a result of the slow penetration rate of 10% NBF fixative (Thekkinkattil, 2017; Thompson et al., 2013). Laboratories should have a written policy on how long they could keep remaining tissues not submitted for clinical use in formalin. Tissues left in formalin for a minimum of 5 years may show good H&E staining but suboptimal staining for histochemical stains such as Mallory's trichrome and Van Giesons' due to potential alteration in protein structure and tissue morphology (Likhithaswamy et al., 2020).

In addition to fixation, bone specimens should usually go through a decalcification process to remove the minerals from calcified tissues. There are many commercially available decalcification solutions, the selection of either strong mineral acids (nitic acid, hydrochloric acid, Von Ebner's solution, Pereny's fluid), weaker organic acids (formic acid), or solutions based on chelating agents (ethylenediaminetetraacetic acid, EDTA) depends on the laboratory's desired turn-around-time or the anticipated downstream ancillary testing of the specimens (Bumalee et al., 2022; Sanjai et al., 2012). For instance, bone specimens meant for either cytogenetics studies, molecular profiling, or immunohistochemistry should be decalcified with EDTA, not strong acid, to preserve the nucleic acids (Ghosh et al., 2020; Miquelestorena-Standley et al., 2020; Sanjai et al., 2012; Singh et al., 2013).

Tissue processing

The goal of tissue processing in histopathology specimens is to remove the free interstitial water molecules from the tissues, leaving the bound water molecules intact, and replace the removed molecules with a supportive medium such as melted paraffin wax (Aziz & Zeman-Pocrnich, 2022; Day, 2014). Specimen processing is an important step in histologic preparation and it involves dehydration with graded alcohols such as ethanol and isopropanol to remove the free water through osmosis; tissue clearing using clearing agents such as xylene or xylene substitutes that could remove the preceding dehydrants and dissolve lipids which could impede paraffin wax penetration; and infiltration with melted paraffin wax (Day, 2014). Using a series of graded alcohols as dehydration agents prevent damage to delicate tissues and avoid tissue shrinkage (Carson & Cappellano, 2020; Lerch et al., 2019). Excessive exposure to the dehydration or clearing agents may lead to tissue shrinkage, distortion, or change in compressibility (Day, 2014; Lerch et al., 2019). To be effective, a good clearing agent should be soluble in the dehydrant and be able to dissolve the lipids in the tissue that could affect the paraffin infiltration. Tissue infiltration displaces the clearing agent and provides support for proper tissue hardening necessary for sectioning. Insufficient infiltration may lead to crumbly and soft tissue while excessive infiltration may result in tissue shrinkage and difficulty in sectioning (Day, 2014). It is best practice to process biopsies separate from large surgical resection specimens to avoid suboptimal and overprocessing of the tissues. Proper tissue processing is crucial in rendering an accurate histologic evaluation of histopathology specimens (Yousif & Qasem, 2016b). Histology laboratories may employ different methods of tissue processing including routine or rapid manual processing, and automated tissue processor with or without microwave technology (Aziz & Zeman-Pocrnich, 2022; Sivapathasundharam et al., 2007). The choice of the processing method is up to the laboratory's leadership, after considering the pros and cons of each of the available methods.

Table 4.1 shows a sample automated tissue processor protocol for processing smaller tissues and the short formalin time in the protocol assumes that the sample has been adequately fixed, prior to processing. Table 4.2 is a sample automated tissue processor protocol for large and fatty tissues and the extended formalin fixation time ensures additional fixation, after the specimen is grossed and sectioned into standard cassettes. Although most larger specimens are placed in their fixatives for an ample amount of time, the center of the tissues may not

Table 4.1 Sample protocol for biopsy processing.

Step	Reagent	Time (in min)
1	Formalin	5
2	70% Alcohol	5
3	80% Alcohol	5
4	95% Alcohol	5
5	95% Alcohol	15
6	Absolute alcohol	15
7	Absolute alcohol	30
8	Xylene	10
9	Xylene	15
10	Xylene	20
11	Paraffin	10
13	Paraffin	15
14	Paraffin	15

Table 4.2 Sample protocol for large resection and fatty tissues.

Step	Reagent	Time (in min)
1	Formalin	240
2	70% Alcohol	20
3	80% Alcohol	20
4	95% Alcohol	20
5	95% Alcohol	30
6	Absolute alcohol	40
7	Absolute alcohol	80
8	Xylene	20
9	Xylene	35
10	Xylene	60
11	Paraffin	30
13	Paraffin	40
14	Paraffin	50

be adequately fixed due to the slow fixation rate of some of the fixatives. It is therefore important to leave sectioned tissues in the fixative for some time, prior to processing. The processing protocols shown below assume 10% NBF as the fixative.

Tissue embedding

After infiltration, processed tissues are embedded by orienting them in a mold and surrounded with a molten medium. Tissue embedding provides support to the tissue, prevents tissue distortion, and preserves the tissue for long-term storage (Dey, 2018). Depending on the type of tissue, microtome, and microscope, a laboratory may choose one of the following embedding media: paraffin wax, epoxy resin, acrylic medium, agar gel, or gelatin. Tissues from core biopsies must be stretched and placed flat at the bottom of the mold whereas the cut surfaces of other tissues (if identified) are usually placed down at the bottom of the mold containing a small amount of the embedding medium. After careful orientation, an additional embedding medium is added to the mold, and then the cassette bearing the unique identity of the tissue is placed at the top of the mold and then allow to solidify on a cold plate. The solidified embedded medium containing the tissue is carefully popped out from the mold. If the tissue had been fixed with formalin and embedded with paraffin wax, the block is usually referred to as formalin-fixed paraffin-embedded (FFPE) tissue block. The embedding quality may be affected by sample thickness, tissue orientation in the mold, the composition of the paraffin wax, mold size, and cassette type, and lighting at the station (Dey, 2018). It is particularly important to appropriately orient certain tissue types such as tubular structures and skin punches or shaves in order not to destroy the architecture of the relevant histology structures. For instance, tissues such as the ovary, testes, lymph nodes, thymus, and mammary glands must be embedded flat; bisected skin punches should be embedded on the edge, and tubular structures must be embedded to show their cross sections. The number of pieces embedded into a single cassette can affect the embedding quality. For multiple pieces of tissue in one cassette, all the pieces must be placed on the same plane and firmly pressed to the bottom of the mold (Taqi et al., 2018). Without placing all pieces into the same plane, some of the pieces may be depleted, prior to getting to the pieces embedded deep into the block. Similarly, larger tissue sections must be placed into appropriate-sized molds to allow all the margins to be visible, on the glass slide, during microtomy (Bryant et al., 2019; Zhanmu et al., 2020).

Microtomy

During microtomy, the paraffin block is mounted unto a microtome that uses sharp blades to section the tissue at the desired thickness. The thickness of each slice could be 2–10 μm, with an average slice thickness of 4–5 μm (Dey, 2018; Yousif & Qasem, 2016a). The cut tissue sections are floated over a warm water

Table 4.3 Recommended tissue thickness.

Tissue	Thickness (μm)
Routine paraffin	4–5
Renal sections	1–3
Bone marrow	2–3
Nerve histochemical staining	6–15
Amyloid demonstration	6–12

(temperature set at 5°C–10°C below the melting point of the paraffin) floatation bath to remove wrinkles and any distortion to the tissue. The histotechnologist must skim the water bath after each block to prevent tissue contamination. Similarly, any tissue debris around the microtome block holder must be cleaned with a microtome brush. Since the processed tissue is transitioned from a paraffin-embedded block to a glass slide during microtomy, the technician or technologist needs to pay the ultimate attention to ensure that the label information on the glass slides exactly matches the information on the block. Factors such as section thickness, water temperature, sectioning techniques, placement of tissue on a glass slide (charged or uncharged), number of tissue sections placed on each glass slide, and microtome blade sharpness could have an adverse impact on the slide review and subsequently on whole slide digital imaging.

Different tissue types may be cut at different thicknesses, based on the laboratory's standard operating procedure. Highly cellular tissues may be cut at 4 μm, neurological tissues at 6–20 μm, and myelinated nerves at 15–20 μm (Shields & Heinbockel, 2019). The CAP standard ANP.11716 gives the following recommendations as shown in Table 4.3, at the discretion of the laboratory director.

Staining

The tissue staining process starts by reversing the hydrophobic tissue environment through deparaffinization with the clearing agent to allow water-soluble dyes to penetrate the tissues (Shields & Heinbockel, 2019). The basic stain in the histopathology laboratory is the H&E stain. This simple, easy-to-use, and relatively cheap stain is the most widely used histological stain and the gold standard for almost all histopathologic diagnoses of solid tumors (Almangush et al., 2021; Bancroft & Layton, 2019a). The two dyes, hematoxylin and eosin, provide the necessary contrast to demonstrate cellular changes in the tissue.

The hematoxylin, a basic dye, binds to elements in the nucleus, specifically the chromatin within the nucleus and the nuclear membrane. This dye, however, does not bind to the nucleoplasm of the nucleus (NSH, 2001). The active ingredient of hematoxylin, a red-brown material called hematein, shows a reddish color at pH below 1, yellowish at pH between 1 and 1.5, and purple at pH 6 or higher—the reason for using tap water during histological staining (Kumar & Kiernan, 2010). Hematoxylins can be categorized as progressive—Mayer's, Delafield's, Gill's, and Harris' hematoxylin or regressive—Ehrlich's, Harris', and Delafield's hematoxylins, each category with different formulations that provide a variety of staining colors and patterns. Regressive hematoxylins overstain the tissue and require an acidic solution to pull out the excess color. In contrast, progressive hematoxylins stain the chromatin to the desired intensity, and do not overstain. Thus progressive hematoxylins which are less concentrated take a slightly longer time (5–10 minutes) to reach the desired intensity slowly and selectively. Regressive hematoxylins which require a differentiation step, on the other hand, is more concentrated and rapidly overstain both the nuclear elements and cytoplasm, prior to the differentiation step with dilute acid alcohol (0.5%–1% hydrochloric acid in 70% alcohol), and it may be ideal for frozen sections (Sohal, 2016).

Progressive hematoxylins can be used in a regressive method by using weak acids such as acetic acid as a differentiator, to remove nonspecific hematoxylin staining from the glass slides (NSH, 2001). Laboratories are supposed to follow the manufacturer's direction for the recommended differentiating agent. Bluing agents which are usually dilute alkaline solutions such as ammonia solutions, lithium carbonate solutions, and Scott's solution with a pH value of approximately 8 help change the less permanent reddish-purple hematoxylin to a more permanent insoluble blue or purple–blue color. Although the bluing agents cannot overblue the stains, leaving slides in the solution for more than 1 minute may lead to severe tissue loss. Some commercially available bluing agents are purposely diluted to prevent tissue loss when slides are left for up to 2 minutes.

Eosin is an acidic dye and is the most suitable dye to combine with a hemalum to demonstrate the general histological structures of tissues (Bancroft & Layton, 2019b). Eosin, an artificial derivative of fluorescein, consists of two closely related compounds—a tetrabromo derivative with a slight yellowish cast (Eosin Y) and a dibromo dinitro derivative with a faint bluish cast (Eosin B), and the two can be used interchangeably (Lai & Lü, 2012). Eosin Y can be aqueous or alcoholic and is the most widely used in histology laboratories as it binds to the basic cytoplasmic elements of the cell containing eosinophilic compounds with positive charges to provide the needed contrast of the stain (Bancroft & Layton, 2019b). Additionally, eosin produces distinct hues for red blood cells (dark reddish orange), collagen (lighter pastel pink), and smooth muscle fibers (bright pink) and it may be used to differentiate various tissue elements (Lai & Lü, 2012; NSH, 2001). Since eosin is soluble in water, a higher amount of water in the alcohols following the eosin will remove more eosin from the tissue. Factors such as elevated pH (greater than 5.0), inefficient rinsing after bluing, excessive alcohol rinse or longer time in aqueous solution after eosin, inadequate staining time, contamination in the alcohol rinse, and thin sections may lead to weak eosin staining. Excessive eosin staining, on the other hand, could result from low pH values of less than 4.0, thicker sections, longer staining time, and

Table 4.4 Sample protocol for automated H&E staining.

Step	Reagent	Time (mm: ss)
Default: Include oven step	Oven @ 70°C	15:00
1	Xylene	5:00
2	Xylene	5:00
3	100% Alcohol	2:00
4	100% Alcohol	2:00
5	95% Alcohol	2:00
6	Tap water	2:00
7	Hematoxylin	9:00
8	Tap water	2:00
9	Differentiator	1:00
10	Tap water	2:00
11	Bluing agent	2:00
12	Tap water	2:00
13	95% Alcohol neutralizing	2:00
14	Eosin	4:30
15	95% Alcohol	2:00
16	95% Alcohol	2:00
17	100% Alcohol	2:00
18	100% Alcohol	2:00
19	Xylene	2:00
20	Xylene	2:00

the use of isopropanol as the rinsing agent (NSH, 2001). The use of clearants such as xylene or xylene substitutes displace alcohol from the tissue sections and provide a compatible solvent with the desired mounting medium. Inadequate clearing will leave the anhydrous alcohol on the tissue and may cause the eosin to bleed from the tissue, after the coverslips (NSH, 2001). Table 4.4 shows a sample H&E staining protocol.

Digital pathology in histology laboratory

Prior to the introduction of digital pathology, the quality management systems (QMS) of most histology laboratories mainly focused on generating quality stains that could be read and accurately interpreted by the immediate pathology group, for the right patients (Adyanthaya & Jose, 2013). However, digital workflows require the implementation of additional quality control checks to ensure quality-scanned images to prevent diagnostic errors with digital images. All factors that affect the quality of physical glass slides constitute the preanalytical variables of whole slide images (WSIs). For instance, factors such as the size and thickness of tissues submitted during grossing, quality of the cassettes and slide label barcodes, location of barcodes on the glass slide labels, tissue section placement, and the number of sections placed on each glass slide may significantly impact the WSI. Other slide preparation factors including the type of mounting media used, viscosity of the medium, volumes dispensed on each glass slide, proper placement of the coverslip on the glass slide, and how well the glass slide needs to dry prior to scanning do have an effect on the WSI quality. While these factors are important in the conventional pathology QMS, each of these variables could independently play a crucial role in affecting the quality of the digital images representing the material on the glass slide. Traditional QMS mainly focus on ensuring the quality and consistency of the laboratory processes to obtain accurate laboratory results (Adyanthaya & Jose, 2013; Mohammedsaleh, 2014). The workflow for whole slide imaging, however, requires detailed attention to many variables that we overlooked, under the traditional QMS. Thus digital pathology workflow requires the incorporation of additional quality improvement projects into the existing traditional laboratory workflows. For instance, the quality of the slide label does not only have to be human-readable, the information embedded in the barcode may hold the primary key to link the scanned image to the rest of the patient's information and hence, absolute care must be taken to design the slide label, clean the slide label printer head, and maintain the printing quality at all times.

The type of coverslip (glass, film, or liquid) and method of cover-slipping (manual or automated) could easily affect the number of air bubbles, scanning time, and the average size of the WSI file (Ferreira et al., 2022). Unlike the conventional microscopes which could easily be adjusted to navigate around air bubbles, the presence of such air bubbles creates uneven slide surfaces on the digital scanner, leading to multiple focus points of the images (Atallah et al., 2022). Although the glass coverslip is found to generate more air bubbles, compared to the film tape, and the latter also minimizes the rate of scanning errors (Ferreira et al., 2022; Retamero et al., 2020), some laboratories have already built in specific coverslip methods in their traditional workflows and have concerns of changing them due to pathologists preference or effects of long-term storage. Additionally, coverslips that slightly overhang either to the sides or to the bottom of the microscope glass slide could easily interfere with the digital scanner's mechanics and cause the slide to break or jam the scanning operations.

In terms of tissue placement on the glass slides, pathologists are trained to review every tissue section on a given glass slide, including the ones close to the edge of the glass slides. Different tissue finder algorithms employed by different WSI platforms, however, may ignore peripheral tissues on the glass slide in order to speed up the scanning process (Atallah et al., 2022). Tissue sections meant for digital pathology must therefore be placed at the center of the slide and probably limit the number of tissue sections placed on a single slide. Any extraneous tissue on the slide could also affect the scanned image quality, especially when the extraneous tissue is folded, found at the corner of the slide, or is more prominent than the actual tissue that needs to be captured.

The microscope slide labels which contain the patient information and the slide's identification information such as accession number, barcode, type of stain, slide level, and patient name may be printed directly on the slide or on an adhesive paper label. These labels must be resistant to all the staining reagents (alcohols, xylenes, and staining kits), smudges, and abrasions. For laboratories using adhesive paper microscope slide labels, the histotechnicians ensure that no portion of the label hangs off the glass slide as that may interfere with the mechanics of the digital scanner. Although it may be a little time-consuming to perfectly align the slide labels during microtomy, preventing any label from overhanging will save time during digital scanning. Slide printers that directly print the slide identification on the glass slide avoid any potential hanging artifact, however, the laboratory must work with its IT team to ensure high printing quality, at least a resolution of 12 dot/mm (300 dpi), and also maintain frequent monitoring of the printing quality. A suboptimal resolution could lead to failure to associate the scanned images with the right patient in the laboratory's information system. Slide label barcodes that are found at the extreme edge of the label or partially cut off could fail to link to the appropriate patient.

Artifacts

Artifacts that directly affect the tissue on the glass slide could easily distort the quality of the digital image. Artifacts are unwanted structures that should not be present on the prepared glass slides and their presence could either distort the histology of the tissue or mislead the pathologists during slide reading and interpretation of physical glass slides or digital images. Specimen handling and processing involve multiple steps and at each stage of the process, an artifact could be introduced and potentially affect the quality of the slide. Artifacts in the histopathology laboratory could occur prior to fixation, during fixation, specimen sectioning at the grossing, tissue processing, embedding, microtomy, floatation and mounting, staining, and coverslipping (Komura & Ishikawa, 2018; Taqi et al., 2018). The artifacts could originate either from the slide preparation or from the specimens themselves and could adversely affect the scanning time as well as the quality of the scanned images. Slide preparation artifacts such as contaminants from other tissue samples, foreign objects, tissue folds, ink markers, dusty or dirty glass slides, air bubbles, overused staining reagents, and unfiltered staining solutions could potentially lead to color variations of the stains and affect the biological structures to be interpreted (Brixtel et al., 2022; Komura & Ishikawa, 2018). Specimen-dependent artifacts may result from specific inherent characteristics or the specimen preparations which include thick-an-thin sections, calcification, tissue tearing, and markings on glass slides (Haghighat et al., 2022). Most of the artifacts could easily be confused with out-of-focused regions and may require unnecessary re-scans. Bautista and Yagi (2010) concluded that the presence of tissue artifacts could have an adverse effect on the quality of whole slide scanned images. Although the presence of artifacts is inherent in the histological slide preparation, the necessary process and procedures must be implemented to prevent or minimize such artifacts, when adopting digital pathology.

To minimize histological artifacts that affect stained slides and digital image quality, histology laboratories should standardize their specimen handling and processes to avoid variations in their daily activities. Laboratories must assign qualified pathologists, pathologists' assistants, or histotechnologists to prepare all fresh specimens and place them into their respective fixatives, within the shortest possible time; and also minimize freeze-thaw cycles for specimens retrieved from biobanking (Compton et al., 2019; Fan et al., 2019; Khoury, 2018; Pekmezci et al., 2012). Larger specimens should be regularly sliced to increase the surface areas exposed to the fixative and gross tissue sections should not be thicker than 3–4 mm per section that could easily fit into a standard processing cassette (Lott et al., 2015). For needle core biopsies, not more than four pieces should be placed in a single cassette. Lesser than five pieces of tissue in a cassette makes it easier to orient all the pieces on the same plane during embedding and enhance the scanning time, during slide digitization. Larger tissue sections or multiple pieces of tissue on a glass slide not only extend beyond the length and breadth of the coverslip area, they also affect the focus points of the digital scanners, significantly increase the scanning time, and generate larger file sizes (Bertram & Klopfleisch, 2017; Farahani et al., 2015). For needle core biopsies, not more than four pieces should be placed in a single cassette. Lesser than five pieces of tissue in a cassette makes it easier to orient all the pieces on the same plane during embedding and enhance the scanning time, during slide digitization. Larger tissue sections or multiple pieces of tissue on a glass slide not only extend the length and width of the coverslip but also affect the focus points of the digital scanners and significantly increase the scanning time (Farahani et al., 2015).

Another quality improvement process that can minimize histological artifacts is to establish a standardized procedure for tissue embedding and microtomy. Laboratories without automated tissue embedders need to implement specific routine processes such as using forceps with teeth, regularly cleaning the forceps wells, selecting of appropriate mold size, and cleaning molds after every single use (Chatterjee, 2014; Nelson et al., 2020; Taqi et al., 2018). Artifacts at microtomy such as venetian blind, scratch lines, zones of compressed tissue, and chattering could be minimized by using new blades or changing them more often, avoiding knives with a small nick at the edges, firmly securing the microtome blades, and using paraffin of appropriate hardness of embedding medium for specific tissues (Bindhu et al., 2013; Taqi et al., 2018).

Immunohistochemistry and special stains

For laboratories with automated immunohistochemistry and histochemical staining platforms, the type and design of the microscope slide labels required by the staining platforms must be taken into consideration when planning the digital pathology workflow. Some staining platforms may require linear barcodes or optical character recognition to run their staining protocols. These barcodes on the staining platform, including those with 2-D barcodes, may be external to the laboratory's information system so slides from these staining platforms may not be able to go on to the digital scanner, without any additional re-labeling. One way of mitigating these issues is to adopt a global slide label throughout the laboratory processes. Thus the same label type should be used to run on the automated Immunohistochemistry (IHC) and special stain staining platforms, regardless of the platform vendor, and the information on those barcode labels should be read and accepted into the LIS. Double labeling that involves the use of separate labels—one for running the stains and the other for scanning the same slide not only creates double work but could also easily lead to a laboratory error with an adverse impact on patient care. Weak counterstain in IHC could pose a challenge to evaluating scanned images of IHC slides as a result of hazy areas around fatty tissue (Atallah et al., 2022). Tissue sections for IHC are usually thinner or of variable thickness and the rigorous process of epitope retrieval renders the sections more susceptible to tissue loss and folding. The tissue disruption on the slides and variability in tissue section thickness present different parts of the tissue to be largely at different planes that lead to many out-of-focus areas in the scanned image (Atallah et al., 2022; Basak et al., 2023). Weak immunohistochemical staining intensity and weak counterstains can be problematic and increase the rescan rates in a digital workflow (Lujan et al., 2021; Rojansky et al., 2023).

Logistical challenges

The type of slide racks used by most histology laboratories also plays a significant role when implementing a digital pathology workflow. Currently, there is no standardized slide rack that could be used across the available WSI scanning platforms. For instance, the Aperio GT 450 scanner only accepts the Leica Universal rack or Sakura Prisma and cover slipper rack; IntelliSite Ultra Fast Scanner works with Winlab LS-20/Winlab LSM-20 or Sakura 4768 20-slide basket; NanoZoomer S20 accepts Hamamatsu Slide cassette or Sakura 4768 20-slide basket; 3DHistech uses Leica Universal rack or Sakura 4768 20-slide basket; and both Roche Ventana DP 200 and DP 600 accept specific slide trays—they do not use any slide rack. Unfortunately, some of the slide racks or trays on the H&E stainer models and cover slippers such as the Roche Ventana HE 600 and Symphony slide trays, slide racks on Leica's ST5010, ST5020, CV5030, and Tissue-Tek 24 slide used in the histology laboratories are not compatible to any of the available whole slide digital scanners (as shown in Fig. 4.1). Stained slides from the H&E stainers and cover slippers sometimes must be manually transferred to the preferred slide racks acceptable to the given slide scanner.

FIGURE 4.1

Histology slide racks that do not support WSI platforms.

Laboratories using stainers and cover slippers whose staining or output racks are not compatible with their installed WSI scanners must employ the time-consuming manual processes to transfer the slides to the appropriate slide racks, before loading to the WSI scanner. The manual transfer is not only time-consuming, but it also introduces fingerprints and smudges that could interfere with the image quality of the scanned slides, resulting in suboptimal image quality (Chen et al., 2021; Ghaznavi et al., 2013; Patel et al., 2022). Fig. 4.1 shows images of histology racks that are still in use but not compatible with any of the available digital scanning platforms.

One other challenge faced in the histology laboratory is making the decision on whether to scan all slides prospectively or retrospectively, after the case is signed out. Retrospective digital slide scanning is good for slide archival and image retrieval; it does not provide a good histological workflow for digital pathology meant for digital sign-out. A complete digital pathology workflow requires all slides to be scanned prospectively, at the laboratory, prior to releasing the glass slides from the laboratory. However, laboratories without barcoded labels may not be able to link digitally scanned images to a particular case. Thus the laboratory would only depend on a manual process to assign those scanned microscope glass slides to a particular case and such a tedious workflow is not feasible for large-volume histology laboratories. Another logistical challenge is with immunohistochemically stained slides where each slide must be reviewed for appropriate reactivity of the known positive control or the internal positive control. Even if the coverslip output rack is compatible with the WSI scanning platform, the histotechnicians would have to unrack each slide to review the known positive controls (or the internal controls of tissue types with expected internal controls) before the slides are re-racked and digitized. Depending on the laboratory's policy, every IHC slide may be run with a known positive control tissue. Ideally, this known positive control tissue is placed on the same slide as that of the patient tissue to ensure equal treatment during processing (Hewitt et al., 2014; Torlakovic et al., 2010). Laboratories without global slide labels (one label that could be used and read on all the laboratory's staining platforms) may have to adopt double labeling—starting with a slide label for the primary purpose such as immunohistochemical staining or special stain platforms, and then re-label with a slide label that could be read on the whole slide imaging scanner and link the scanned image to the appropriate patient. The process of re-labeling and lack of uniformity in material labeling is tedious and prone to medical errors (Brown et al., 2015; Layfield & Anderson, 2010; Morelli et al., 2013; Tabatabaee et al., 2023). Some of the labeling errors could lead to digital images that are assigned to the wrong patient or immunohistochemistry stains being re-labeled with the wrong antibody markers. Thus such errors may easily mislead the reviewing pathologist and lead to a wrong diagnosis, a major breach in patient safety (Nakhleh, 2008; Tang et al., 2018).

Adoption of whole slide imaging workflow

The quality of the WSI depends on how well the histology slides were prepared. Histological artifacts which may be introduced at any point of the workflow process—tissue handling and processing, embedding, microtomy, staining, and coverslipping can easily compromise the quality of WSI. Even before the image compression phase in WSI, histological slide preparation factors such as variations in section thickness, tissue section placement on the glass slides, presence of chatter artifacts, and tissue folds may negatively affect the focus of the virtual images of the tissue (Ghaznavi et al., 2013). Similarly, the thickness of tissue sections and the quality of slide preparation have a significant impact on WSI capture speed and image quality (Yagi & Gilbertson, 2008). The printing quality of the slide label, viscosity of the mounting media, thickness, and length of the cover glass, and quality of the staining reagents contribute to the overall quality of the scanned images and how well the images are linked to the appropriate corresponding data in the LIS. Thus each factor affecting histological slide preparation, prior to whole slide imaging, plays a vital role in digital image quality.

Each of the preanalytical factors which include tissue handling and processing, starting from the specimen grossing to tissue processing, has a potential impact on digital image quality. Tissue-marking dyes such as Indian ink, colored gelatin, and other commercially available dyes used for tissue margin evaluation and orientation could affect the whole slide digital images. A study by Hon et al. (2019) showed diminished sensitivity of various colors of tissue-dye markers on digitized images. According to their study, the sensitivity of chromatic dyes on digital images ranged from 29% (red) to 75% (blue). Thus instead of maintaining their true colors, some digital scanners may misread red, green, and blue as black (Hon et al., 2019). The darker regions introduced by the tissue-marker dyes may affect the autoexposure mechanism of the digital scanners and ultimately affects the quality of the digital image. Histological slide preparation could also lead to color variations in the scanned images (Badano et al., 2015). Ideally, to ensure accurate and reproducible morphometric quantification of tumor areas on a digitized glass slide, Shankar et al. (2022) suggest that the freshly stained H&E slides should be digitized within a month since fewer tumor areas may be detected as the stained slides age. Slide aging degrades color saturation and luminosity which ultimately increases the average pathologists' scoring time of the digitized images (Avanaki et al., 2015). Quality histological slide preparation plays a key role in whole slide imaging so additional quality measures must be implemented when adopting WSI laboratory workflow.

As part of the decision-making process, laboratories must consider the technical functionality and reliability as well as the ease of operability of the WSI system (Scott Campbell et al., 2012). The consideration must include sizes of slides (standard 75 mm × 25 mm microscope slides or large format 200 mm × 150 mm slides), types of stains (standard H&E, Frozen section H&E, special stains, and immunohistochemical stains), WSI device slide capacity, and how quickly to scan the slides (Evans et al., 2017; Farahani et al., 2015; Patel et al., 2021). Some WSI platforms such as Leica's Aperio and GT450, Roche's DP 200, and DP 600 are only designed to scan the standard 3″ × 1″ glass slides, not in large format. Similarly, not all scanners are designed to handle wet slides or slides without coverslips (Farahani et al., 2015). WSI scanners like Vision Tek can manage wet slides, FNA smears, biopsy touch preparations, and noncover slipped slides. Ideally, WSI devices with horizontal loading trays instead of vertical racks are recommended for wet slides such as frozen sections (Evans et al., 2021). The type of coverslip instrument and the coverslips at the lab must be taken into consideration when adopting WSI workflow as some scanners may have specific coverslip requirements to enhance slide thickness (Farahani et al., 2015; Patel et al., 2021). Another crucial factor is the loading mechanisms of the WSI device. Since WSI workflow extends the traditional histology workflow, devices that allow automatic or continuous loading is preferred for efficient laboratory workflow.

Laboratories that intend to integrate WSI into their workflow need to fully validate the new workflow as well as the WSI devices. The validation of WSI workflow validation must include staff training on new policies and procedures that have been implemented. Such training and competency assessment of staff must be documented. The laboratory also must find ways to provide continuous education on WSI to the staff. Besides the clinical validation of the WSI workflow that is appropriate and applicable for the intended clinical use and clinical setting of the application (Pantanowitz et al., 2013), each laboratory should validate the WSI devices as a laboratory instrument, following their written policies on new instrument validation. All documents related to the instrument validation and preventive maintenance must be kept for the duration specified by the regulatory standards.

References

Adyanthaya, S., & Jose, M. (2013). Quality and safety in histopathology laboratory. *Journal of Oral and Maxillofacial Pathology*, *17*.

Alenazi, S. M., & Bugis, B. A. (2022). The role of laboratory information system in improving the delivery of laboratory services: A recent systematic review. *Combinatorial Chemistry and High Throughput Screening*, *26*(8), 1451–1460. Available from https://doi.org/10.2174/1386207325666220914112713, PMID:36111772.

Almangush, A., Mäkitie, A. A., & Leivo, I. (2021). Back to basics: Hematoxylin and eosin staining is the principal tool for histopathological risk assessment of oral cancer. *Oral Oncology*, *115*. Available from https://doi.org/10.1016/j.oraloncology.2020.105134.

Atallah, N. M., Toss, M. S., Verrill, C., Salto-Tellez, M., Snead, D., & Rakha, E. A. (2022). Potential quality pitfalls of digitalized whole slide image of breast pathology in routine practice. *Modern Pathology*, *35*(7), 903–910. Available from https://doi.org/10.1038/s41379-021-01000-8.

Avanaki, A. R. N., Espig, K. S., Sawhney, S., Pantanowitzc, L., Parwani, A. V., Xthona, A., & Kimpe, T. R. L. (2015). Aging display's effect on interpretation of digital pathology slide. *Medical Imaging 2015: Digital Pathology*, *9420*, 942006. Available from https://doi.org/10.1117/12.2082315.

Aziz, S. J., & Zeman-Pocrnich, C. E. (2022). *Immunohistochemistry and immunocytochemistry: Methods and protocols*. Springer Science + Business Media, LLC. Available from http://www.springer.com/series/7651.

Badano, A., Revie, C., Casertano, A., Cheng, W. C., Green, P., Kimpe, T., Krupinski, E., Sisson, C., Skrøvseth, S., Treanor, D., Boynton, P., Clunie, D., Flynn, M. J., Heki, T., Hewitt, S., Homma, H., Masia, A., Matsui, T., Nagy, B., … Yokoi, H. (2015). Consistency and standardization of color in medical imaging: A consensus report. *Journal of Digital Imaging*, *28*(1), 41–52. Available from https://doi.org/10.1007/s10278-014-9721-0.

Bancroft, J. D., & Layton, C. (2019a). The hematoxylins and eosin. In S. K. Suvarna, C. Layton, & J. D. Bancroft (Eds.), *Bancroft's theory and practice of histological techniques* (8th ed., pp. 126–138). Elsevier.

Bancroft, J. D., & Layton, C. (2019b). *The hematoxylins and eosin, Editor(s): S. Kim Suvarna, Christopher Layton, John D. Bancroft,Bancroft's Theory and Practice of Histological Techniques (Eighth Edition), Elsevier*. 12–138.

Basak, K., Ozyoruk, K. B., & Demir, D. (2023). Whole slide images in artificial intelligence applications in digital pathology: Challenges and pitfalls. *Turk Patoloji Dergisi*, *39*(2), 101–108. Available from https://doi.org/10.5146/tjpath.2023.01601.

Bautista, P. A., & Yagi, Y. (2010). Improving the visualization and detection of tissue folds in whole slide images through color enhancement. *Journal of Pathology Informatics*, *1* (1), 25. Available from https://doi.org/10.4103/2153-3539.73320.

Bertram, C. A., & Klopfleisch, R. (2017). The pathologist 2.0: An update on digital pathology in veterinary medicine. *Veterinary Pathology*, *54*(5), 756–766. Available from https://doi.org/10.1177/0300985817709888.

Bindhu, P., Krishnapillai, R., Thomas, P., & Jayanthi, P. (2013). Facts in artifacts. *Journal of Oral and Maxillofacial Pathology*, *17*(3), 397–401. Available from https://doi.org/10.4103/0973-029X.125206.

Brixtel, R., Bougleux, S., Lézoray, O., Caillot, Y., Lemoine, B., Fontaine, M., Nebati, D., Renouf, A., Bougleux, S., & Lemoine, B. (2022). Whole slide image quality in digital pathology: Review and perspectives. *IEEE Access*, *10*. Available from https://doi.org/10.1109/ACCESS.2022.3227437i.

Brown, R. W., Speranza, V. D., Alvarez, J. O., Eisen, R. N., Frishberg, D. P., Rosai, J., Santiago, J., Tunnicliffe, J., Colasacco, C., Lacchetti, C., & Thomas, N. E. (2015). Uniform labeling of blocks and slides in surgical pathology: Guideline from the college of American pathologists pathology and laboratory quality center and the national society for histotechnology. *Archives of Pathology and Laboratory Medicine*, *139*(12), 1515−1524. Available from https://doi.org/10.5858/arpa.2014-0340-SA.

Bryant, P., Haine, N., Johnston, J., & Ntiamoah, P. (2019). Application of large format tissue processing in the histology laboratory. *Journal of Histotechnology*, *42*(3), 150−162.

Bumalee, D., Lapthanasupkul, P., Songkampol, K., Srimaneekarn, N., Kitkumthorn, N., & Arayapisit, T. (2022). Qualitative histological evaluation of various decalcifying agents on human dental tissue. *European Journal of Dentistry*. Available from https://doi.org/10.1055/s-0042-1755615.

Carson, F., & Cappellano, C. (2020). *Histotechnology: A Self-Instructional Text* (5th ed.). Chicago: ASCP Press.

Chatterjee, S. (2014). Artefacts in histopathology. *Journal of Oral and Maxillofacial Pathology*, *18*(5), 111−116. Available from https://doi.org/10.4103/0973-029X.141346.

Chen, Y., Zee, J., Smith, A., Jayapandian, C., Hodgin, J., Howell, D., Palmer, M., Thomas, D., Cassol, C., Farris, A. B., Perkinson, K., Madabhushi, A., Barisoni, L., & Janowczyk, A. (2021). Assessment of a computerized quantitative quality control tool for whole slide images of kidney biopsies. *Journal of Pathology*, *253*(3), 268−278. Available from https://doi.org/10.1002/path.5590.

Compton, C. C., Robb, J. A., Anderson, M. W., Berry, A. B., Birdsong, G. G., Bloom, K. J., Branton, P. A., Crothers, J. W., Cushman-Vokoun, A. M., Hicks, D. G., Khoury, J. D., Laser, J., Marshall, C. B., Misialek, M. J., Natale, K. E., Nowak, J. A., Olson, D., Pfeifer, J. D., Schade, A., ... Yohe, S. L. (2019). Preanalytics and precision pathology: Pathology practices to ensure molecular integrity of cancer patient biospecimens for precision medicine. *Archives of Pathology and Laboratory Medicine*, *143*(11), 1346−1363. Available from https://doi.org/10.5858/arpa.2019-0009-SA.

Cotter, M. B., & Loda, M. (2017). Introduction to Pathology. In M. Loda, L. A. Mucci, M. L. Mittelstadt, M. Van Hemelrijck, & M. B. Cotter (Eds.), *Pathology and Epidemiology of Cancer* (pp. 27−38). Switzerland: Springer International Publishing AG Switzerland.

Day, C. E. (2014). Histopathology: Methods and protocols. *Methods in Molecular Biology*, *1180*. Available from http://www.springer.com/series/7651.

Dey, P. (2018). *Embedding of tissue in histopathology. Basic and advanced laboratory techniques in histopathology and cytology*. Singapore: Springer. Available from https://doi.org/10.1007/978-981-10-8252-8_3, https://doi.org/10.1007/978-981-10-8252-8.

Evans, A. J., Depeiza, N., Allen, S. G., Fraser, K., Shirley, S., & Chetty, R. (2021). Use of whole slide imaging (WSI) for distance teaching. *Journal of Clinical Pathology*, *74*(7), 425−428. Available from https://doi.org/10.1136/jclinpath-2020-206763.

Evans, A. J., Salama, M. E., Henricks, W. H., & Pantanowitz, L. (2017). Implementation of whole slide imaging for clinical purposes issues to consider from the perspective of early adopters. *Archives of Pathology and Laboratory Medicine*, *141*(7), 944–959. Available from https://doi.org/10.5858/arpa.2016-0074-OA.

Fan, X. J., Huang, Y., Wu, P. H., Yin, X. K., Yu, X. H., Fu, X. H., Feng, L. L., Wang, Y. L., Yi, H. J., Chen, Z. T., Yin, J. X., Zhang, D. L., Feng, W. X., Bai, S. M., Kim, T., Mills, G. B., Lu, Y. L., Wan, X. B., & Wang, L. (2019). Impact of cold ischemic time and freeze-thaw cycles on RNA, DNA and protein quality in colorectal cancer tissues biobanking. *Journal of Cancer*, *10*(20), 4978–4988. Available from https://doi.org/10.7150/jca.29372.

Farahani, N., Parwani, A. V., & Pantanowitz, L. (2015). Whole slide imaging in pathology : advantages, limitations, and emerging perspectives. *Pathology and Laboratory Medicine International*, *7*, 23–33.

Ferreira, D., Vale, J., Curado, M., Polónia, A., & Eloy, C. (2022). The impact of different coverslipping methods in the quality of the whole slide images used for diagnosis in pathology. *Journal of Pathology Informatics*, *13*. Available from https://doi.org/10.1016/j.jpi.2022.100098.

Ghaznavi, F., Evans, A., Madabhushi, A., & Feldman, M. (2013). *Digital imaging in pathology: Whole-slide imaging and beyond*. *Annual review of pathology: Mechanisms of disease* (Vol. 8, pp. 331–359). Annual Reviews Inc. Available from https://doi.org/10.1146/annurev-pathol-011811-120902.

Ghosh, S., Lewis, M. B., & Walters, B. J. (2020). Comparison of ethylenediaminetetraacetic acid and rapid decalcificier solution for studying human temporal bones by immunofluorescence. *Laryngoscope Investigative Otolaryngology*, *5*(5), 919–927. Available from https://doi.org/10.1002/lio2.449.

Haghighat, M., Browning, L., Sirinukunwattana, K., Malacrino, S., Khalid Alham, N., Colling, R., Cui, Y., Rakha, E., Hamdy, F. C., Verrill, C., & Rittscher, J. (2022). Automated quality assessment of large digitised histology cohorts by artificial intelligence. *Scientific Reports*, *12*(1), 1–16. Available from https://doi.org/10.1038/s41598-022-08351-5.

Hanna, M. G., & Pantanowitz, L. (2015). Bar coding and tracking in pathology. *Surgical Pathology Clinics*, *8*(2), 123–135. Available from https://doi.org/10.1016/j.path.2015.02.017.

Hanna, M. G., Ardon, O., Reuter, V. E., Sirintrapun, S. J., England, C., Klimstra, D. S., & Hameed, M. R. (2022). Integrating digital pathology into clinical practice. *Modern Pathology*, *35*(2), 152–164. Available from https://doi.org/10.1038/s41379-021-00929-0.

Hewitt, S. M., Baskin, D. G., Frevert, C. W., Stahl, W. L., & Rosa-Molinar, E. (2014). Controls for immunohistochemistry: The Histochemical Society's standards of practice for validation of immunohistochemical assays. *Journal of Histochemistry and Cytochemistry*, *62*(10), 693–697. Available from https://doi.org/10.1369/0022155414545224.

Hon, J. D., Chen, W., Minerowicz, C., Thomas, S., Barnard, N., Gilbert, N., & Fyfe, B. (2019). Analysis and comparison of tissue-marking dye detection via light microscopy, telemicroscopy, and virtual microscopy. *American Journal of Clinical Pathology*, *151*(1), 95–99. Available from https://doi.org/10.1093/ajcp/aqy117.

Jalali, S. F., Jadidi-niaragh, F., Enderami, S. E., & Sattar, A. (2023). Characterization of fixatives and their application in histopathology. *Biomedical Journal of Scientific and Technical Research*, 38945–38953. Available from https://doi.org/10.26717/BJSTR.2023.47.007563.

Khoury, T. (2018). Delay to formalin fixation (cold ischemia time) effect on breast cancer molecules. *American Journal of Clinical Pathology*, *149*(4), 275–292. Available from https://doi.org/10.1093/AJCP/AQX164.

Komura, D., & Ishikawa, S. (2018). Machine learning methods for histopathological image analysis. *Computational and Structural Biotechnology Journal*, *16*, 34–42. Available from https://doi.org/10.1016/j.csbj.2018.01.001.

Kumar, G. L., & Kiernan, J. A. (2010). *Education guide: Special Stains and H&E* (2nd Edition). Carpinteria, CA: Dako North America.

Lai, M., & Lü, B. (2012). *Tissue Preparation for Microscopy and Histology. Comprehensive Sampling and Sample Preparation: Analytical Techniques for Scientists* (pp. 53–93). Elsevier. Available from https://doi.org/10.1016/B978-0-12-381373-2.00070-3.

Layfield, L. J., & Anderson, G. M. (2010). Specimen labeling errors in surgical pathology: An 18-month experience. *American Journal of Clinical Pathology*, *134*(3), 466–470. Available from https://doi.org/10.1309/AJCPHLQHJ0S3DFJK.

Lerch, M. L., Bauer, D. R., Theiss, A., Chafin, D., Otter, M., & Baird, G. S. (2019). Monitoring dehydration and clearing in tissue processing for high-quality clinical pathology. *Biopreservation and Biobanking*, *17*(4), 303–311. Available from https://doi.org/10.1089/bio.2018.0122.

Li, X., Deavers, M. T., Guo, M., Liu, P., Gong, Y., Albarracin, C. T., Middleton, L. P., & Huo, L. (2013). The effect of prolonged cold ischemia time on estrogen receptor immunohistochemistry in breast cancer. *Modern Pathology*, *26*(1), 71–78. Available from https://doi.org/10.1038/modpathol.2012.135.

Likhithaswamy, H., Madhushankari, G., Selvamani, M., Kokila, G., Mohan Kumar, K., & Chethana, K. (2020). Comparison of staining adequacy between tissues stored in formalin and paraffin embedded blocks for prolonged duration. *Journal of Oral and Maxillofacial Pathology*, *24*(3), 586–587. Available from https://doi.org/10.4103/jomfp.JOMFP_49_20.

Lott, R., Tunnicliffe, J., Sheppard, E., Santiago, J., Hladik, C., Nasim, M., Zeitner, K., Haas, T., Kohl, S., & Movahedi-lankarani, S. (2015). Practical guide to specimen handling in surgical pathology. *College of American Pathologists*, *1*, 53.

Lujan, G. M., Savage, J., Shana'ah, A., Yearsley, M., Thomas, D., Allenby, P., Otero, J., Limbach, A. L., Cui, X., Scarl, R. T., Hardy, T., Sheldon, J., Plaza, J. A., Whitaker, B., Frankel, W., Parwani, A. V., & Li, Z. (2021). Digital pathology initiatives and experience of a large academic institution during the coronavirus disease 2019 (COVID-19) pandemic. *Archives of Pathology and Laboratory Medicine*, *145*(9), 1051–1061. Available from https://doi.org/10.5858/arpa.2020-0715-SA.

Miquelestorena-Standley, E., Jourdan, M. L., Collin, C., Bouvier, C., Larousserie, F., Aubert, S., Gomez-Brouchet, A., Guinebretière, J. M., Tallegas, M., Brulin, B., Le Nail, L. R., Tallet, A., Le Loarer, F., Massiere, J., Galant, C., & de Pinieux, G. (2020). Effect of decalcification protocols on immunohistochemistry and molecular analyses of bone samples. *Modern Pathology*, *33*(8), 1505–1517. Available from https://doi.org/10.1038/s41379-020-0503-6.

Mohammedsaleh, M. Z. (2014). The role of technical quality control in histology laboratories. *Journal of Cytology and Histology*, *05*(05). Available from https://doi.org/10.4172/2157-7099.1000264.

Morelli, P., Ruspini, M., & Banfi, G. (2013). Analysis of errors in histology by root cause analysis: A pilot study. *Journal of Preventive Medicine and Hygiene*, *54*.

Nakhleh, R. E. (2006). What is quality in surgical pathology? *Journal of Clinical Pathology*, *59*(7), 669–672. Available from https://doi.org/10.1136/jcp.2005.031385.

Nakhleh, R. E. (2008). Patient safety and error reduction in surgical pathology. *Patient Safety-Nakhleh, 132*.

Nakhleh, R. E., & Zarbo, R. J. (1996). Surgical pathology specimen identification and accessioning: A College of American Pathologists Q-Probes Study of 1 004 115 cases from 417 institutions. *Archives of Pathology & Laboratory Medicine, 120*(3), 227–233.

Nelson, B. G., Patel, E., Arth, D., & Nelson, P. T. (2020). Embedding brain tissue for routine histopathology: A processing step worthy of consideration in the digital pathology era. *Applied Immunohistochemistry and Molecular Morphology, 28*(10), 791–793. Available from https://doi.org/10.1097/PAI.0000000000000832.

NSH. (2001). *Guidelines for hematoxylin and eosin staining*. https://nsh.org/sites/default/files/Guidelines_For_Hematoxylin_and_Eosin_Staining.pdf.

O'Rourke, M. B., & Padula, M. P. (2016). Analysis of formalin-fixed, paraffin-embedded (FFPE) tissue via proteomic techniques and misconceptions of antigen retrieval. *Biotechniques, 60*(5), 229–238. Available from https://doi.org/10.2144/000114414.

Pantanowitz, L., Sinard, J. H., Henricks, W. H., Fatheree, L. A., Carter, A. B., Contis, L., Beckwith, B. A., Evans, A. J., Otis, C. N., Lal, A., & Parwani, A. V. (2013). Validating whole slide imaging for diagnostic purposes in pathology: Guideline from the College of American pathologists Pathology and Laboratory Quality Center. *Archives of Pathology and Laboratory Medicine, 137*(12), 1710–1722. Available from https://doi.org/10.5858/arpa.2013-0093-CP.

Patel, A. U., Shaker, N., Erck, S., Kellough, D. A., Palermini, E., Li, Z., Lujan, G., Satturwar, S., & Parwani, A. V. (2022). Types and frequency of whole slide imaging scan failures in a clinical high throughput digital pathology scanning laboratory. *Journal of Pathology Informatics, 13*. Available from https://doi.org/10.1016/j.jpi.2022.100112.

Patel, A., Balis, U. G. J., Cheng, J., Li, Z., Lujan, G., McClintock, D. S., Pantanowitz, L., & Parwani, A. (2021). Contemporary whole slide imaging devices and their applications within the modern pathology department: A selected hardware review. *Journal of Pathology Informatics, 12*(1), 50. Available from https://doi.org/10.4103/jpi.jpi_66_21.

Pekmezci, M., Szpaderska, A., Osipo, C., & Erşahin, Ç. (2012). The effect of cold ischemia time and/or formalin fixation on estrogen receptor, progesterone receptor, and human epidermal growth factor receptor-2 results in breast carcinoma. *Pathology Research International, 2012*. Available from https://doi.org/10.1155/2012/947041.

Retamero, J. A., Aneiros-Fernandez, J., & del Moral, R. G. (2020). Complete digital pathology for routine histopathology diagnosis in a multicenter hospital network. *Archives of Pathology and Laboratory Medicine, 144*(2), 221–228. Available from https://doi.org/10.5858/arpa.2018-0541-OA.

Rojansky, R., Jhun, I., Dussaq, A. M., Chirieleison, S. M., Nirschl, J. J., Born, D., Fralick, J., Hetherington, W., Kerr, A. M., Lavezo, J., Lawrence, D. B., Lummus, S., Macasaet, R., Montine, T. J., Ryan, E., Shen, J., Shoemaker, J., Tan, B., Vogel, H., ... Folkins, A. (2023). Rapid deployment of whole slide imaging for primary diagnosis in surgical pathology at stanford medicine: Responding to challenges of the COVID-19 pandemic. *Archives of Pathology and Laboratory Medicine, 147*(3), 359–367. Available from https://doi.org/10.5858/arpa.2021-0438-OA.

Sanjai, K., Kumarswamy, J., Patil, A., Papaiah, L., Jayaram, S., & Krishnan, L. (2012). Evaluation and comparison of decalcification agents on the human teeth. *Journal of*

Oral and Maxillofacial Pathology, *16*(2), 222–227. Available from https://doi.org/10.4103/0973-029X.99070.

Scott Campbell, W., Lele, S. M., West, W. W., Lazenby, A. J., Smith, L. M., & Hinrichs, S. H. (2012). Concordance between whole-slide imaging and light microscopy for routine surgical pathology. *Human Pathology*, *43*(10), 1739–1744. Available from https://doi.org/10.1016/j.humpath.2011.12.023.

Shankar, V., Grewal, D., Sunder, S., Fossen, B., Peters, K., & Agarwal, A. (2022). Digital marketing communication in global marketplaces: A review of extant research, future directions, and potential approaches. *International Journal of Research in Marketing*, *39*(2), 541–565. Available from https://doi.org/10.1016/j.ijresmar.2021.09.005.

Shields, V. D. C., & Heinbockel, T. (2019). *Introductory chapter: Histological microtechniques. Histology*. IntechOpen. Available from https://doi.org/10.5772/intechopen.82017.

Singh, V. M., Salunga, R. C., Huang, V. J., Tran, Y., Erlander, M., Plumlee, P., & Peterson, M. R. (2013). Analysis of the effect of various decalcification agents on the quantity and quality of nucleic acid (DNA and RNA) recovered from bone biopsies. *Annals of Diagnostic Pathology*, *17*(4), 322–326. Available from https://doi.org/10.1016/j.anndiagpath.2013.02.001.

Sivapathasundharam, B., Panja, P., Sriram, G., & Saraswathi, T. (2007). Comparison of three different methods of tissue processing. *Journal of Oral and Maxillofacial Pathology*, *11*(1), 15. Available from https://doi.org/10.4103/0973-029X.33958.

Snyder, S. R., Favoretto, A. M., Derzon, J. H., Christenson, R. H., Kahn, S. E., Shaw, C. S., Baetz, R. A., Mass, D., Fantz, C. R., Raab, S. S., Tanasijevic, M. J., & Liebow, E. B. (2012). Effectiveness of barcoding for reducing patient specimen and laboratory testing identification errors: A laboratory medicine best practices systematic review and meta-analysis. *Clinical Biochemistry*, *45*(13–14), 988–998. Available from https://doi.org/10.1016/j.clinbiochem.2012.06.019.

Sohal, S. (2016). *CXA121 Tutorial: Staining*. https://www.utas.edu.au/__data/assets/pdf_file/0004/955813/CXA121Tutorial3StainingAnswers.pdf.

Tabatabaee, S. S., Ghavami, V., Kalhor, R., Amerzadeh, M., & Zomorrodi-Niat, H. (2023). Evaluation of errors related to surgical pathology specimens of different hospital departments with a patient safety approach: A case study in Iran. *Patient Safety in Surgery*, *17*(1). Available from https://doi.org/10.1186/s13037-023-00360-1.

Tang, D., Dowbeus, P. A., Firstenberg, M. S., & Papadimos, T. J. (2018). *Patient safety issues in pathology: From mislabeled specimens to interpretation errors. Vignettes in patient safety - volume 3*. InTech. Available from https://doi.org/10.5772/intechopen.79634.

Taqi, S. A., Sami, S. A., Sami, L. B., & Zaki, S. A. (2018). A review of artifacts in histopathology. *Journal of Oral and Maxillofacial Pathology*, *22*(2), 279. Available from https://doi.org/10.4103/jomfp.JOMFP_125_15.

Thekkinkattil, D. (2017). Evaluation of discrepancies in weights of fresh and fixed specimens in breast surgery: A retrospective cohort study. *Gland Surgery*, *6*(1), 27–34. Available from https://doi.org/10.21037/gs.2016.09.10.

Thompson, S. M., Craven, R. A., Nirmalan, N. J., Harnden, P., Selby, P. J., & Banks, R. E. (2013). Impact of pre-analytical factors on the proteomic analysis of formalin-fixed paraffin-embedded tissue. *Proteomics - Clinical Applications*, *7*(3–4), 241–251. Available from https://doi.org/10.1002/prca.201200086.

Tolonen, T. T., Isola, J., Kaipia, A., Riikonen, J., Koivusalo, L., Huovinen, S., Laurila, M., Porre, S., Tirkkonen, M., & Kujala, P. (2015). Length of prostate biopsies is not necessarily compromised by pooling multiple cores in one paraffin block: An observational study Histopathology. *BMC Clinical Pathology*, *15*(1). Available from https://doi.org/10.1186/s12907-015-0001-9.

Torlakovic, E. E., Riddell, R., Banerjee, D., El-Zimaity, H., Pilavdzic, D., Dawe, P., Magliocco, A., Barnes, P., Berendt, R., Cook, D., Gilks, B., Williams, G., Perez-Ordonez, B., Wehrli, B., Swanson, P. E., Otis, C. N., Nielsen, S., Vyberg, M., & Butany, J. (2010). Canadian Association of Pathologists-Association canadienne des pathologistes National Standards Committee/Immunohistochemistry: Best practice recommendations for standardization of immunohistochemistry tests. *American Journal of Clinical Pathology*, *133*(3), 354–365. Available from https://doi.org/10.1309/AJCPDYZ1XMF4HJWK.

Westra, W. H., Phelps, T. H., Hruban, R. H., & Isacson, C. (2003). Surgical Pathology Dissection: An Illustrated Guide. In *Pathology (Philadelphia, Pa.)*. (Second Edition) (Vol. 3, Issue 2). New York: Springer-Verlag.

Wisse, E., Duimel, H., Vreuls, C., Frederik, P., Braet, F., Koek, G., Olde Damink, S. W., van den Broek, M. A., & Dejong, C. H. (2010). Fixation methods for electron microscopy of human and other liver. *World Journal of Gastroenterology*. Available from https://doi.org/10.3748/wjg.v16.i23.

Woollons, A., Holmes, G. J., Gratian, M. J., Bhogal, B. S., & Black, M. M. (1999). Michel's medium: A potential alternative to cryoprotection for tissue transport in the investigation of genetic skin disease. *Clinical and Experimental Dermatology*, *24*(6), 487–489. Available from https://doi.org/10.1046/j.1365-2230.1999.00540.x.

Yagi, Y., & Gilbertson, J. R. (2008). A relationship between slide quality and image quality in whole slide imaging (WSI). *Diagnostic Pathology*, *3*(Suppl. 1). Available from https://doi.org/10.1186/1746-1596-3-S1-S12.

Yildiz-Aktas, I. Z., Dabbs, D. J., & Bhargava, R. (2012). The effect of cold ischemic time on the immunohistochemical evaluation of estrogen receptor, progesterone receptor, and HER2 expression in invasive breast carcinoma. *Modern Pathology*, *25*(8), 1098–1105. Available from https://doi.org/10.1038/modpathol.2012.59.

Yousif, M. Q., & Qasem, S. A. (2016a). Tissue processing and staining for histological analyses. In M. Z. Albanna, & J. H. Holmes IV (Eds.), *Skin tissue engineering and regenerative medicine* (pp. 49–59). Academic Press. Available from https://doi.org/10.1016/B978-0-12-801654-1.00003-6.

Yousif, M. Q., & Qasem, S. A. (2016b). *Tissue processing and staining for histological analyses*. Skin tissue engineering and regenerative medicine (pp. 49–59). Elsevier Inc. Available from https://doi.org/10.1016/B978-0-12-801654-1.00003-6.

Yu, M. H., Lee, T. T., & Mills, M. E. (2019). The effect of barcode technology use on pathology specimen labeling errors. *AORN Journal*, *109*(2), 183–191. Available from https://doi.org/10.1002/aorn.12585.

Zhanmu, O., Yang, X., Gong, H., & Li, X. (2020). Paraffin-embedding for large volume bio-tissue. *Scientific Reports*, *10*(1). Available from https://doi.org/10.1038/s41598-020-68876-5.

CHAPTER

Quality management in digital pathology: analytic and postanalytic phases

5

Orly Ardon

Department of Pathology and Laboratory Medicine, Memorial Sloan Kettering Cancer Center, New York, NY, United States

Introduction

Digital pathology (DP) is destined to change the practice of pathology and is now being adopted in clinical, education, and research use cases. As pathologists became increasingly familiar with using DP for primary diagnostic use, we have witnessed an increase in pathologist buy-in and comfort using the technology (Hanna et al., 2022). Current clinical applications in pathology include primary diagnosis, second opinions, intra-operative consultations, molecular pathology workflows, and multidisciplinary team sharing of pathology digital images. Additional use cases are expected to be implemented in upcoming years.

In order to enable successful clinical implementation, a suitable, multicomponent DP infrastructure must be in place (Ardon, Klein, et al., 2023; Hanna et al., 2022). Each laboratory and organization will require foundational components, including hardware, software, and network components to support the DP system. The exact infrastructure blueprint will depend on the laboratory use cases, resources, and deployment strategy.

In 2013 the College of American Pathologists (CAP) published a guideline on whole slide imaging (WSI) validation for diagnostic purposes. This guideline included 12 statements to guide pathology laboratories in their DP adoption and clinical validation requirements (Pantanowitz et al., 2013). In 2021 an updated guideline that included strong recommendations and good practice statements was introduced, providing additional validation guidance (Evans et al., 2022). Another 2021 publication from the European Society for Digital and Integrative Pathology detailed guidelines including quality checkpoints for clinical DP workflows (Fraggetta, L'Imperio et al., 2021). The DP systems validation will include adequate training and support for the pathologists and laboratory staff to ensure proper use of the DP systems in providing patient care. The clinical implementation will also need to include a quality management system (QMS) as with any other laboratory operation (Ardon, Labasin, et al., 2023; Williams et al., 2019).

Digital Pathology. DOI: https://doi.org/10.1016/B978-0-443-13809-6.00007-5

The World Health Organization Quality Management Handbook summarized quality control/quality assurance (QC/QA) as the set of planned and systematic activities focused on providing confidence that the quality requirements will be fulfilled; QC are the measures that must be included in each assay to verify that the test is working properly, while QA is defined as the overall program that ensures that the final results reported by a laboratory are as accurate and correct as possible (Laboratory Quality Management System: Handbook. World Health Organization, 2011). QC/QA in any pathology laboratory extend to any clinically validated test and should be standard practice in anatomic pathology laboratories (Qihui et al., 2017).

There is a substantiated need for ensuring quality WSI workflows and ensuring the quality of the resulting digital images for all downstream use cases (Fraggetta, L'Imperio et al., 2021; Hanna et al., 2022; Hartman et al., 2017; Williams et al., 2019; Wright et al., 2020). As with other disruptive healthcare technologies, the need for consistent, reliable quality images is essential for pathologists' accuracy and accurate diagnoses as well as for building trust and adoption of digital pathology workflows. The level of image quality required would be dependent on the use case of the digitized image and the potential future downstream applications of the data derived from the slide.

Digital pathology process: preanalytical, analytical, and postanalytical phases

A successful process of tissue digitization includes preanalytical, analytical, and postanalytical steps with multiple checkpoints to monitor quality and ensure successful availability of a quality image for all downstream applications (Fig. 5.1).

The preanalytical step in DP refers to the steps taken before the digitization of the glass slide. This phase includes processing of tissue samples, embedding into paraffin blocks, sectioning, staining, coverslipping. and drying the glass slides. This process is crucial for the production of a quality glass slide. The analytic step refers to the process of WSI where glass slides are scanned using WSI equipment, resulting in digital slide images. The postanalytic phase refers to the processes that occur after the digital image is produced by the scanner.

Ensuring image quality: multicheckpoint process

Ensuring the availability of quality pathology images for pathologist review and accurate diagnostic results relies on multiple quality control checkpoints that should be implemented throughout the DP workflow in pathology laboratories. These checkpoints are comprised of two glass slide level checkpoints in the preanalytical phase, two additional digital image checkpoints in the analytical phase, and one checkpoint in the postanalytical phase (Fig. 5.2).

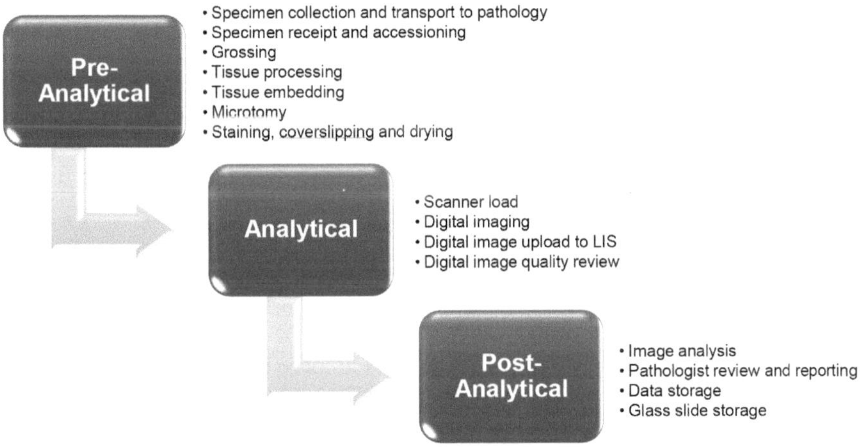

FIGURE 5.1

Workflow steps identified in the preanalytical, analytical, and postanalytical steps in a DP laboratory operation.

The preanalytic QC process requires high quality, specific specimen preparation, and careful macro-evaluation of slide artifacts that have to be resolved before scanning whenever possible (Fraggetta, L'Imperio et al., 2021; Hanna et al., 2022; Patel et al., 2022; Schüffler et al., 2021; Shaker et al., 2022). This QC step is especially important for glass slides submitted for consultation from other hospitals and laboratories as these have varying, inconsistent degrees of slide preparation artifacts which have to be minimized as much as possible before scanning (Hanna, Parwani, et al., 2020; Hartman & Pantanowitz, 2017; Hartman et al., 2017). Consultation slides submitted from other institutions may also have multiple barcode labels and these may cause barcode errors on detection and cause the digitized slide not to be linked or available within the laboratory information system (LIS). Including a barcoding and tracking station to scan each glass slide ensures each barcode is readable by handheld barcode readers prior to placing the slides on the whole slide scanner and can also be configured to track the glass slide in the LIS as scanned or marked for scanning. Glass slides without barcodes, or defective barcodes, should have new patient labels with barcodes printed and affixed on the glass slide. Initial training of pathology staff to apply slide labels with barcode in appropriate areas of the slide to keep essential data (patient identifiers, outside accession number, stain name, etc.) visible on WSI may be needed (Hanna et al., 2022).

In addition, the handling and transportation of glass slides poses a risk of introducing environmental-derived quality issues for the glass slides. Fig. 5.3 shows examples of consultation slides submitted from external institutions after arrival at Memorial Sloan Kettering (MSK) and relabeling with secondary MSK labels.

These include cracked or chipped glass or coverslip, dried mounting media, peeling labels, and accumulated dust and debris. Another challenging category of

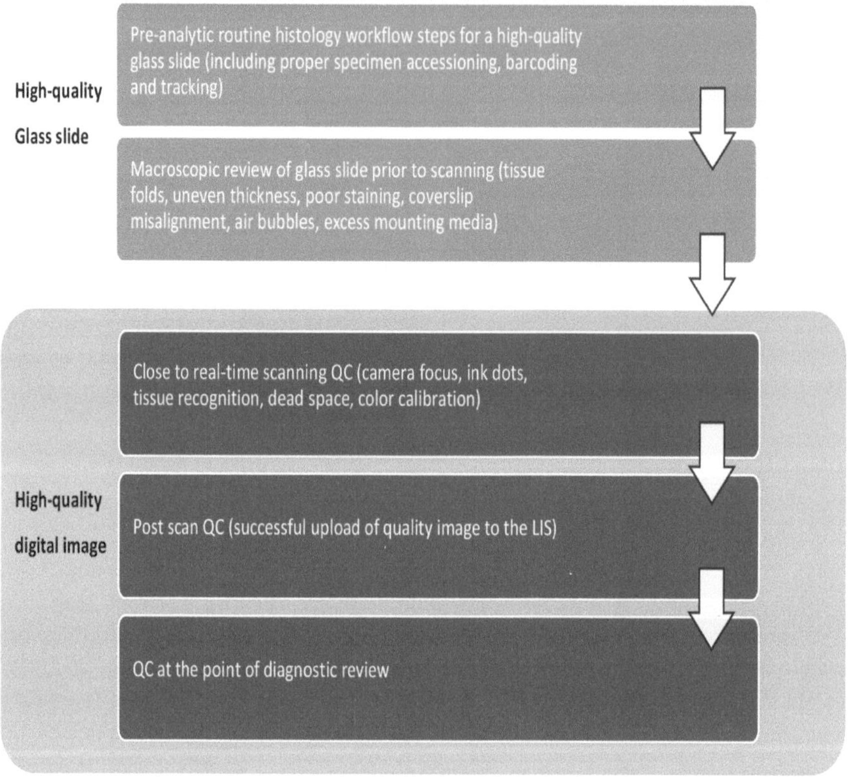

FIGURE 5.2

DP slide and image quality checkpoints introduced in the preanalytic, analytic, and post phases of whole slide imaging at Memorial Sloan Kettering Cancer Center.

slides are older, archival slides that may have been affected by years of storage in suboptimal conditions as well as require the addition of newer, barcoded labels. Slides with glass defects cannot get successfully scanned on most high-throughput scanners and laboratories can experience potential damage to the scanner if attempting to scan in some of these compromised slides (Pantanowitz et al., 2015).

Analytic step

Scanner level quality checkpoint

Once a well-stained and prepared histologic glass slide is prepared, it is ready to be loaded to the WSI. This scanning process of the glass slide resulting in a whole slide digital image is known as the analytical step. This step relies on a combination of equipment identified error codes and manual inspections of the resulting images.

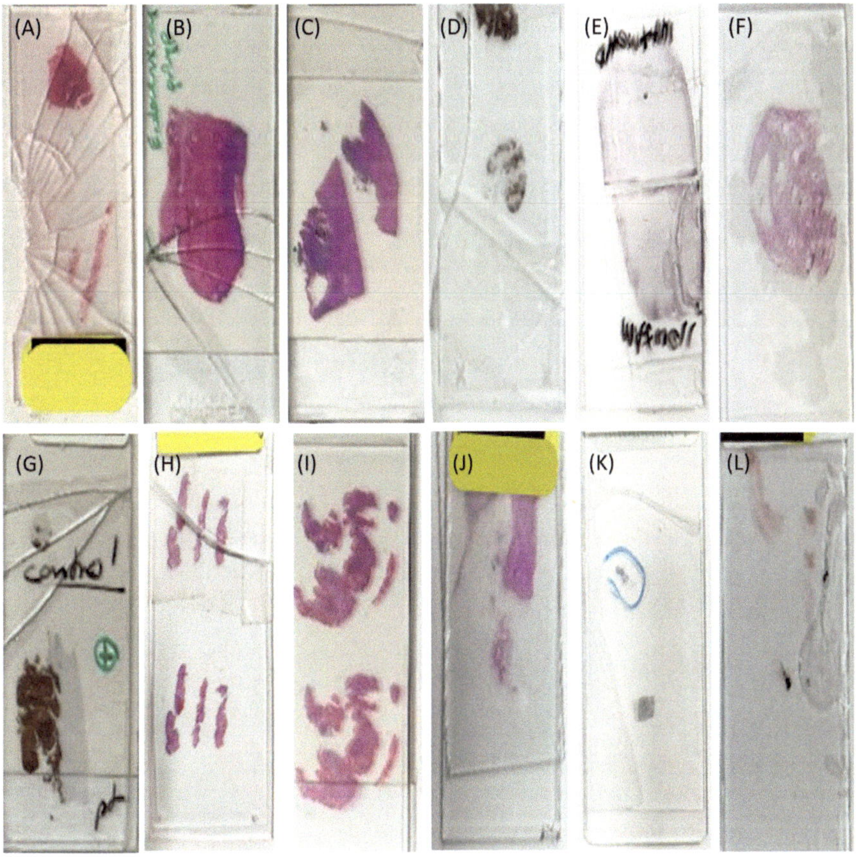

FIGURE 5.3

Examples of slide level quality artifacts are typical of consultation slides sent to Memorial Sloan Kettering from external institutions. These slides were collected within a few hours in a routine daily workflow and require additional preparation work in order to allow placement on the whole slide scanner and obtain a satisfactory quality image. Artifacts include: cracked, broken glass (A–D, G, H), ink marking (B, C, E, G, K), excessive, dry mounting media (D, E, K, L), air bubbles (F, J), tissue outside glass (G, I), misaligned coverslip (D, E, J), slides with tape that needs removal (D, H).

The key parameters that are typically used to evaluate WSI quality in the analytic phase include:

1. High resolution and clear images with no pixelated areas or stitching (i.e., misalignment of tissue) (Fig. 5.4D, J, K).
2. Out of focus regions: Images should be uniformly focused across the entire slide, with no areas that are out of focus or blurry (e.g., Fig. 5.4A, J).

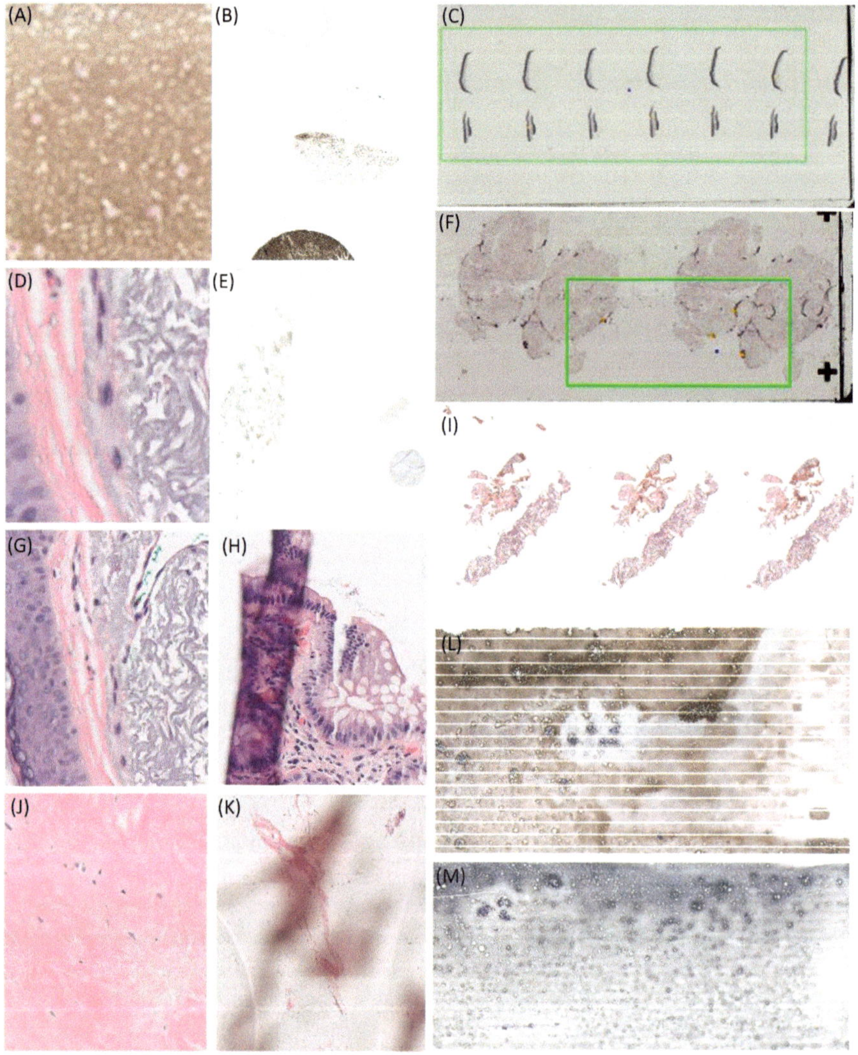

FIGURE 5.4

Examples of common digital image artifacts detected in the analytic phase by the digital scan team at MSK: blurry image (A, J), slide containing contaminating fiber leading to blurry image (B), missing tissue (C, F), raised coverslip with tissue outside coverslip (E), stitching errors with out of focus regions (D, G), tissue fold (H), image banding (I, L, M), stitching due to excessive mounting media (K).

3. Tissue detection capture completeness (i.e., tissue outside coverslip, under coverslip, or no tissue detected) (Fig. 5.4C, F, E).
4. Color accuracy: The images should accurately reflect the colors of the original tissue sample, which is crucial for identifying cellular structures and abnormalities.

5. Banding on the image which occurs when there is debris on the objective (Fig. 5.4I, L, M).
6. Glass slide artifacts: WSI images should be free from artifacts such as cover slip stains, scratches, or dust (Fig. 5.3B, C, E, G, K).
7. Uniformity: Images should be homogenous across the entire slide, with no areas that are over- or underexposed, or with variations in color or contrast.

Fig. 5.4 shows some examples of frequent artifacts detected in postscan manual review of digital slides in the MSK DP operation.

Digital scanner identified error codes will depend on scanner hardware but typically include barcode detection failures, failures to detect all tissue on glass slide (i.e., usually slides with scant tissue, fatty tissue, or faint immunohistochemical stains) and image quality error. It is preferable to perform the QC checks on the scanned images, such as checking for artifacts or inconsistencies in the image before the slide is removed from the scanner. The mitigation for these errors will include a second full slide scan that should not add a significant delay in the turnaround time of the availability of a slide to the pathologist (Ardon, Labasin, et al., 2023; Atallah et al., 2022; Hanna et al., 2022; Hanna, Parwani, et al., 2020; Wright et al., 2020).

The internal quality control procedures of most of the WSI systems are not sufficient as incorrect focus points, scanning and glass slide artifacts or missing tissue are not always detected by the different scanners and require manual post-processing after scanning (Schüffler et al., 2021). At some institutions, an additional manual QC check of the digital image quality is performed by the scanning team after the slides have been unloaded from the scanner. The scanned image from the DP system database or the uploaded image in the LIS is manually evaluated for quality issues. Slides with identified image quality deficiencies require retrieval of affected slides so they can be re-scanned in an attempt to generate an improved, adequate digital image (Ardon, Labasin, et al., 2023; Hanna, Reuter, et al., 2020; Hanna, Parwani, et al., 2020; Patel et al., 2022).

Scanning operations can be delegated to histologic technicians/technologists or nontechnical laboratory staff with the appropriate training and demonstrated competencies. This trained staff is responsible for quality scanning operations in the preanalytical and analytical phases and minimize WSI defects prior to reaching the pathologist for postanalytic, diagnostic review. The team size will be determined by the size of the operation, use case image quality needs, and the number of representative slides chosen for routine review.

Manual QC review of glass and digital slides is laborious, qualitative, subject to intra- and interreader variability (Hanna et al., 2022; Janowczyk et al., 2019) and can add significant cost to digital scanning operations (Ardon, Klein, et al., 2023). All errors and troubleshooting should be recorded and monitored to establish quality metrics and performance thresholds as well as to identify educational opportunities for staff (Ardon, Klein, et al., 2023; Williams et al., 2017). An alternative to the manual QC review would be a computational automated focus quality assessment technique if that could be compatible with widely available scanning platforms, without additional custom hardware. It can also be a cost-effective solution for DP departments (Ardon, Labasin, et al., 2023). Automated QC software is being introduced in some organizations and may be the future standard for DP operations (Janowczyk et al., 2019).

Scanner quality variability

There are many digital glass scanners available in the market that vary in technical features and suitability for user-specific scanning needs (Hanna et al., 2022; Hanna, Parwani, et al., 2020; Patel et al., 2021). Scanner preference and selection considerations vary greatly according to the use cases, volume, turnaround time, specimen specifics, magnification required, and the available space and other infrastructure. Digital scanner selection criteria to be considered before instrument purchase and onboarding will include features such as scanner capacity, image quality continuous/batch load options, glass slide size, magnification, brightfield/fluorescence needs, scan speed and time for image availability, file format and viewer, ergonomics, cost, lab footprint, expected downtime, reliability, and availability of vendor service (Ardon, Labasin, et al., 2023; Hanna et al., 2022).

Scanner considerations will also depend on the laboratory's available budget (Ardon, Klein, et al., 2023; Hanna et al., 2022; Patel et al., 2021). Laboratories need to consider the cost and reliability of vendor service contracts necessary to support digital scanning equipment and systems as part of their overall decision making (Ardon, Klein, et al., 2023). Additional budget considerations include the costs associated with maintaining digital file storage including a back-up archive for digital image access during DP system downtime. The selection of digital scanning equipment/systems will vary among institutions and is also likely to change over time with improved technology, evolving scanning needs of the users and improved pricing and financing options (Ardon, Klein, et al., 2023; Hanna et al., 2022).

The different technical features of the WSI system will determine quality and image resolution such as objective lens magnification, sensor pixel size, and the monitor used for viewing the digital slide image (Patel et al., 2021, 2022; Sellaro et al., 2013; Zarella et al., 2018). Most modern scanners come equipped with autofocus optics system to select focal planes to accurately capture the three-dimensional tissue morphology similar to a two-dimensional digital image (Zarella et al., 2018). One of the main causes of image quality degradation is a scanning failure, which introduces focus blur or noise in the WSI scanner. With these types of scanners, focus points are selected automatically or manually for each glass slide and if focus points are selected from a region of different focus depth than the desired tissue area, the neighboring tissue region will be out of focus and the slide will need to be re-scanned (Shakhawat et al., 2020).

Some WSI scanners can generate a prescan image that can be manually inspected to assess focus quality prior to full glass slide scanning. Newer WSI scanner models have a built-in QC algorithm to allow detection of image distortion including blurred image, banding, stitching, and missing tissue scans thereby bypassing the need for time-consuming manual QC (Zarella et al., 2018). For laboratories with high-throughput scanning operations, it is impractical to manually perform focus QC checks for each individual slide (Hosseini et al., 2020). Farahani et al. (Pantanowitz et al., 2015) and later Patel et al. (Patel et al., 2021) have published extensive reviews of the hardware capabilities of different scanner models and the resolution and quality of their individual components. Scanner features can have implications on any subsequent image analysis procedure. For example, Khan et al. (2022) presented a study that assessed the impact of scanner variability on lymph node segmentation. By employing the same glass slides scanned across different scanners, they removed all nonscanner variabilities from downstream consideration. Their results demonstrated that scanner variabilities alone can severely negatively affect both traditional and deep learning algorithms. They recommend calibrating all scanners to the same slide printed with a color scheme according to the International Color Consortium (ICC) standards (Khan et al., 2022). Ogura et al. observed 23.1% of discordant classification results between digital slides obtained from two independent scans of the same glass slide on the same NanoZoomer scanner (Hamamatsu Photonics K. K., Shizuoka, Japan).

Differences between scanners and their quality image capabilities should be considered by pathology departments and accounted for the intended volume and use cases of the WSI efforts. The scanner features as well as the manual labor involved should be part of an extensive prepurchase evaluation study of the scanner adding to the purchase decision of the institution and their specific needs (Ardon, Klein, et al., 2023).

Barcode quality considerations

Barcodes are essential for modern pathology operations and DP prerequisites include having glass slides barcoded (Atallah et al., 2022; Ferreira et al., 2023; Fraggetta, Caputo et al., 2021; Hanna et al., 2019; Hanna, Reuter, et al., 2020; Lujan et al., 2021; Montezuma et al., 2022; Ramaswamy et al., 2021; Rao et al., 2021; Retamero et al., 2019; Schüffler et al., 2021). The barcodes should enable tracking of patient specimen and slides across the entire path of workflow and interface across the different technologies and IT systems used including the LIS. Requiring bar-code scanning at key steps (e.g., microtomy, staining, WSI, slide distribution, sign-out, and transcription) prior to advancing to the next step allows auditing and tracking of each step in the DP laboratory workflow (Hanna & Pantanowitz, 2015).

There are various glass slide barcoding printing technologies available including printing barcodes directly on glass slides or printing barcodes on adhesive labels using label printers that can be applied to corresponding slides (Fraggetta et al., 2021; Hanna & Pantanowitz, 2015). Most newer DP scanners can detect 1D and/or 2D barcodes (Hanna et al., 2022). 2D barcodes are preferred to 1D barcodes because of their smaller size, their ability to fit on the surface of both tissue blocks and glass slides, their ability to adhere to convex surfaces of tissue specimen containers, and are generally less prone to scanning issues (Fraggetta et al., 2021). Once the whole slide scanners decode barcodes captured in the glass slide label during the digitization process, messaging (e.g., HL7, XML) and high speed network communication protocols allow seamless integration with DP system databases as well as the LIS (Hanna & Pantanowitz, 2015).

Barcoding needs for secondary labeling of outside pathology slides received for secondary consultation must also be considered if such slides will be digitally scanned. One solution is to place a bar-coded secondary label below the primary label to avoid obstruction of the originating laboratory label data. However, this placement can result in multiple barcodes detected by the scanner thus preventing successful scanning. In these situations, another label may need to be placed on the glass slide to temporarily mask the outside institutional label information and allow the scanner to only recognize and decode the desired internal lab barcode. There are some DP systems with software capabilities to automatically suppress noninstitutional barcodes without the need to mask the outside institutional barcodes (Hanna & Pantanowitz, 2015; Hanna et al., 2022).

Barcode failures can occur at the digital slide scanner or at the LIS level and a failure rate associated with the successful barcode detection and subsequent digital image upload into the laboratory LIS can be monitored. Barcodes printed on glass slide labels should be of high-quality ink, legible with no cutoff or smeared regions. Pro-active barcode scanning of barcoded glass slide labels at the preloading stage can minimize WSI scanner errors downstream. Codifying these scanning steps within the LIS (e.g., "marked for scanning," "scanned") allows for scanning workflow tracking and glass slide tracking. Barcodes that failed successful reading at the preloading stage will require the creation and adherence of new patient labels with readable barcodes prior to digital scanning. This may contribute to delays in digital scanning turnaround time (Ardon, Labasin, et al., 2023; Hanna & Pantanowitz, 2015). DP staff should be trained in managing barcode scanning deficiencies and mitigation steps (Ardon, Labasin, et al., 2023).

Label printer maintenance is essential to avoid printing of defective labels. In one study, Hanna and Pantanowitz (2015) reported that proper slide label printer maintenance reduced barcode scanning errors by up to 20%. In a WSI quality failure study published in 2019, Hanna et al. reported that out of 2221 glass slide scanned, 4 out of 148 glass slides needed to be rescanned, representing 5.9% or 0.2% of total slides scanned failed due to barcode errors preventing image upload into the LIS (Hanna et al., 2019). Barcode failure issues are inherent and prevalent across institutions and scanner models used but can be proactively addressed to minimize errors and ensure efficient DP scanning operations.

Postanalytic step quality factors

The postanalytical phase of DP involves the interpretation and reporting of results by a pathologist from the analysis of DP images. Pathologist digital slide workflow requires the pathologists to view and evaluate digitized slides using a computer workstation instead of evaluating glass slides using a traditional microscope. This pathologist review of digitized slides requires a standardized pathologist workstation setup, including validated displays/monitors, viewer software, and input devices. Some of the infrastructure required can be tailored to a pathologist's individual preferences (e.g., monitors), yet there are basic technical requirements for the pathology digital review workstations (Clarke et al., 2019; Hanna & Ardon, 2023; Kim et al., 2020; Williams et al., 2017). Additional requirements include high speed internet connections for the efficient transmission and quick retrieval of digital images from digital storage systems (Ardon, Klein, et al., 2023; Eloy et al., 2021; Fraggetta et al., 2017; Fraggetta, L'Imperio et al., 2021; Hanna, Parwani, et al., 2020; Montezuma et al., 2022; Schüffler et al., 2021). DP has been shown to slow down the efficiency of pathologists due to technical constraints of image visualization on a computer screen and a computer mouse as input device (Hanna et al., 2019).

Quality control at the postanalytical phase includes ensuring DP images are of sufficient quality to enable the evaluation and diagnosis by the pathologist (Hanna et al., 2022). Some institutions skip the routine QC in the analytic phase as described earlier before delivery to the pathologists as they find the QC errors infrequent. Patel et al. report analytic QC error rates of less than 5% and that affect only small parts of the slide, thus rarely causing diagnostic problems (Patel et al., 2022). They suggest that assigning the manual check of each WSI scan to the scanning team technicians would lead to a significant time-consuming process for less than 5% of rescan. As an alternative solution, some institutions recommend deferring the QC step to the pathologists who can request, if needed, a rescan of slide directly from the LIS (Ferreira et al., 2023; Fraggetta, L'Imperio et al., 2021; Patel et al., 2022). Other institutions choose to invest in extensive QC image review to prevent the delay in the pathologist workflow, and increase the comfort level of the pathologist with DP (Ardon, Klein, et al., 2023; Hanna et al., 2019, 2020).

WSI diagnosis can be audited in a similar way to existing departmental glass slide diagnostic audit, with a random sample representing a proportion of the diagnostic workload reviewed by a second pathologist (Williams et al., 2020).

Pathologists need to understand and adjust to digital workflows and there are several factors that could impact digital image presentation and evaluation by the pathologist. Kim et al. reported different variables that influence high power field (HPF) in WSI review including the slide scanner device, scan resolution, software image viewer, monitor display resolution, monitor size, and aspect ratio (Kim et al., 2020). The findings overall showed a wide variability in the HPF area generated by different hardware and software and that such differences in tissue area presented to a pathologist are divergent from those seen using a light microscope. Understanding these differences and validating these alterations are necessary to ensure that the image being presented and evaluated by the pathologist is optimal and equivalent to the quality visualized microscopically using glass slide. Pathologists who are unaware of the impact different variables may have on the

HPF can lead to an over- or underestimation of mitotic counts, eosinophil quantification, lymphocyte quantitation, and so on when rendering pathology reports. The authors recommend additional software development to ensure comparable tissue presentation of a single HPF to a pathologist using WSI or re-evaluation of formal pathologic criteria/scoring systems to accommodate DP (Kim et al., 2020).

Whole slide image viewer software

For wide adoption of DP, digital slide viewers must be able to efficiently open cases and display and navigate digital slides obtained from whole slide scanner. The image viewer software used by the pathologist plays a critical role in DP workflows as it enables pathologists to pan, zoom, and focus on an image in a similar sequence as compared to a brightfield microscope while examining the digital slides. While microscopes enable instantaneous glass slide visualization, digital viewers may be delayed by slide loading times, image rendering times, and mosaic effects. Viewer errors affecting the postanalytic phase of DP may be derived from failures in launching cases from the LIS or image management system, loading and rendering digital slide file/image, and field of view navigation (e.g., rendering of images during pan and zoom functions) (Hanna et al., 2022).

Different whole slide scanner image viewer software is offered by the hardware vendors but is limited to a specified vendor format. Viewers have different slide visualization performance metrics, including time to open a case, time to change a slide, and time to change a field of view (Schüffler et al., 2022). When multiple whole slide scanning systems from different vendors are integrated into a single clinical operation, a vendor-agnostic viewer should be considered so that all digital slides can be viewed on a single image viewer with consistent digital tools and user interface (Kim et al., 2020; Schüffler et al., 2021).

Image viewer software was found to result in differences in HPF area though these were minor in comparison to the reported differences seen with display resolution and monitor size (Kim et al., 2020). These viewer differences were attributed to variations with each viewer application interface (e.g., toolbars and navigation panes) and various elements presented on the screen that may obscure tissue being viewed (Fig. 5.5).

Image viewers should include the ability for a pathologist to document QC rescan requests and other identified QC findings at the postanalytical phase. Some digital image viewers show a macro image for every slide and enable pathologists to easily report poor quality slides or missed tissue to the scanning group, such that those slides can be quickly rescanned. While viewing a digital slide in the whole slide image viewer software, pathologists may identify missing tissue or poor-quality digital slides and require prompt rescanning of the glass slide or need to defer to review the corresponding glass slide. An example of this type of pathologist QC documentation tool is described in a publication by Schüffler et al. who describe the creation of a "Report Slide Problem" feature in their viewing software (Schüffler et al., 2021). This feature provides the capability of the

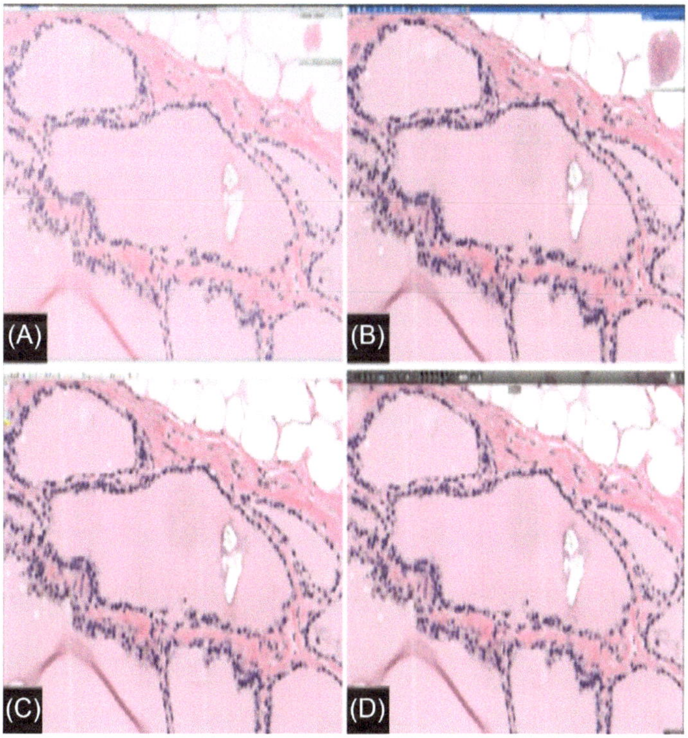

FIGURE 5.5

High-power field tissue area shown on four WSI viewer software displayed on the same monitor (Philips PS27QHDCR). (A) Philips IMS; (B) MSK Viewer; (C) Aperio ImageScope; (D) 3DHistech CaseViewer. The same high-power field area of tissue seen on four different whole slide image viewer software viewed on the same monitor size and resolution with very minor differences in the tissue size area represented as a result of user interface variations (e.g., toolbars) (Kim et al., 2020).

pathologist to document their QC findings or send re-scan request via email to the scanning team. The functionality links the request to the case accession number and includes the details of the reported issue, such that the slide can quickly be retrieved and rescanned (Fig. 5.6). Similar features are now available in several commercially available image viewers.

Display monitors

Display monitors enable pathologists to view and analyze the digital slides. The reliability and quality of the monitors are essential for accurate pathologist review and diagnosis and must be optimized and validated to ensure consistent and

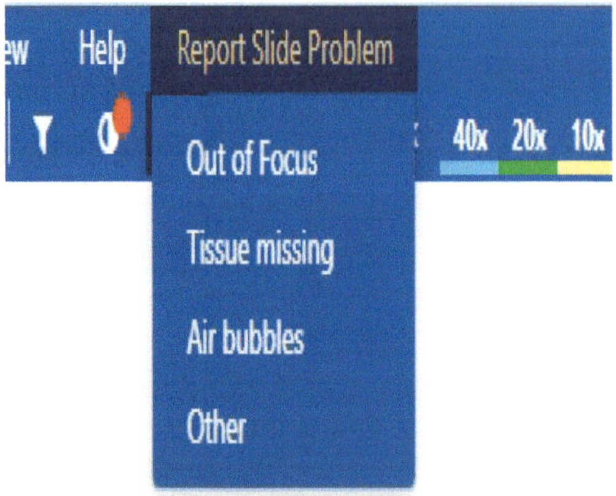

FIGURE 5.6

Example of a pathologist level image quality checkpoint from the MSKViewer. The functionality allows pathologists to report specific QC findings such as focus issues, missing tissue, and presence of air bubbles as specific to the case being reviewed. A link to the accession number and slide is sent to the scanning team via email to quickly retrieve the relevant glass slide and rescan successfully using appropriate steps to mitigate the specific reported QC finding.

reliable results (Clarke & Treanor, 2017; Clarke et al., 2019; Hanna, Parwani, et al., 2020; Randell et al., 2015; Sellaro et al., 2013). The amount and detail of image information visible on monitors will depend on monitor size and resolution and was found to affect time to detect areas of interest, time to diagnosis, and user confidence (Randell et al., 2015; Rojo & Bueno, 2015).

Kim et al. studied differences between glass slide visualization on a traditional light microscope versus a digital representation of the WSI, and reported that monitor resolution and size affected the HPF (Kim et al., 2020). These variables should be considered and all DP system components should be appropriately validated to ensure reliability (Fraggetta et al., 2021; Hanna, Reuter, et al., 2020; Pantanowitz et al., 2013). The need to optimize DP display screen environments for making reliable observations led to the introduction quality assurance tools such as the Point of Use Quality Assurance Remote Image (POUQA) tool. This tool was originally developed to assess monitors for radiologists working remotely (Clarke et al., 2019; Wright et al., 2020). In 2019 the tool was modified for testing monitors for digital microscopy and pathology remote use as the display screen equipment and environmental lighting may impact the visual quality of digital slides, and therefore patient diagnoses (Wright et al., 2020). The software tool is clinical use (Lujan et al., 2021; Wright et al., 2020) and is available for online for use remote pathology work and reporting (Point of Use QA & Point of Use Device Validation for Digital Pathology, 2023).

Automated digital slide quality tools

The growth of DP adoption in clinical use and the current practice of most laboratory teams, to manually review digital images to ensure digital slide quality, make the case for automated solutions for detecting quality issues encountered in the DP workflow checkpoints. A major requirement from an automated clinical grade QC tool is a seamless, speedy analysis that will not result in a delay in availability of the images to the pathologist.

Automated QC solutions are described in the literature. Kanwal et al. (2022) reviewed QC tools based on artifact detection (e.g., blur, tissue folds, blood, and damaged regions) and overall quality evaluation. Brixtel et al. listed tools based on the pathology domain (histology and cytology), the type of artifact (air bubble, marker, ink, tissue folds foreign objects, biological contaminant), and type of tool method (image processing, machine learning, and deep learning) (Brixtel et al., 2022). At the time of this publication, it is unknown if these tools are available for use in clinical DP operations.

The following section cites some automated quality tools that address scanner level errors in the chronological order in which they were developed and published. Not all research contributions to this field were included due to publication limitations.

Walkowski and Szymas (2011) developed an algorithm dedicated to the comparison of the quality of digital slides generated by different scanner models. Their resulting algorithm that was developed can also be used to identify blurry regions (Senaras et al., 2018; Walkowski & Szymas, 2011).

The same year, Zerbe et al. developed a distributed image analysis system to calculate the amount of sharpness of image patches, classifying each patch into one of the four sharpness categories (Zerbe et al., 2011). Hashimoto et al. (2012) proposed a method for assessing blur and noise by an evaluation index.

Lopez et al. (2013) developed a model to characterize tile sharpness using a thresholding step. Although these and other earlier studies have shown promising results, Seneras et al. (2018) suggested that they were not validated to justify how they generalize to unseen data, or how well they perform for varying amounts of blurriness.

Avanaki et al. (2016) developed a tool for images with blur, stitching, folded tissue, or air bubble artifacts and reported that their results were correlated with the ratings given by a pathologist.

Campanella et al. (2018) developed a blur detection algorithm that produces fine-grained sharpness assessment maps. This tool was tested on three datasets and against human experts (Campanella et al., 2018).

Hossain et al. (2018) developed a machine learning-based tool but reported that some artifacts (i.e., tissue-fold, air-bubble) were detected as false positives. A later 2020 paper from the authors proposed discriminating between quality failure slide preparation or scanning causes. Their algorithm relies on low resolution WSI for artifact detection and offered a speedy quality assessment classifying images into good, average, or poor quality groups (Shakhawat et al., 2020).

Senaras et al. (2018) describe the development of DeepFocus, a convolutional neural network (CNN)-based tool for identification of blurry regions. The authors acquired data at different focal planes to introduce controlled blurring and performed validation on independent test sets. The time requirement for the algorithm to analyze a full digital slide (16 GB) at 40× magnification was 643 seconds, which may be impractical for high-throughput clinical workflows.

Another CNN-based quality detection tool, HistoQC, was developed by Janowczyk et al. (2019). It is a modular open-source QC application that allows assessment of color histograms, brightness, and contrast of individual digital slides as well as identifications of cohort-level outliers (e.g., darker or lighter stain than other slides in the cohort) (Janowczyk et al., 2019).

Hosseini et al. (2020) developed FQPath, for analysis of blur levels and generation of a local slide-level focus quality heatmap. Smit and Cigéhn (2021) describe the development of a tool to identify six semantic artifact types in an abstract, claiming to outperform other solutions. The authors also proposed a multidecision computational quality control system (no action, re-prepare slide, and re-scan) for routine clinical practice (Smit & Cigéhn, 2021). Haghighat et al. (2022) developed a QC tool, PathProfiler, that provides three slide-level scores to predict the WSI usability, focus, and staining qualities.

Generative adversarial network (GAN)-based approaches in eliminating noise to produce normalized hematoxylin and eosin (H&E)-stained images have also been proposed (Barua et al., 2023; Shimizu & Nakayama, 2020; Zanjani et al., 2018). GAN class of machine learning models are built with two competing subnetworks: a generator network and a discriminator network (Goodfellow et al., 2020). In the training stage, the generator network learns to generate fake images that are hard to distinguish from real images by the discriminator. The discriminator network aims to separate the generated fake images from real images. Once the discriminator and the generator reach equilibrium, the generator is able to produce like-real images that cannot be distinguished by the discriminator (Rong et al., 2023). Rong et al. (2023) described the development of GAN to improve image quality issues such as blurred regions, enhance low resolution images, and normalize stain colors.

CNN- and GAN-based quality tools are promising to play an essential role in the automated and quantifiable QC of whole-slide images and allow the standardization of the quality of images in computational pathology.

Wang et al. (2021) proposed that new tools are needed for understanding where machine learning tools in pathology models might fail and had developed a tool for generating simulated artifacts such as simulated artifact, bubbles, tissue folds, uneven illumination, pen marks, sectioning artifacts, altered staining, and tissue tears for testing and building trust in machine learning models.

Brixtel et al. (2022) made three recommendations for future global quality assessment of WSI:

1. Ability to simultaneously detect different quality problems and to globally assess the slide quality in terms of analyzability for a final diagnosis.

2. Ability to detect out-of-the-scope artifacts to avoid their analysis by an artificial inteligence (AI).
3. Methods have to be robust to the different real life quality situations, including generalization abilities for different datasets coming from multiple institutions.

The importance of automation in the QC processes is likely to increase with the adoption WSI in pathology, and the need to minimize manual QC processes and to standardize workflows (Boyd & Hawker, 2012). As laboratories will need to address their QC software gaps, vendor-provided tools that could integrate in the image viewer system will be essential to help ensuring the image quality of the WSI process.

Image quality and computational pathology tools

Image quality is essential to the creation of large pathology data sets for the development and validation of AI-based research and computational pathology tools (Janowczyk et al., 2019; Khan et al., 2022; Michielli et al., 2022; Swiderska-Chadaj et al., 2020; Tizhoosh & Pantanowitz, 2018). Immunohistochemistry (IHC)/H&E staining of tissue sections can vary significantly across laboratories and at intra-laboratory level, with different scanners and workflows. Minor variation in the WSI, which represents no problem for the human eye, can be enough to significantly affect the performance of these tools (Janowczyk et al., 2019; Khan et al., 2022; Michielli et al., 2022; Swiderska-Chadaj et al., 2020). H&E and IHC slides that are produced in different clinical operations also have unique characteristics (e.g., variation in tissue processing, tissue stain protocols, stain quality, color intensity) and other confounding factors that may be introducing batch effect bias leading to differences in image properties between the batches (Brixtel et al., 2022; Clarke et al., 2019; Dehkharghanian et al., 2023; Khan et al., 2022; Kothari et al., 2013). Source data quality is especially challenging in collaborative repositories, such as the Cancer Genome Atlas (TCGA) where a wide spectrum of artifacts and image qualities can be seen in many of the 30,000 archived digitized slides. This occurs despite the fact that these slide images undergo manual QC before being introduced into TCGA (Janowczyk et al., 2019; Kothari et al., 2013). Analysis performed on low-quality tissue, histology slides, or staining will compromise the validity of data (Parwani et al., 2023). It is therefore important to remember that high-quality, well-annotated, and large datasets composed from high-quality histology yield high performing deep learning models (Janowczyk et al., 2019; Khan et al., 2022; O'Hurley et al., 2014).

Published quality error rates

Detected image quality error rates vary among institutions, use cases, scanner models, subjectivity of staff members involved in manual QC and selected QC activities. Several publications reported site-specific error rates. The range of rescan rates is reported to be between 1% and 16% in different studies and suggest a subsequent increase in first time scanning success rate over time with experience and workflow adjustments (Eloy et al., 2021; Ferreira et al., 2023; Fraggetta et al., 2017; Fraggetta et al., 2021; Hanna et al., 2019; Hanna, Reuter, et al., 2020; Lujan et al., 2021; Mukhopadhyay et al., 2018; Patel et al., 2022; Retamero et al., 2019; Schüffler et al., 2021).

Lujan et al. summarized a 3-year large scale scanning operations error rate (for the years 2017–20) that averaged 2138 slides a day (Lujan et al., 2021). Scan failures by the Philips Ultra Fast Scanners (Philips Medical Systems, Nederland B.V.) were uncommon, and reported at a rate of 1.2%. The average pass rate for their daily quality metrics is 97.4%. Failure typically resulted from a failure to scan the entire tissue area on slide, weak H&E or immunohistochemical staining intensity, or incorrect slide labels (Lujan et al., 2021).

In a later publication from the same group, Patel et al. reported on the experience of WSI imaging of close to 2.3 million slides using additional scanners (Patel et al., 2022). The study reported 1.2% of the total slides had scanning issues, most commonly due to focus problems or in the tissue finding algorithm. For example, abundant white adipose tissue that is ignored by the tissue finder, or some parts of the slide that are out of focus (Patel et al., 2022). The rescan requests by the pathologists were only 0.2% for primary diagnostic slides.

Ferreira et al. reported that the main reasons for a rescan in their private laboratory DP validation using Aperio GT450DX (Leica Biosystems, Buffalo Grove, Illinois, USA) were bad barcodes, thick slides, excess glue/mounting medium, and air bubbles resulting in an out-of-focus slide (Ferreira et al., 2023). The total rescan rate of their validation set of 8474 slides was 0.6%.

Mukhopadhyay et al. (2018) studied image quality using the Philips Ultra Fast scanners. In the first scan of 3390 slides, the system was able to automatically detect an issue, such as no tissue or label detection in 2.3% of the slides. A total of 2.1% of the images did not pass the image quality check by the scanning operator for slide-related issues such as prior ink markings, broken slides, or debris on the slide. A total of 1.6% of the slides had an out of focus image, and 0.03% had missing tissue. In a rescan, issues were detected in 0.6% of the slides.

Eloy et al. studied rescan rates on Pannoramic1000 scanner (3DHISTECH Ltd., Budapest, Hungary), and reported a rescan rate of 0.1%–1.8% in a 12-month study period from July 2020 to Feb 2021, with a mean of 1% (Eloy et al., 2021). The reasons reported in the study for rescanning slides were poor focus or incomplete scanning of the fragments and/or difficulties associated with uneven thickness of the tissue. The most frequent cause of scanner failure was the misprinted QR code, leading to failure to scan sections placed in the lower limits of the slide (Eloy et al., 2021).

Fraggetta et al. studied failure rates following go live of their production system with more than >115,000 histology glass slides that were digitized using Aperio AT2 (Leica Biosystems, Buffalo Grove, Illinois, USA). The scan fail rate experienced was around 1% (Fraggetta et al., 2017). In another study, Fraggetta et al. (2021) observed a scan failure rate of approximately 0.5% when scanning on the Pannoramic 250 flash III (3DHISTECH Ltd., Budapest, Hungary), mostly due to problems in the recognition of the 2D barcode label, and occasionally due to network connectivity.

Hanna et al. studied the quality of 2221 glass slide scans using Leica Aperio AT2, including 148 glass slides that needed to be rescanned (7%). Missing tissue accounted for 3.6% rescans, where the entire bounds of tissue present on the slide were not scanned and present in the whole slide image. True scanner technical error necessitated 3.1% rescans (45 H&E slides and 23 IHC slides). These errors resulted from faint slides, barcode recognition errors, focus point errors, or incomplete tissue detection. Of the 68 total rescans in the study, 64 slides were regarded as the true rescan rate, representing 2.9% of all glass slides (Hanna et al., 2019). In a later validation study of 2119 slides scanned on a Aperio GT450 (Leica Biosystems, Buffalo Grove, Illinois, USA) scanner, QC review revealed an overall rescan rate of 1.8%. Rescanning was due to barcode detection failures, no tissue detected, partial tissue detection failure, no macro image focus, and image quality (Hanna, Reuter, et al., 2020). The total rescan rate reported from the same team is less than 1.5% for prospective slide scanning on Aperio GT450 (Ardon, Labasin, et al., 2023). All error types are routinely tracked and grouped into four categories: barcode errors, missing tissue, out of focus regions, and stitching/banding. The highest weekly error rate is consistently due to missing scanned tissue in the digital slide, followed by barcode errors and by blurry/out of focus images (Ardon, Labasin, et al., 2023).

A recent study evaluated a Pramana 2-head scanning system (Pramana.ai, Cambridge, MA) and demonstrated a scanner rejection rate of 1.5% using submitted slides of mixed slide quality. The errors were a result of scan quality and missing tissue. Most error rates occurred in extremely faint slides or ones with minimal tissue or significant slide artifacts. Focus errors were detected algorithmically in 9.3% of the slides, with 2.3% of slides impacted over more than 1% of the slide pixel area. Manual review verified the algorithm-detected errors (Pathology Visions 2022 Overview, 22 Overview, 2023).

Currently, the error and rescan rates in different departments rely heavily on manual subjective review by digital scan staff or pathologists, highlighting the need for automated, standard review tools to speed process, improve quality, and minimize cost.

Digital pathology quality management systems

Digital pathology operations should be incorporated in the laboratory's overall laboratory QMS. A QMS is a systematic, integrated set of activities to establish and control the work processes from preanalytical through postanalytical phases, manage resources, conduct evaluations, and make continual improvements to ensure consistent quality results. QMS has been described by the International Organization for Standardization (ISO) and the Clinical Laboratory Standards Institute (CLSI) as "coordinated activities to direct and control an organization with regard to quality" (Daley & Berte, 2023; Wayne, 2013). In the United States, a QMS is a CLIA requirement for all clinical laboratories that test human specimens for health assessment or to diagnose, prevent, or treat disease. Implementing a QMS is critical to providing quality reliable results for clinical decision making for all laboratory operations (Andea et al., 2017).

Key QMS practices include periodic audits and quality monitors to assess compliance with regulatory-related requirements and laboratory-defined expectations. Quality practices and metrics are in place to ensure quality throughout the preanalytic, analytic, and postanalytic phases of testing. The QMS includes measures to identify and evaluate errors, defects, incidents, and other problems that may interfere with patient care services.

As previously described, the preanalytic steps of DP include histology laboratory processes that are already covered in most histology laboratories' QMS. The QMS should be expanded to include the novel processes associated with WSI/DP operations and laboratories should develop quality assurance essentials for their digital pathology following published guidelines (Andea et al., 2017; Daley & Berte, 2023; Wayne, 2013).

A recommended framework for QMS considerations for DP operations was developed (Ardon, Labasin, et al., 2023) and is summarized in Fig. 5.7. This DP QMS prompts staff to develop and follow workflow-specific standard operating procedures (SOPs) based on the specific operation and to document defects when expected scanning activities outcomes do not occur. Due to the disruptive nature of DP, this will be an ongoing process, and as digital scanning activities expand, new and revised SOPs should be linked to additional policies, instructions, and other appendices.

As laboratories adopt DP workflows, quality practices need to be in place to ensure quality throughout the preanalytic, analytic, and postanalytic phases of the DP operations, and the QMS will allow to control and manage these quality processes.

Establishing quality and performance metrics

Quality metrics can be used to assess the performance of a digital scan system or process in the laboratory and identify areas of improvement. QC documentation should include all aspects of the DP workflow, from slide preparation defects as well as QC scanner-related errors. Monitoring the quality and performance allows for detection of unforeseen test system drift, identifying areas of improvement, and confirmation of a stable testing system (Wayne, 2013). Action is taken when the metrics reveal unacceptable performance or a trend in that direction. A pathology laboratory's quality assurance plan should include quality and performance metrics specific to DP workflow processes. Data should be collected routinely on

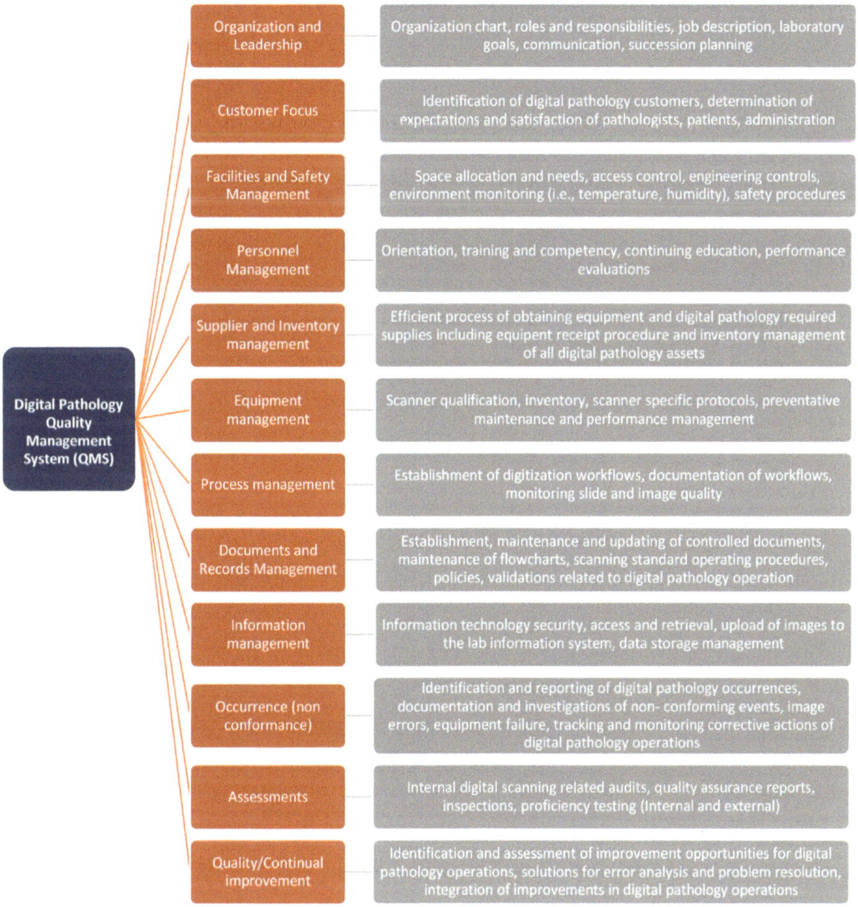

FIGURE 5.7

Summary of digital pathology quality management system (QMS), including digital pathology quality essentials (DPQE) at MSK (orange) and their digital pathology-specific measures (gray).

the different types and frequency of DP workflow deficiencies. Investigation of root cause and factors that contribute to poor quality images, out of focus images, presence of image artifacts, and deferrals to glass slide review based on image quality should be collected and monitored. Table 5.1 outlines some recommended metrics as were developed for the analytical and postanalytical DP.

As with other quality factors outlined in this chapter, the specific use cases, geographical location, and size of the institution will dictate the quality and performance metrics that are identified by the laboratory for the DP operations monitoring.

Table 5.1 Recommended quality and performance metrics for the analytical and postanalytical phases of a digital pathology operations.

Phase	Metric	Rationale
Analytic: Slides loaded into racks and into scanner and scan run	Number of cases, slides, and type of slides scanned Quality defects of slides Timing of scanner loading Scanner-specific errors Time for successful scan of slides	Scanner capacity and turnaround time requirements Instrument reliability and performance Preanalytic specimen quality issues Laboratory production and scanning capacity
Analytic: Image review	Qualitative and quantitative metrics of all image errors Tracking of all nonconformance events	Repeat scan workflow needs for prevention of delays in turnaround time Monitoring image quality errors for process improvement Understanding nonconformance events for education and training
Postanalytic: Image upload into the LIS	Postscan metrics: Quantify image size and timing of image upload	Monitor if storage of digital images becomes a constriction of operations Ability to correlate digital scan input to populating in the LIS
Postanalytic: Case review by pathologist	Time to open image, quality, and missing tissue errors	Monitoring for servers and network performance Use data to maintain expected image quality and turnaround needs

The cost of quality

The DP QMS, as part of a laboratory's overall operation, aims to monitor and evaluate all relevant data to promote accuracy and safety, and mitigate risk to patients while reducing waste and controlling costs.

In order to maintain its QMS and quality essentials, the laboratory will incur costs that are associated with internal and external quality studies, validations, compliance, and regulatory needs (Ardon, Klein, et al., 2023). In designing quality systems, one needs to determine what it will cost to ensure that financial resources are available yet must justify the effort in improved service to patients (AuBuchon, 1999). Investing in a quality may redirect money from other healthcare initiatives. In economic terms, this is called an "opportunity cost," or a lost opportunity for alternative investment (Danzon et al., 2018; Thusini et al., 2022).

The quality checkpoints detailed for the DP phases as well and the DP performance metrics allow quality improvements including maintaining turnaround time and minimizing cost and waste. Clinically, slides rejected on quality grounds represent a strain on the clinical pathology workflow, because these slides need to be recut or rescanned, in turn causing additional delays and unnecessary costs (Janowczyk et al., 2019).

In general, the earlier the quality issue is detected, the less costly the process is in terms of turnaround time disruption and financial cost of generating a new digital image. Ideally, quality slide errors are minimized during the production of the slide in the lab, detected before digital scanning is performed or while the slide is still located on the scanner. Every image failure after the slide is removed from the scanner will incur additional costs and potential turnaround time (TAT) delays that are associated with the retrieval of the slide for repeated WSI. If the image quality fails the pathologist checkpoint, additional delay and cost is incurred while a new slide may need to get recut or scanned.

Conclusions

DP is revolutionizing the practice of pathology and relies on high quality digital images for all successful DP applications. High-quality images need to be seamlessly integrated with the different institutional and laboratory information systems to facilitate clinical workflows, research, development of novel computational tools, and for educational applications (Ardon, Labasin, et al., 2023; Fraggetta et al., 2021; Hanna, Parwani, et al., 2020; Janowczyk et al., 2019; Montezuma et al., 2022; Williams et al., 2017). Reported WSI error and rescan rates are highly variable among teams, specific workflows and technologies used and reporting institutions, highlighting the need for the development and use of objective, automated QC review. The need for establishing new quality measures for DP operations has been documented by several early adopters of DP (Ardon, Labasin, et al., 2023; Evans et al., 2021; Evans et al., 2022; Fraggetta, L'Imperio et al., 2021; Pantanowitz et al., 2013; Retamero et al., 2019; Vodovnik, 2016). Slides rejected on image quality grounds represent a strain on the clinical pathology workflow, because these slides need to be recut or rescanned, in turn causing additional delays and unnecessary costs (Ardon, Klein, et al., 2023; Janowczyk et al., 2019).

As DP becomes increasingly adopted in clinical use, ensuring clinical reproducibility is necessary to allow pathologists to migrate fully to primary diagnosis using digital workflows. Institutions need to understand and incorporate DP operations as part of their overall QMS with this addition to standard pathology practice. The need for a DP QMS, as for other laboratory operations, has the goals of promoting accuracy and safety, and mitigating risk to patients and employees while reducing waste and controlling costs (Ardon, Labasin, et al., 2023).

References

Andea, A. A., Myers, J. L., Owens., & Scott, R. (2017). *Designing a quality management plan, . Quality management in anatomic pathology* (2017). CAP Press.

Ardon, O., Klein, E., Manzo, A., Corsale, L., England, C., Mazzella, A., Geneslaw, L., Philip, J., Ntiamoah, P., Wright, J., Sirintrapun, S. J., Lin, O., Elenitoba-Johnson, K., Reuter, V. E., Hameed, M. R., & Hanna, M. G. (2023). Digital pathology operations at a tertiary cancer center: Infrastructure requirements and operational cost. *Journal of Pathology Informatics*, *14*, 100318. Available from https://doi.org/10.1016/j.jpi.2023.100318.

Ardon, O., Labasin, M., Friedlander, M., Manzo, A., Corsale, L., Ntiamoah, P., Wright, J., Elenitoba-Johnson, K., Reuter, V. E., Hameed, M. R., & Hanna, M. G. (2023). Quality management system in clinical digital pathology operations at a tertiary cancer center. *Laboratory Investigation*, *103*(11). Available from https://doi.org/10.1016/j.labinv.2023.1002462023.

Atallah, N. M., Toss, M. S., Verrill, C., Salto-Tellez, M., Snead, D., & Rakha, E. A. (2022). Potential quality pitfalls of digitalized whole slide image of breast pathology in routine practice. *Modern Pathology*, *35*(7), 903–910. Available from https://doi.org/10.1038/s41379-021-01000-8, http://www.nature.com/modpathol/index.html.

AuBuchon, J. P. (1999). Optimizing the cost-effectiveness of quality assurance in transfusion medicine. *Archives of Pathology and Laboratory Medicine*, *123*(7), 603–606. Available from https://doi.org/10.5858/1999-123-0603-otceoq.

Avanaki, A. R. N., Espig, K. S., Xthona, A., Lanciault, C., & Kimpe, T. R. L. (2016). Automatic image quality assessment for digital pathology. *Lecture Notes in Computer Science*, *9699*, 431–438. Available from https://doi.org/10.1007/978-3-319-41546-8_54, http://springerlink.com/content/0302-9743/copyright/2005/.

Barua, B., Bora, K., Das, A. K., Ahmed, G. N., & Rahman, T. (2023). Stain color translation of multi-domain OSCC histopathology images using attention gated cGAN. *Computerized Medical Imaging and Graphics*, *106*, 102202. Available from https://doi.org/10.1016/j.compmedimag.2023.102202.

Boyd, J. C., & Hawker, C. D. (2012). *Automation in the clinical laboratory* (pp. 469–485). Elsevier BV. Available from 10.1016/b978-1-4160-6164-9.00019-6.

Brixtel, R., Bougleux, S., Lezoray, O., Caillot, Y., Lemoine, B., Fontaine, M., Nebati, D., & Renouf, A. (2022). Whole slide image quality in digital pathology: Review and perspectives. *IEEE Access*, *10*, 131005–131035. Available from https://doi.org/10.1109/ACCESS.2022.3227437, http://ieeexplore.ieee.org/xpl/RecentIssue.jsp?punumber = 6287639.

Campanella, G., Rajanna, A. R., Corsale, L., Schüffler, P. J., Yagi, Y., & Fuchs, T. J. (2018). Towards machine learned quality control: A benchmark for sharpness quantification in digital pathology. *Computerized Medical Imaging and Graphics*, *65*, 142–151. Available from https://doi.org/10.1016/j.compmedimag.2017.09.001, http://www.elsevier.com/locate/compmedimag.

Clarke, E. L., Brettle, D., Sykes, A., Wright, A., Boden, A., & Treanor, D. (2019). Development and evaluation of a novel point-of-use quality assurance tool for digital pathology. *Archives of Pathology and Laboratory Medicine*, *143*(10), 1246–1255. Available from https://doi.org/10.5858/arpa.2018-0210-oa.

Clarke, E. L., & Treanor, D. (2017). Colour in digital pathology: A review. *Histopathology*, *70*(2), 153–163. Available from https://doi.org/10.1111/his.13079, http://www3.interscience.wiley.com/journal/118511588/home.

Daley, A. T., & Berte, L. M. (2023). QMS01: QMS - A Model for Laboratory Services - CLSI. *Clinical & Laboratory Standards Institute, 2023.*

Danzon, P. M., Drummond, M. F., Towse, A., & Pauly, M. V. (2018). Objectives, budgets, thresholds, and opportunity costs—a health economics approach: An ISPOR special task force report. *Value in Health*, *21*(2), 140–145. Available from https://doi.org/10.1016/j.jval.2017.12.008, http://www.elsevier.com/wps/find/journaldescription.cws_home/724501/description#description.

Dehkharghanian, T., Bidgoli, A. A., Riasatian, A., Mazaheri, P., Campbell, C. J. V., Pantanowitz, L., Tizhoosh, H. R., & Rahnamayan, S. (2023). Biased data, biased AI: Deep networks predict the acquisition site of TCGA images. *Diagnostic Pathology*, *18* (1). Available from https://doi.org/10.1186/s13000-023-01355-3, https://diagnostic-pathology.biomedcentral.com.

Eloy, C., Vale, J., Curado, M., Polónia, A., Campelos, S., Caramelo, A., Sousa, R., & Sobrinho-Simões, M. (2021). Digital pathology workflow implementation at IPATIMUP. *Diagnostics*, *11*(11), 2111. Available from https://doi.org/10.3390/diagnostics11112111.

Evans, A. J., Brown, R. W., Bui, M. M., Chlipala, E. A., Lacchetti, C., Milner, D. A., Pantanowitz, L., Parwani, A. V., Reid, K., Riben, M. W., Reuter, V. E., Stephens, L., Stewart, R. L., & Thomas, N. E. (2022). Validating whole slide imaging systems for diagnostic purposes in pathology guideline update from the College of American Pathologists in collaboration with the American Society for Clinical Pathology and the Association for Pathology Informatics. *Archives of Pathology and Laboratory Medicine*, *146*(4), 440–450. Available from https://doi.org/10.5858/arpa.2020-0723-CP, https://meridian.allenpress.com/aplm/article/146/4/440/464968/Validating-Whole-Slide-Imaging-Systems-for.

Evans, A. J., Vajpeyi, R., Henry, M., & Chetty, R. (2021). Establishment of a remote diagnostic histopathology service using whole slide imaging (digital pathology. *Journal of Clinical Pathology*, *74*(7), 421–424. Available from https://doi.org/10.1136/jclinpath-2020-206762, http://jcp.bmj.com/.

Ferreira, I., Montenegro, C. S., Coelho, D., Pereira, M., da Mata, S., Carvalho, S., Araújo, A. C., Abrantes, C., Ruivo, J. M., Garcia, H., & Oliveira, R. C. (2023). Digital pathology implementation in a private laboratory: The CEDAP experience. *Journal of Pathology Informatics*, *14*, 100180. Available from https://doi.org/10.1016/j.jpi.2022.100180.

Fraggetta, F., Caputo, A., Guglielmino, R., Pellegrino, M. G., Runza, G., & L'imperio, V. (2021). A survival guide for the rapid transition to a fully digital workflow: The "caltagirone example.". *Diagnostics*, *11*(10). Available from https://doi.org/10.3390/diagnostics11101916, https://www.mdpi.com/2075-4418/11/10/1916/pdf.

Fraggetta, F., Garozzo, S., Zannoni, G. F., Pantanowitz, L., & Rossi, E. D. (2017). Routine digital pathology workflow: The Catania experience. *Journal of Pathology Informatics*, *8*(1), 51. Available from https://doi.org/10.4103/jpi.jpi_58_17.

Fraggetta, F., L'Imperio, V., Ameisen, D., Carvalho, R., Leh, S., Kiehl, T.-R., Serbanescu, M., Racoceanu, D., Mea, V. D., Polonia, A., Zerbe, N., & Eloy, C. (2021). Best practice recommendations for the implementation of a digital pathology workflow in the anatomic pathology laboratory by the European Society of Digital and Integrative Pathology (ESDIP). *Diagnostics*, *11*(11), 2167. Available from https://doi.org/10.3390/diagnostics11112167.

Goodfellow, I., Pouget-Abadie, J., Mirza, M., Xu, B., Warde-Farley, D., Ozair, S., Courville, A., & Bengio, Y. (2020). Generative adversarial networks. *Communications of the ACM*, *63*(11), 139–144. Available from https://doi.org/10.1145/3422622.

Haghighat, M., Browning, L., Sirinukunwattana, K., Malacrino, S., Alham, N. K., Colling, R., Cui, Y., Rakha, E., Hamdy, F. C., Verrill, C., & Rittscher, J. (2022). Automated quality assessment of large digitised histology cohorts by artificial intelligence. *Scientific Reports*, *12*(1). Available from https://doi.org/10.1038/s41598-022-08351-5.

Hanna, M. G., & Ardon, O. (2023). Digital pathology systems enabling quality patient care. *Genes, Chromosomes and Cancer*, 1045–2257. Available from https://doi.org/10.1002/gcc.23192.

Hanna, M. G., Ardon, O., Reuter, V. E., Sirintrapun, S. J., England, C., Klimstra, D. S., & Hameed, M. R. (2022). Integrating digital pathology into clinical practice. *Modern Pathology*, *35*(2), 152–164. Available from https://doi.org/10.1038/s41379-021-00929-0, http://www.nature.com/modpathol/index.html.

Hanna, M. G., & Pantanowitz, L. (2015). Bar coding and tracking in pathology. *Surgical Pathology Clinics*, *8*(2), 123–135. Available from https://doi.org/10.1016/j.path.2015.02.017, http://www.elsevier.com/wps/find/journaldescription.cws_home/716205/descriptio.

Hanna, M. G., Parwani, A., & Sirintrapun, S. J. (2020). Whole slide imaging: Technology and applications. *Advances in Anatomic Pathology*, *27*(4), 251–259. Available from https://doi.org/10.1097/PAP.0000000000000273, http://www.anatomicpathology.com.

Hanna, M. G., Reuter, V. E., Ardon, O., Kim, D., Sirintrapun, S. J., Schüffler, P. J., Busam, K. J., Sauter, J. L., Brogi, E., Tan, L. K., Xu, B., Bale, T., Agaram, N. P., Tang, L. H., Ellenson, L. H., Philip, J., Corsale, L., Stamelos, E., Friedlander, M. A., … Hameed, M. (2020). Validation of a digital pathology system including remote review during the COVID-19 pandemic. *Modern Pathology*, *33*(11), 2115–2127. Available from https://doi.org/10.1038/s41379-020-0601-5, http://www.nature.com/modpathol/index.html.

Hanna, M. G., Reuter, V. E., Hameed, M. R., Tan, L. K., Chiang, S., Sigel, C., Hollmann, T., Giri, D., Samboy, J., Moradel, C., Rosado, A., Otilano, J. R., England, C., Corsale, L., Stamelos, E., Yagi, Y., Schüffler, P. J., Fuchs, T., Klimstra, D. S., & Sirintrapun, S. J. (2019). Whole slide imaging equivalency and efficiency study: Experience at a large academic center. *Modern Pathology*, *32*(7), 916–928. Available from https://doi.org/10.1038/s41379-019-0205-0, http://www.nature.com/modpathol/index.html.

Hartman, D. J., & Pantanowitz, L. (2017). The intersection between quality assurance and informatics. Quality management in anatomic pathology. *Strategies for Assessment*, *2017*, 177–184.

Hartman, D. J., Pantanowitz, L., McHugh, J. S., Piccoli, A. L., OLeary, M. J., & Lauro, G. R. (2017). Enterprise implementation of digital pathology: Feasibility, challenges, and opportunities. *Journal of Digital Imaging*, *30*(5), 555–560. Available from https://doi.org/10.1007/s10278-017-9946-9.

Hashimoto, N., Bautista, P. A., Yamaguchi, M., Ohyama, N., & Yagi, Y. (2012). Referenceless image quality evaluation for whole slide imaging. *Journal of Pathology Informatics*, *3*(1), 9. Available from https://doi.org/10.4103/2153-3539.93891.

Hossain, M.S., Nakamura, T., Kimura, F., Yagi, Y., & Yamaguchi, M. (2018). SPIE Japan Practical image quality evaluation for whole slide imaging scanner. Proceedings of SPIE - the international society for optical engineering. http://spie.org/x1848.xml.

Hosseini, M. S., Brawley-Hayes, J. A. Z., Zhang, Y., Chan, L., Plataniotis, K., & Damaskinos, S. (2020). Focus quality assessment of high-throughput whole slide imaging in digital pathology. *IEEE Transactions on Medical Imaging*, *39*(1), 62–74. Available from https://doi.org/10.1109/TMI.2019.2919722, http://ieeexplore.ieee.org/xpl/RecentIssue.jsp?punumber = 42.

Janowczyk, A., Zuo, R., Gilmore, H., Feldman, M., & Madabhushi, A. (2019). HistoQC: An open-source quality control tool for digital pathology slides. *JCO Clinical Cancer Informatics*, *3*(3), 1–7. Available from https://doi.org/10.1200/cci.18.00157.

Kanwal, N., Perez-Bueno, F., Schmidt, A., Engan, K., & Molina, R. (2022). The devil is in the details: Whole slide image acquisition and processing for artifacts detection, color variation, and data augmentation: A review. *IEEE Access*, *10*, 58821–58844. Available from https://doi.org/10.1109/access.2022.3176091.

Khan, A., Janowczyk, A., Müller, F., Blank, A., Nguyen, H. G., Abbet, C., Studer, L., Lugli, A., Dawson, H., Thiran, J.-P., & Zlobec, I. (2022). Impact of scanner variability on lymph node segmentation in computational pathology. *Journal of Pathology Informatics*, *13*, 100127. Available from https://doi.org/10.1016/j.jpi.2022.100127.

Kim, D., Pantanowitz, L., Schüttler, P., Yarlagadda, D. V. K., Ardon, O., Reuter, V. E., Hameed, M., Klimstra, D. S., & Hanna, M. G. (2020). (Re) Defining the high-power field for digital pathology. *Journal of Pathology Informatics*, *11*(1), 33. Available from https://doi.org/10.4103/jpi.jpi_48_20.

Kothari, S., Phan, J. H., Stokes, T. H., & Wang, M. D. (2013). Pathology imaging informatics for quantitative analysis of whole-slide images. *Journal of the American Medical Informatics Association*, *20*(6), 1099–1108. Available from https://doi.org/10.1136/amiajnl-2012-001540United, http://jamia.bmj.com/content/20/6/1099.full.pdf.

Laboratory quality management system: Handbook. World Health Organization. (2011).

Lopez, X. M., D'Andrea, E., Barbot, P., Bridoux, A. S., Rorive, S., Salmon, I., Debeir, O., & Decaestecker, C. (2013). An automated blur detection method for histological whole slide imaging. *PLoS ONE*, *8*(12). Available from https://doi.org/10.1371/journal.pone.0082710Belgium, http://www.plosone.org/article/fetchObject.action?uri = info%3Adoi%2F10.1371%2Fjournal.pone.0082710&representation = PDF.

Lujan, G. M., Savage, J., Shana'ah, A., Yearsley, M., Thomas, D., Allenby, P., Otero, J., Limbach, A. L., Cui, X., Scarl, R. T., Hardy, T., Sheldon, J., Plaza, J. A., Whitaker, B., Frankel, W., Parwani, A. V., & Li, Z. (2021). Digital pathology initiatives and experience of a large academic institution during the coronavirus disease 2019 (COVID-19) pandemic. *Archives of Pathology and Laboratory Medicine*, *145*(9), 1051–1061. Available from https://doi.org/10.5858/arpa.2020-0715-SA, https://allen.silverchair-cdn.com.

Michielli, N., Caputo, A., Scotto, M., Mogetta, A., Pennisi, O. A. M., Molinari, F., Balmativola, D., Bosco, M., Gambella, A., Metovic, J., Tota, D., Carpenito, L., Gasparri, P., & Salvi, M. (2022). Stain normalization in digital pathology: Clinical multi-center evaluation of image quality. *Journal of Pathology Informatics*, *13*, 100145. Available from https://doi.org/10.1016/j.jpi.2022.100145.

Montezuma, D., Monteiro, A., Fraga, J., Ribeiro, L., Gonçalves, S., Tavares, A., Monteiro, J., & Macedo-Pinto, I. (2022). Digital pathology implementation in private practice: Specific challenges and opportunities. *Diagnostics*, *12*(2), 529. Available from https://doi.org/10.3390/diagnostics12020529.

Mukhopadhyay, S., Feldman, M. D., Abels, E., Ashfaq, R., Beltaifa, S., Cacciabeve, N. G., Cathro, H. P., Cheng, L., Cooper, K., Dickey, G. E., Gill, R. M., Heaton, R. P.,

Kerstens, R., Lindberg, G. M., Malhotra, R. K., Mandell, J. W., Manlucu, E. D., Mills, A. M., Mills, S. E., ... Taylor, C. R. (2018). Whole slide imaging versus microscopy for primary diagnosis in surgical pathology. *American Journal of Surgical Pathology*, *42*(1), 39–52. Available from https://doi.org/10.1097/pas.0000000000000948.

O'Hurley, G., Sjöstedt, E., Rahman, A., Li, B., Kampf, C., Pontén, F., Gallagher, W. M., & Lindskog, C. (2014). Garbage in, garbage out: A critical evaluation of strategies used for validation of immunohistochemical biomarkers. *Molecular Oncology*, *8*(4), 783–798. Available from https://doi.org/10.1016/j.molonc.2014.03.008.

Pantanowitz, L., Farahani, N., & Parwani, A. (2015). Whole slide imaging in pathology: Advantages, limitations, and emerging perspectives. *Pathology and Laboratory Medicine International*, *23*, 1179–2698. Available from https://doi.org/10.2147/PLMI.S59826.

Pantanowitz, L., Sinard, J. H., Henricks, W. H., Fatheree, L. A., Carter, A. B., Contis, L., Beckwith, B. A., Evans, A. J., Otis, C. N., Lal, A., & Parwani, A. V. (2013). Validating whole slide imaging for diagnostic purposes in Pathology: Guideline from the College of American pathologists Pathology and Laboratory Quality Center. *Archives of Pathology and Laboratory Medicine*, *137*(12), 1710–1722. Available from https://doi.org/10.5858/arpa.2013-0093-CPUnitedStates, http://www.archivesofpathology.org/doi/pdf/10.5858/arpa.2013-0093-CP.

Parwani, A. V., Patel, A., Zhou, M., Cheville, J. C., Tizhoosh, H., Humphrey, P., Reuter, V. E., & True, L. D. (2023). An update on computational pathology tools for genitourinary pathology practice: A review paper from the Genitourinary Pathology Society (GUPS. *Journal of Pathology Informatics*, *14*, 100177. Available from https://doi.org/10.1016/j.jpi.2022.100177.

Patel, A., Balis, U. G. J., Cheng, J., Li, Z., Lujan, G., McClintock, D. S., Pantanowitz, L., & Parwani, A. (2021). Contemporary whole slide imaging devices and their applications within the modern pathology department: A selected hardware review. *Journal of Pathology Informatics*, *12*(1), 50. Available from https://doi.org/10.4103/jpi.jpi_66_21.

Patel, A. U., Shaker, Na, Erck, S., Kellough, D. A., Palermini, E., Li, Z., Lujan, G., Satturwar, S., & Parwani, A. V. (2022). Types and frequency of whole slide imaging scan failures in a clinical high throughput digital pathology scanning laboratory. *Journal of Pathology Informatics*, *13*, 100112. Available from https://doi.org/10.1016/j.jpi.2022.100112.

Pathology Visions 2022 Overview. (2023). Journal of Pathology Informatics. Available from https://doi.org/10.1016/j.jpi.2023.100310.

Point of Use QA | Point of use device validation for digital pathology. (2023). Available from: https://www.virtualpathology.leeds.ac.uk/research/systems/pouqa/.

Qihui., Zhai, M. D., & Gene, P. (2017). *Quality management in anatomic pathology - Pub125* (2017). CAP Press.

Ramaswamy, V., Tejaswini, B. N., & Uthaiah, S. B. (2021). Remote reporting during a pandemic using digital pathology solution: Experience from a tertiary care cancer center. *Journal of Pathology Informatics*, *12*(1), 20. Available from https://doi.org/10.4103/jpi.jpi_109_20.

Randell, R., Ambepitiya, T., Mello-Thoms, C., Ruddle, R. A., Brettle, D., Thomas, R. G., & Treanor, D. (2015). Effect of display resolution on time to diagnosis with virtual pathology slides in a systematic search task. *Journal of Digital Imaging*, *28*(1), 68–76. Available from https://doi.org/10.1007/s10278-014-9726-8.

Rao, V., Kumar, R., Rajaganesan, S., Rane, S., Deshpande, G., Yadav, S., Patii, A., Pai, T., Menon, S., Shah, A., Rabade, K., Ramadwar, M., Panjwani, P., Mittal, N., Sahay, A., Rekhi, B., Bal, M., Sakhadeo, U., Gujral, S., & Desai, S. (2021). Remote reporting from home for primary diagnosis in surgical pathology: A tertiary oncology center experience during the COVID-19 pandemic. *Journal of Pathology Informatics*, *12*(1), 3. Available from https://doi.org/10.4103/jpi.jpi_72_20.

Retamero, J. A., Aneiros-Fernandez, J., & del Moral, R. G. (2019). Complete digital pathology for routine histopathology diagnosis in a multicenter hospital network. *Archives of Pathology and Laboratory Medicine*, *144*(2), 221–228. Available from https://doi.org/10.5858/arpa.2018-0541-oa.

Rojo, M. G., & Bueno, G. (2015). Analysis of the impact of high-resolution monitors in digital pathology. *Journal of Pathology Informatics*, *6*(1), 57. Available from https://doi.org/10.1016/s2153-3539(22)00512-0.

Rong, R., Wang, S., Zhang, X., Wen, Z., Cheng, X., Jia, L., Yang, D. M., Xie, Y., Zhan, X., & Xiao, G. (2023). Enhanced pathology image quality with restore–generative adversarial network. *The American Journal of Pathology*, *193*(4), 404–416. Available from https://doi.org/10.1016/j.ajpath.2022.12.011.

Schüffler, P. J., Geneslaw, L., Yarlagadda, D. V. K., Hanna, M. G., Samboy, J., Stamelos, E., Vanderbilt, C., Philip, J., Jean, M. H., Corsale, L., Manzo, A., Paramasivam, N. H. G., Ziegler, J. S., Gao, J., Perin, J. C., Kim, Y. S., Bhanot, U. K., Roehrl, M. H. A., Ardon, O., ... Fuchs, T. J. (2021). Integrated digital pathology at scale: A solution for clinical diagnostics and cancer research at a large academic medical center. *Journal of the American Medical Informatics Association*, *28*(9), 1874–1884. Available from https://doi.org/10.1093/jamia/ocab085, http://jamia.oxfordjournals.org/content/22/e1.

Schüffler, P. J., Stamelos, E., Ahmed, I., Yarlagadda, D. K. V., Ardon, O., Hanna, M. G., Reuter, V. E., Klimstra, D. S., & Hameed, M. (2022). Efficient visualization of whole slide images in web-based viewers for digital pathology. *Archives of Pathology and Laboratory Medicine*, *146*(10), 1273–1280. Available from https://doi.org/10.5858/arpa.2021-0197-OA, https://meridian.allenpress.com/aplm/article/146/10/1273/476350/Efficient-Visualization-of-Whole-Slide-Images-in.

Sellaro, T. L., Filkins, R., Hoffman, C., Fine, J. L., Ho, J., Parwani, A. V., Pantanowitz, L., & Montalto, M. (2013). Relationship between magnification and resolution in digital pathology systems. *Journal of Pathology Informatics*, *4*(1), 21. Available from https://doi.org/10.4103/2153-3539.116866.

Senaras, C., Niazi, M. K. K., Lozanski, G., Gurcan, M. N., & Lo, C.-M. (2018). DeepFocus: Detection of out-of-focus regions in whole slide digital images using deep learning. *PLoS One*, *13*(10), e0205387. Available from https://doi.org/10.1371/journal.pone.0205387.

Shaker, N., Sardana, R., Hamasaki, S., Nohle, D. G., Ayers, L. W., & Parwani, A. V. (2022). Accuracy of whole slide image based image analysis is adversely affected by preanalytical factors such as stained tissue slide and paraffin block age. *Journal of Pathology Informatics*, *13*. Available from https://doi.org/10.1016/j.jpi.2022.100121, https://www.sciencedirect.com/journal/journal-of-pathology-informatics/about/aims-and-scope.

Shakhawat, H. M., Nakamura, T., Kimura, F., Yagi, Y., & Yamaguchi, M. (2020). Automatic quality evaluation of whole slide images for the practical use of whole slide imaging scanner. *ITE Transactions on Media Technology and Applications*, *8*(4), 252–268. Available from https://doi.org/10.3169/mta.8.252.

Shimizu, H., & Nakayama, K. I. (2020). Artificial intelligence in oncology. *Cancer Science*, *111*(5), 1452–1460. Available from https://doi.org/10.1111/cas.14377, http://onlinelibrary.wiley.com/journal/10.1111/(ISSN)1349-7006.

Smit, G., & Cigéhn, M. (2021). Proceedings of the international conference on medical imaging with deep learning (MIDL). pp. 1–3.

Swiderska-Chadaj, Z., de Bel, T., Blanchet, L., Baidoshvili, A., Vossen, D., van der Laak, J., & Litjens, G. (2020). Impact of rescanning and normalization on convolutional neural network performance in multi-center, whole-slide classification of prostate cancer. *Scientific Reports*, *10*(1). Available from https://doi.org/10.1038/s41598-020-71420-0.

Thusini, S., Milenova, M., Nahabedian, N., Grey, B., Soukup, T., Chua, K.-C., & Henderson, C. (2022). The development of the concept of return-on-investment from large-scale quality improvement programmes in healthcare: an integrative systematic literature review. *BMC Health Services Research*, *22*(1). Available from https://doi.org/10.1186/s12913-022-08832-3.

Tizhoosh, H. R., & Pantanowitz, L. (2018). Artificial intelligence and digital pathology: Challenges and opportunities. *Journal of Pathology Informatics*, *9*(1), 38. Available from https://doi.org/10.4103/jpi.jpi_53_18.

Vodovnik, A. (2016). Diagnostic time in digital pathology: A comparative study on 400 cases. *Journal of Pathology Informatics*, *7*(1), 4. Available from https://doi.org/10.4103/2153-3539.175377.

Walkowski, S., & Szymas, J. (2011). Quality evaluation of virtual slides using methods based on comparing common image areas. *Diagnostic Pathology*, *6*(1). Available from https://doi.org/10.1186/1746-1596-6-S1-S14Poland, http://www.diagnosticpathology.org/content/6/S1/S14.

Wang, N. C., Kaplan, J., Lee, J., Hodgin, J., Udager, A., & Rao, A. (2021). Stress testing pathology models with generated artifacts. *Journal of Pathology Informatics*, *12*(1), 54. Available from https://doi.org/10.4103/jpi.jpi_6_21, http://www.jpathinformatics.org/browse.asp?sabs = n.

Wayne, P. A. (2013). –, . (6th ed.). *Quality management system: Development and management of laboratory documents; approved guideline*, (2013). Clinical Laboratory and Standards Institute.

Williams, B. J., Bottoms, D., & Treanor, D. (2017). Future-proofing pathology: The case for clinical adoption of digital pathology. *Journal of Clinical Pathology*, *70*(12), 1010–1018. Available from https://doi.org/10.1136/jclinpath-2017-204644, http://jcp.bmj.com/.

Williams, B. J., Knowles, C., & Treanor, D. (2019). Maintaining quality diagnosis with digital pathology: A practical guide to ISO 15189 accreditation. *Journal of Clinical Pathology*, *72*(10), 663–668. Available from https://doi.org/10.1136/jclinpath-2019-205944, http://jcp.bmj.com/.

Williams, B. J., Williams, B. J., Treanor, D., & Treanor, D. (2020). Practical guide to training and validation for primary diagnosis with digital pathology. *Journal of Clinical Pathology*, *73*(7), 418–422. Available from https://doi.org/10.1136/jclinpath-2019-206319, http://jcp.bmj.com/.

Wright, A. I., Clarke, E. L., Dunn, C. M., Williams, B. J., Treanor, D. E., & Brettle, D. S. (2020). A point-of-use quality assurance tool for digital pathology remote working. *Journal of Pathology Informatics*, *11*(1), 17. Available from https://doi.org/10.4103/jpi.jpi_25_20.

Zanjani, F.G., Zinger, S., Bejnordi, B.E., Van Der Laak, J.A. W.M., & De With, P.H. N. (2018). Stain normalization of histopathology images using generative adversarial networks. Proceedings - international symposium on biomedical imaging. http://ieeexplore.ieee.org/xpl/conferences.jsp2018-.

Zarella, M. D., Bowman, D., Aeffner, F., Farahani, N., Xthona, A., Absar, S. F., Parwani, A., Bui, M., & Hartman, D. J. (2018). A practical guide to whole slide imaging: A white paper from the digital pathology association. *Archives of Pathology and Laboratory Medicine*, *143*(2), 222–234. Available from https://doi.org/10.5858/arpa.2018-0343-ra.

Zerbe, N., Hufnagl, P., & Schlüns, K. (2011). Distributed computing in image analysis using open source frameworks and application to image sharpness assessment of histological whole slide images. *Diagnostic Pathology*, *6*(1). Available from https://doi.org/10.1186/1746-1596-6-S1-S16Germany, http://www.diagnosticpathology.org/content/6/S1/S16.

CHAPTER

Use cases in surgical pathology

6

Meera Hameed

Department of Pathology and Laboratory Medicine, Memorial Sloan Kettering Cancer Center, New York, NY, United States

Introduction

For many decades, histological diagnosis was based on pattern recognition under the microscope of disease processes based on a hematoxylin and eosin-stained glass slide created from a tissue block obtained from formalin-fixed paraffin-embedded patient tissue. Accurate recognition of diseases in this manner is based on the human brain's ability not only to recognize patterns but also to be able to logically connect and integrate these patterns in order to make appropriate decisions. While the practice of disease recognition with an analog microscope for >150 years is fundamental to the practice of pathology, there are limitations that include subjectivity in cancer grading, quantitation of biomarker expression, counting of mitoses, and assessment of necrosis percentage. These are essential elements for the estimation of both tumor prognosis and guide therapy. However, the images on glass slides have petabytes worth of data and only a small percentage of this is used by the human brain. Digital pathology and whole slide imaging (WSI) have gained momentum in the last decade largely due to improvements in the technology for decoding barcodes on glass slides and the ability to integrate them into laboratory information system (LIS), thus enabling a workflow conducive to daily pathology sign-out. In recent years, the advancement of technology has allowed the visualization of images for not only pattern recognition but to allow for image analysis through computer-assisted techniques, which enables us to overcome the above-mentioned limitations. In addition, there is an opportunity to explore the petabytes of data on the images at cellular and molecular levels. In this chapter, I will discuss the use cases in oncologic surgical pathology where digital pathology, machine learning (ML), and artificial intelligence (AI) can aid in improving accuracy, efficiency, and prognosis and outcome prediction. The following areas will be discussed with examples: telepathology; tumor detection and classification; metastases detection in lymph nodes; quantitative biomarker detection; prediction of genetic mutation on digital images; and image-based prognostication and outcome.

Digital Pathology. DOI: https://doi.org/10.1016/B978-0-443-13809-6.00008-7

Telepathology

Telepathology, where a pathologist can remotely view pathology images, is one of the earliest implementations of digital pathology predominantly for intraoperative consultations as static, dynamic, robotic, or as virtual slides. Although this approach was not digital pathology based, digital pathology has vastly expanded its utility. The concept was proposed in 1986 by Weinstein (1986) to integrate robotic microscopy, video imaging, and broadband telecommunications to develop an infrastructure for telepathology. Even in those early days, his group demonstrated using real-time video imaging, diagnostic accuracy can be achieved on frozen sections (Weinstein et al., 1997). Over the years, telepathology using dynamic systems gained popularity over the static viewing due to higher level of diagnostic accuracy rendered by the telepathologist with the opportunity to view all the fields, albeit being driven by the consulting pathologist (Dunn et al., 1997; Oberholzer et al., 1993; Shimosato et al., 1992). Taking this one step further, robotic microscopy allows the consulting pathologist to assume control of the microscope stage, objective, and focus, still allowing live dynamic image projection. Currently, WSI where large images, which are composed of multiple small images which are stitched together are more readily available, and the digital images can be transmitted remotely, without the need for two pathologists to simultaneously manipulate a microscope. With the advancement of WSI technology, telepathology usage has extended to primary diagnosis, second opinion consultation, education, quality assurance, and research. There are also hybrid systems which are combinations of robotic/WSI or dynamic/WSI, which are available for use now, but ultimately, WSI-based telepathology will be the most widely used technology.

Diagnostic accuracy and feasibility

It is imperative that diagnostic performance is measured before implementing telepathology in clinical practice. In this regard, multiple feasibility, validation, and clinical studies have been performed and reported in the literature (Weinstein et al., 1997). Between 2005 and 2016, nearly 5000 feasibility studies have been performed worldwide, with sample sizes ranging from 10 to 1862 for a multitude of telepathology applications, including intraoperative consultations, histopathological diagnosis, cytology, and tumor banking (Bashshur et al., 2017). Overall, the results of the feasibility and diagnostic concordance of these studies using a variety of telepathology systems and technologies for the various applications showed that telepathology is a beneficial and useful modality for the practice of pathology with good general diagnostic accuracy with diagnostic concordance in more than 90% of cases, ready availability of expertise, worldwide data-sharing capabilities, improved workflow and data integration with efficient use of time, and good resource for quality assurance programs, education, and tumor banking (Bashshur et al., 2017).

Intraoperative consultation

As indicated above, remote reviewing of frozen sections/intraoperative consultation was the early use case for telepathology. Dietz et al. (2020) conducted a systematic review of intraoperative consultation by telepathology versus conventional microscopy studies from 2000 to 2020. A total of 56 studies with 13,996 cases were identified and the concordance of telepathology diagnosis to light microscopy was excellent with a mean of 96.9% (Dietz et al., 2020). This concordance was much improved compared to prior studies where mean concordance was 91.0% (Wellnitz et al., 2000). This improvement is attributed to technological advances such as image transmission, speed, quality, as well as the use of robotic microscopy and WSI rather than static images. Dietz et al. (2020) also noted that deferrals were inconsistently reported in these studies and ranged from none to 30%. Generally, these were attributed to image quality, unusual cases, nature of specimen, and expertise of the pathologist. Contributing factors to discrepancies included poor tissue sampling, staining and slide preparation, inadequate knowledge of gross specimen, and technical errors including image focusing and lack of images in diagnostic area. Many of these are also areas encountered in discrepant results using conventional light microscopy. It is also worth noting that telepathology diagnosis by a subspecialist pathologist resulted in a better diagnosis than a general pathologist using conventional light microscopy (Liang et al., 2008; Vitkovski et al., 2015). Similar to all technologies, the various telepathology systems have their advantages and disadvantages. The static/dynamic systems are generally low cost and allow for Z-pane for focus, however, need a host to drive the fields, whereas the robotic system comes with a higher cost but does not need a host to drive and allow Z-plane focus, and WSIs come at a high cost, larger footprint, longer scan time but have shorter viewing time (Dietz et al., 2018). There is a perceived notion that the telepathology platforms are slow for intraoperative consultations; however, Evans et al. (2009) and Rohr et al. (2023) have shown that the desired turnaround time of 20 minutes can be readily achieved by telepathology intraoperative consultations. Regarding cost and financial barriers, while the initial cost and maintenance can be higher especially for robotic and WSI systems, one needs to consider time wasted in having an in-house pathologist for low-volume frozen sections and the advantages to get subspecialty consultations for difficult cases as means of return for investment. Before implementation, validation studies should be performed as per published guidelines for use of telepathology in clinical practice (Pantanowitz et al., 2014). It is also imperative gross images are made available similar to telemicroscopic images. As storage and scanners improve, it will be possible to do Z-plane imaging using WSI systems.

Primary diagnosis

While the use of WSI for primary diagnosis has been in practice in Europe and Canada for many years, in the United States, the Food and Drug Administration granted its first approval for WSI use in primary diagnosis in 2017 (US Food and

Drug Administration. FDA News Release: FDA allows marketing of first whole slide imaging system for digital pathology, 2017. Available at: https://www.fda.gov/drugs/informationondrugs/approveddrugs/ucm553358.htm). In alignment, over the years, multiple validation studies conducted at several institutions comparing digital pathology diagnosis to light microscopy in various subspecialties have shown excellent concordance between the two modalities (Bauer et al., 2013; Buck et al., 2014; Cheng et al., 2016; Goacher et al., 2017; Mukhopadhyay et al., 2018). The noninferiority of digital pathology diagnosis to conventional glass slide diagnosis was further reinforced by College of American Pathologists (CAP) who put forth guidelines for clinical validation of WSI for the practice of primary diagnosis (Pantanowitz et al., 2013). In 2019 and 2022 two more WSI systems have attained approval of Food and Drug Administration (US Food and Drug Administration. FDA News Release: Digital Pathology and Medical Devices [https://www.fda.gov/medical-devices/medical-device-regulatory-science-research-programs-conducted-osel/digital-pathology-program-research-digital-pathology-medical-devices]). While some institutions have transitioned to a digital workflow and sign-out, the adoption is slow in the United States (Betmouni, 2021) due to multitude of factors, which include not only financial barriers such as cost, maintenance, storage, more importantly, integration and interoperability of technologies with LIS and across organizations, thus making it difficult to implement a seamless workflow within the laboratories. However, one must have a long-term vision with the possibilities of moving toward a pathology future. Apart from the immediate use cases of glass slide management, transport, tumor boards, education, efficient interinstitutional and international consultations, and quality assurance, there is a real opportunity to change century-old practices of subjective pathological assessment (grading, mitoses, semiquantitation) into objective data and incorporate computer-aided diagnosis/AI modules into daily practice. The Digital Pathology Association (DPA) has released a white paper as a foundational guide for adoption of digital pathology considering the business and monetary considerations (Lujan et al., 2021). The value of digital pathology and remote diagnosis was also accentuated in 2019 during COVID-19 pandemic, necessitating pathology departments to find innovative ways to safeguard their faculty. While Clinical Laboratory Improvement Amendments (CLIA) required pathologists to issue reports in a CLIA-certified facility (Clinical Laboratory Improvement Amendments of 1988 Title 42: The Public Health and Welfare. Subpart 2: Clinical Laboratories (42 U.S.C. 263a). https://www.govinfo.gov/content/pkg/USCODE-2011-title42/pdf/USCODE-2011-title42-chap6A-subchapII-partF-subpart2-sec263a.pdf.), the public health emergency-induced circumstances led to relaxation of government enforcement of the regulation and allowed pathologists to review and report pathology slides from a remote non-CLIA certified facility. At Memorial Sloan Kettering Cancer Center, we successfully validated a digital pathology system (first of its kind) in non-CLIA certified facility through a secure connection (Hanna et al., 2020) and obtained approval from New York State Department of Health. As of May 2023, Centers for Medicare and Medicaid Services (CMS) will continue to exercise enforcement discretion to permit pathologists to review digital images remotely, without obtaining a separate CLIA certificate for the remote testing site (https://www.cms.gov/files/document/qso-23-15-clia.pdf). With rapid advancement of digital pathology ecosystems, a more widespread application of digital pathology and remote sign-out for primary diagnosis is expected in the coming years.

Tumor detection and classification

The advances in scanning technology and the ready availability of whole slide images have allowed for multiple exploratory studies using ML models to segment, detect, diagnose, and classify various tumor types. These deep learning models essentially provide computer-assisted diagnosis to pathologists by analyzing patterns, structural changes, and characteristics of tumor cells and can potentially lead to better turnaround time and increased efficiency in day-to-day pathology workload. Some illustrative examples are listed below:

Prostate Cancer: In the last few years, there have been several studies using various deep learning algorithms for detection and grading of prostate cancer in needle core biopsy specimens (Arvaniti et al., 2018; Bulten et al., 2020; Campanella et al., 2019; Huang et al., 2021; Lucas et al., 2019; Mun et al., 2021; Nagpal et al., 2019; Ström et al., 2020). These studies applied various ML algorithms using a training and validation set, with or without an external testing set, and achieved an accuracy for benign versus cancer in the range of area under the curve (AUC) 0.92–0.99. The Gleason grading accuracy in available studies ranged from 70% to 90%. In the study by Campanella et al. (2019), using a weakly supervised ML model (multiple instance learning/MIL) on WSIs from 24,859 prostate needle core biopsies (including internal and external cases), the authors were able to achieve 100% sensitivity in the detection of prostate cancer. This study showed that a manual pixel-level annotation is not necessary, thus allowing for possibilities of compiling large datasets and model training with real data similar to that of a pathology practice. A clinical grade decision support tool formulated in this study led to additional validation studies through a commercial spin-off, Paige-AI (Perincheri et al., 2021; Raciti et al., 2020) and FDA clearance for the Paige Prostate model as an adjunct to detect areas suspicious for carcinoma in digital images of prostate needle core biopsy sections (Paige Receives First-Ever FDA Approval for AI Product in Digital Pathology, 2021. Available at: https://www.businesswire.com/news/home/20210922005369/en/Paige-Receives-First-Ever-FDA-Approval-for-AI-Product-in-Digital-Pathology). While these individual studies were encouraging and showed a promise of AI application in diagnosing and grading prostate cancer, a multinational validation study was needed and was launched as the PANDA challenge (Bulten et al., 2022). PANDA (the Prostate cANcer graDe Assessment) is the largest histopathological competition to date consisting of 1290 investigators who participated to develop a reproducible AI algorithm for Gleason grading using 10,616 prostate needle biopsy whole slide images. This study showed that a pathologist-level performance can be achieved by a diverse set of algorithms on independent cross-continental cohorts. The full development set is publicly available for noncommercial research use (https://panda.grand-challenge.org/). These studies show that AI performance in prostate cancer diagnosis and grading is robust and mature and will benefit from prospective clinical trials.

Breast Cancer: Due to morphologic mimics with benign breast lesions, the interobserver diagnostic concordance for breast biopsy diagnosis can be variable and has been reported to be only around 75%, especially distinguishing DCIS (ductal carcinoma in situ) and ADH (atypical duct hyperplasia) (Longacre et al., 2006; Elmore et al., 2015). There has been much interest in exploring ML and AI-based algorithms in breast cancer to improve reproducibility. In radiology, there have been several studies on AI classification models for breast cancer prediction, with 99% accuracy, 98% specificity, and an AUC of 0.95 (Ozsahin et al., 2022). They have shown the caliber of AI models are continuously improving and it is believed that AI-based applications will serve as a supplement to their practice. In pathology, there has been a growing number of studies utilizing ML and AI algorithms to aid in the diagnosis of various breast lesions. In 2018 there was a grand challenge of breast cancer histology images where participants were challenged to classify clinically relevant breast lesions from an annotated public WSI dataset, which showed that the submitted algorithms improved the automatic classification of breast cancer using histology images to an accuracy of 87% (Aresta et al., 2019). Subsequently, multiple studies have been performed to distinguish invasive from noninvasive cancer (Klimov et al., 2019; Mercan et al., 2019) and benign, preneoplastic lesions, and neoplastic lesions of the breast (Sato et al., 2021). In the study by Mercan et al., the authors analyzed digital WSI of 240 breast biopsies from Breast Cancer Surveillance Consortium Registries with three expert pathologists categorizing the lesions as benign, atypia, DCIS, and invasive carcinoma. Two automated image features used for diagnosis (tissue distribution feature and structure feature) on high-resolution images were compared to diagnoses rendered by 87 practicing US pathologists. The accuracy rate for ML tissue distribution features, structure features, and pathologists was 0.94, 0.91, and 0.98, respectively, for invasive versus noninvasive cancer; for atypia and DCIS versus benign tissue was 0.70, 0.70, and 0.81 and DCIS versus atypia was 0.83, 0.85, and 0.80, respectively (Mercan et al., 2019). The sensitivity of ML algorithm was lower than the pathologists for invasive versus noninvasive classification but higher for DCIS versus atypia cases. Sato et al. (2021) studied 595 images from 125 biopsy cases of DCIS, columnar cell lesion (CCL), usual duct hyperplasia (UDH), and normal mammary glands and showed AI-based image analysis can detect lesions with AUCs of 0.992 (normal vs DCIS); 0.9942 (DCIS vs comedo-DCIS); and 0.9786 (normal vs CCL and UDH vs DCIS). These are proof of principle studies and show that there is a future for using AI algorithm as computer-assisted diagnosis of breast lesions for the pathologists. Additional WSI studies with external validation are needed. Another area of interobserver variability in breast cancer is grading. Many studies have reported significant interobserver variability among pathologists with larger variation on intermediate grade for survival and outcomes (Ginter et al., 2021; van Dooijeweert et al., 2020, 2022; Zhang et al., 2010). Wang et al. (2022) conducted a retrospective observational study, used a deep learning model (DeepGrade) on WSIs of 1567 patients with an external set of 1262 patients focusing on Nottingham histological grade 2 (NHG 2), which is considered low clinical value. They were able to classify NG2 high and NG2 low (Nottingham grade) based on DeepGrade, which provided independent prognostic values for recurrence. It is expected that such studies for improved stratification will come forth in the future to overcome the current subjective estimation of tumor grading.

Skin Cancer: In the dermatology literature, AI research has focused on melanoma and on multiple other classes of diagnoses with subsequent therapeutic recommendations (Young et al., 2020). In a metaanalysis of 70 studies, computer-aided diagnoses of melanoma were found to be comparable to that of human experts (Dick et al., 2019). Many studies have reported classification of cutaneous lesions from dermoscopic and nondermoscopic images at levels equal to dermatologists (Brinker et al., 2019; Liu et al., 2019). In dermatopathology, ML and AI algorithms are also being increasingly explored and have the promise for increasing accuracy and reproducibility. In the same study where Campanella et al. (2019) employed MIL for prostate cancer, they also showed that their algorithm accurately classified basal cell carcinomas from 9962 slides from 5325 patients with AUC of 0.988. In dermatopathology, the diagnosis of melanocytic lesions can be challenging, and studies have shown intra- and interobserver discordance up to 25%–26% (Lodha et al., 2008). In a study by Piepkorn et al. (2019), 240 melanocytic lesions representing full spectrum of melanocytic diagnoses from nevi to invasive melanoma evaluated by 187 US pathologists (113 general and 74 dermatopathologists) and compared to consensus opinion from three expert dermatopathologists, it was shown that discordance rate among pathologists was higher for general pathologists and improved reliability of diagnoses was noted after second opinion from a dermatopathologist. These studies have inspired a wave of publications investigating AI-based algorithms for diagnosis of melanocytic lesions which have shown comparable sensitivities with dermatopathologists (Ba et al., 2021; Brinker et al., 2022). Applying a deep learning algorithm, with a head-to head comparison of classification results from cropped images with 11 histopathologists, Hekler et al. (2019) found that the CNN (convolutional neural network) was able to outperform the pathologists. In another study, using 701 WSI from 583 patients (from Central South University Xiangya Hospital, China and The Cancer Genome Atlas Program (TCGA)), a deep learning model using CNN provided an accurate and robust automated diagnosis of melanoma at WSI level (AUC = 0.971) and provided probabilistic heat maps for location of the lesion (Li et al., 2021). While the deep learning models have shown impressive capabilities for detecting melanocytic lesions, there are limitations which include selection criteria, model variations from one study to next, and the fact that the model can only detect what is being taught (De Logu et al., 2020). While high-quality datasets and multiinstitutional collaborative studies are needed, the potential for deep learning models to serve as computer-assisted diagnosis for pathologists remains high. In fact, a pathology deep learning system capable of classifying WSIs of hematoxylin and eosin-stained skin lesions into diagnostically relevant classes (basaloid, squamous, melanocytic, and other) has been published (Ianni et al., 2020).

Apart from above-mentioned cancers, the availability of WSIs has led to development of promising ML models to aid in the diagnosis and subtype and classification of wide variety of tumor types, which include, bladder, kidney, lung, GI tract, GYN, head and neck, brain tumors, and sarcomas (Chuang et al., 2020; Coudray et al., 2018; Ertosun & Rubin, 2015; Farahani et al., 2022; Fenstermaker et al., 2020; Gertych et al., 2019; Ho et al., 2023; Korbar et al., 2017; Li et al., 2022; Sharma et al., 2017; Wang et al., 2019). An example of a deep learning model (deep multi-magnification network) for tissue segmentation of histopathology images is shown in Fig. 6.1. Fig. 6.2 shows its multiclass segmentation prediction in an osteosarcoma. Other advancing areas include tumor grading and mitoses detection, which have remained subjective and prone to interobserver variation and, nevertheless, play an important role in patient management (Ertosun & Rubin, 2015; Veta et al., 2015; Wang et al., 2022). In a recent study, two open datasets (International Conference on Pattern Recognition [ICPR] 2012 and ICPR 2014 [MITOS-ATYPIA-14]) of breast cancer histopathology were used to assess mitoses on a multistage mitotic-cell-detection method based on two CNNs and showed that the platforms achieved 0.876 and 0.848 precision, respectively, which was higher than prior reported methods (Mahmood et al., 2020). While challenges remain, deep learning-based AI algorithms are appealing and have a future in assisting in various aspects of pathological diagnosis and tumor classification and would likely eliminate subjective areas in grading of tumors.

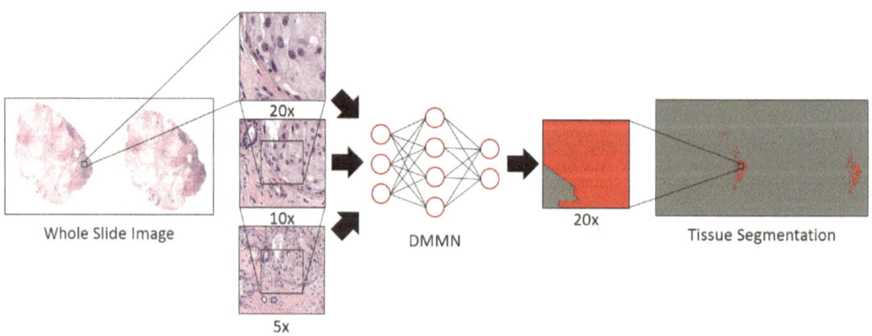

FIGURE 6.1

Deep multimagnification network (DMMN) for tissue segmentation of histopathology whole slide images. The model looks at a set of patches from multiple magnifications centered at the same coordinate to extract morphological patterns from various magnifications to generate more accurate segmentation predictions.

Courtesy David Ho, PhD, Memorial Sloan Kettering Cancer Center.

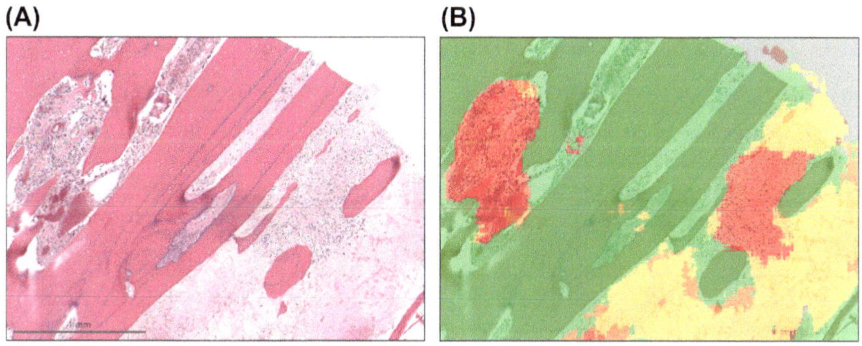

FIGURE 6.2

Osteosarcoma image (A) and (B) its multiclass segmentation prediction where *red* highlights viable tumor, *yellow* highlights necrotic tumor, *green* highlights normal bone, and *gray* highlights blank.

Courtesy David Ho, PhD, Memorial Sloan Kettering Cancer Center.

Metastases detection in lymph nodes

Assessment of metastases in regional lymph nodes is routine and standard of care in pathology practice and is performed either in frozen sections or formalin-fixed paraffin-embedded tissue sections. In breast cancer, the metastatic deposits are measured to determine if they are macrometastases (>2 mm), micrometastases (0.2–2 mm or larger than 200 cells), and isolated tumor cells (single or clusters of tumor cells <0.2 mm or <200 cells). This is a well-established practice and is a component of the American Joint Committee on Cancer Staging System (Amin et al., 2017). The process of detecting metastases, especially micrometastases or isolated tumor cells, is tedious and time-consuming for the pathologists and thus subject to error. With the advancement of digital pathology and WSI, ML models can be trained to perform this task, opening an avenue to develop a test which can be of substantial benefit and practical value. To assess the performance of deep learning algorithms for detecting metastases in sentinel lymph nodes from H&E sections, the CAMELYON challenge (CAncer MEtastases in LYmph nOdes challeNge) was launched in 2016 and 2017 using large well-curated datasets consisting of 1399 annotated WSIs of lymph nodes with and without metastases collected from five medical centers (Litjens et al., 2018). The results showed that the models can achieve comparable results as the pathologist and the AI can aid in improving accuracy for the pathologist (Ehteshami Bejnordi et al., 2017). Subsequent studies have supported these results and affirmed that higher accuracy can be achieved when pathologists used algorithm-assisted detection of lymph node metastases with high sensitivity and significant shorter time with assistance than without (Steiner et al., 2018). An illustrative example is shown in Fig. 6.3. While this is an attractive and useful tool, clinical implementation of the

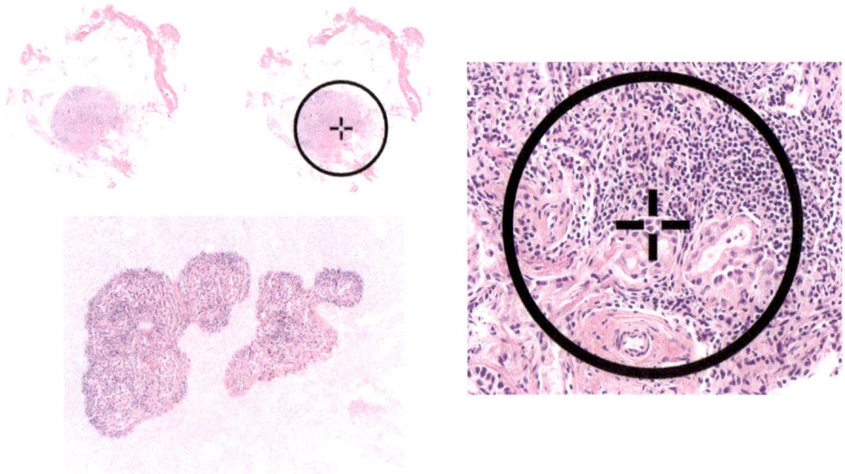

FIGURE 6.3

Micrometastases detection in lymph node: the algorithm detects area of interest which is highlighted for pathologist's review.

Courtesy Matthew Hanna, MD, Memorial Sloan Kettering Cancer Center.

challenge of generalizability in ML models exists, as shown by Jarkman et al. (2022), where a deep learning model trained to detect sentinel lymph node metastases on the CAMELYON multicenter data dropped performance in a new diagnostic setting (local sentinel node data) and change in surgical indication (axillary lymph node dissection), highlighting the necessity of retraining models in new conditions. Other studies have investigated lymph node metastases in cancers such as lung and colorectal carcinomas and have also shown promising results (Chuang et al., 2021; Kindler et al., 2023; Pham et al., 2019).

Quantitative biomarker detection

Quantitative analysis of biomarkers is both prognostic and predictive in cancer treatment. In most routine practice, biomarkers are assessed in a manual and semiquantitative manner, which is prone to interobserver variability. With availability of digital images, quantitative analysis of biomarkers which is more objective is feasible, however, have not been adopted widely into daily practice. Tissue segmentation, stain detection, and quantification are needed for quantitative image analysis. Quantifying immunohistochemical markers, such as estrogen receptor (ER), progesterone receptor (PR), Her2/neu, PDL-1, proliferation markers Ki-67, and PHH3 (phosphorylated histone 3), are the most investigated. AI-based quantitative image analysis can be used to employ additional features such as feature extraction and pattern recognition. Food and Drug Administration has cleared many AI algorithms for quantitative estimation of P53, Ki-67, ER, PR, and Her2 (e.g., Aperio Technologies, Applied Spectral Imaging, Pathiam-Bioimagene, TriPath Imaging-VIAS, Ventana Medical Systems/Roche Diagnostics).

Assessment of ER and PR is routine in diagnostic breast pathology practice with strong predictive value for hormone receptor therapies and also prognosis (Allred et al., 1998; Fitzgibbons et al., 2000). As per ASCO/CAP guidelines, it is recommended that ER and PR status be tested and reported in all primary breast cancer and recurrences (Hammond et al., 2010). A free internet-based application named ImmunoRatio is available to quantify ER, PR, and Ki-67, and the results have been shown to correlate well with manual scoring by pathologists (Tuominen et al., 2010). In this method, a color deconvolution algorithm is used to segment the staining components (diaminobenzidine and hematoxylin) and using an adaptive threshold for nuclear segmentation, the percentage of the positively stained nuclear area is calculated (labeling index). A similar online platform called DeepLIIF is available for reproducible immunohistochemical scoring (Ghahremani et al., 2022), where a virtual multiplex immunofluorescence restaining of the standard IHC stains is done which allows stain deconvolution and cell segmentation/classification simultaneously. This platform can be used for scoring ER/PR, Ki-67, and P53 (https://deepliif.org) (Fig. 6.4). An open-source platform for automated scoring of Ki-67 by a deep learning algorithm on WSIs of Ki-67

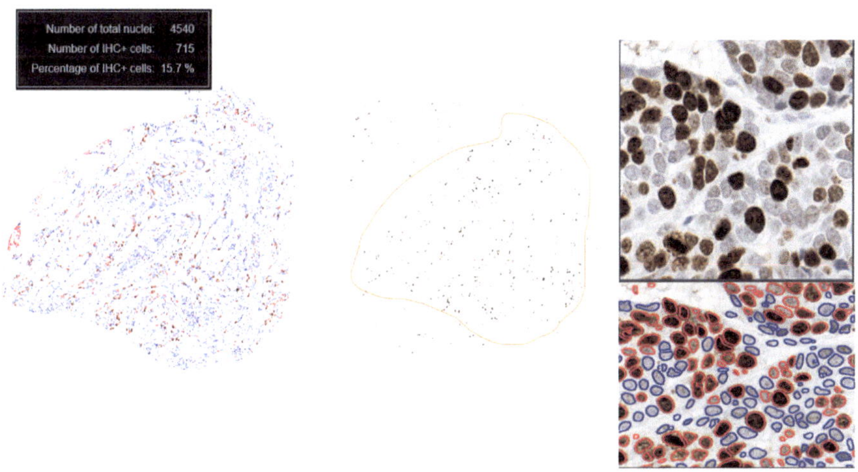

FIGURE 6.4

Automated scoring of Ki-67: https://www.deepliif.org. Ki-67 positive cells are labeled *red* and other cells *blue*. On the marked region of interest (ROI), the program calculates the percentage of IHC-positive cells (see table). High power view (right-hand side IHC and segmented cells). *IHC*, Immunohistochemistry.

Courtesy Matthew Hanna, MD, Memorial Sloan Kettering Cancer Center and Dr. Saad Nadeem, PhD.

stained cases with accuracy, sensitivity, and specificity of 89.44%, 85.05%, and 95.23% has been proposed by Feng et al. (2020). Assessing hormone receptors based on cellular morphology is an area of exploration using ML models and deep neural networks. Naik et al. (2020) have shown that a MIL-based deep learning model can determine ER status from H&E-stained WSIs in a dataset of 3474 patients from two datasets (Australian Breast Cancer Tissue Bank and TCGA) achieving an AUC of 0.92. In this study, the authors designed an attention-based deep neural network called ReceptorNet, which learns to assign weights (high-attention weights to tiles in H&E with maximum discrimination capacity and low-attention weight to tiles that are insignificant for the purpose), thus allowing one to determine which tiles are assigned to make the ER status by the algorithm.

Analysis of Her2/neu by quantitative image analysis is more challenging due to the membranous expression pattern unlike nuclear proteins where hematoxylin or DAPI can be used as counterstains to define nuclear compartment of the cell. In practice, Her2 IHC (immunohistochemical) is evaluated by pathologists in a semiquantitative manner as 0, 1 + , 2 + , and 3 + , based on membranous staining as per recommended ASCO/CAP guidelines (Wolff et al., 2018). The semiquantitative manual assessment can be improved by digital quantitative image analysis which is more objective and reproducible. Several studies (Brügmann et al., 2012; Dobson et al., 2010; Helin et al., 2016; Holten-Rossing et al., 2015) have shown concordance with in situ hybridization (ISH) is better with digital quantitative image analysis, especially for equivocal cases (2 +). These studies have led to the release of guidelines and recommendations by CAP for quantitative image analysis of Her2/neu immunohistochemical expression in breast cancer, which include assay validation, ongoing maintenance, quality control, and quality assurance, with oversight of performance, interpretation, and reporting by pathologists (Bui et al., 2019). Good concordance was also seen in IHC analysis of Her2 in gastric and esophageal adenocarcinoma in comparison with chromogenic ISH for equivocal cases (2 +) using Visiopharm Integrator System (vis) Platform with their HER2CONNECT algorithm (Koopman et al., 2018). Overall digital quantitative image analysis is a valid and feasible alternative tool to manual scoring with increased sensitivity and specificity; however, algorithmic differences exist among different platforms necessitating optical calibration (Koopman et al., 2019).

The field of immune-oncology has been rapidly advancing in the last few years, with improved clinical outcomes in patients treated with immune checkpoint inhibitors targeting programmed death-1 (PD-1)/death ligand-1(PD-L1) in various tumor types (Balar et al., 2017; Herbst et al., 2016; Larkin et al., 2019). Immunohistochemical assays to assess expression of PD-L1 have been approved as a biomarker of response to anti-PD-1/PD-L1 inhibitors and are widely used as companion and/or complementary diagnostic test (Buttner et al., 2017; Cogswell et al., 2017; Emens et al., 2021; Mok et al., 2019; Phillips et al., 2015). IHC assessment of PD-L1 is a time-consuming manual and laborious process with studies showing considerable inter- and intraobserver variability or moderate agreement among pathologists with variable reproducibility in lower cut-off values (Brunnström et al., 2017; Chang et al., 2019; Koomen et al., 2021). Recent studies have shown AI-based methods can reliably show moderate to high correlation with pathologist assessment in several cancer types (Humphries et al., 2018; Taylor et al., 2019). However, validation studies for automated analysis of PD-L1 expression with clinical outcomes are relatively few. In a study by Baxi et al. (2022), the authors developed AI-based algorithms to assess PD-L1 expression by IHC and compared it to manual scoring in nonsmall cell lung carcinoma, melanoma, urothelial carcinoma, and head and neck cancers. In a retrospective analysis of 1746 slides, the AI-based algorithm on whole slide images was able to quantitate more PD-L1 positive cases as compared to manual scoring at cut-offs $\geq 1\%$ and $\geq 5\%$ in most tumor types, with improvements in response and survival by both methods (Baxi et al., 2022). More prospective large-scale clinical trials are warranted to validate the use of AI-based method of quantification of PD-L1 to serve as companion or complementary diagnostics tests in the future.

Prediction of genetic mutation on hematoxylin and eosin whole slide images

Molecular testing of cancer has led to discovery of targetable genetic aberrations (both tumor-specific and agnostic), with successful use of mutation-specific therapies in multiple clinical trials and FDA approval for a variety of targeted agents (Blumenthal et al., 2017; Pérez-Soler et al., 2004; Zhou & Li, 2022). This extraordinary progress in the last several years is due to the development of cutting-edge technologies such as next-generation sequencing, allowing for identification of actionable targets with specific targeted therapeutics and paradigm shift in the treatment of several types of cancers. These technologies require special expertise, destructive tissue usage, high cost, and turnaround time and are largely only available in developed nations and large institutions. Digital pathology which has enabled the availability of whole slide images has opened doors to investigate the large petabytes of data on the H&E-stained images invisible to human eye and perception. Computer-assisted image algorithms have been advancing in the last several years, and with the advent of ML tools/AI, computational tasks can be automated and applied to complex tasks such as deciphering molecular alterations buried in the data on whole slide H&E images. Majority of these tools apply weakly supervised algorithms on whole slide images; however, learn/process the images in patches or tiles which are aggregated for a case or patient-level predictive score. Coudray et al. (2018) trained a CNN (Inceptionv3) on TCGA samples of lung cancer to be able to automatically detect lung adenocarcinoma, squamous cell carcinoma, or normal tissue comparable to pathologists with an AUC of 0.97. Additionally, they were able to train the network to predict six common mutations (*STK11, EGFR, FAT1, KRAS, SETBP1, TP53*) in lung adenocarcinoma with an AUC of 0.733–0.856 on a test dataset (Coudray et al., 2018). This proof-of-concept study generated multiple investigations and publications showing molecular predictions on H&E images across many tumor types (Chen et al., 2020; Fu et al., 2020; Kather, Pearson, et al., 2019; Velmahos et al., 2021). These studies have explored gene mutations, microsatellite instability (MSI), homologous recombination deficiency (HRD), gene expression, and copy number alterations (Cui et al., 2020; Diao et al., 2021; He et al., 2020; Velmahos et al., 2021; Yamashita et al., 2021).

In breast cancer, prediction of germline *BRCA* mutation, an important biomarker for DNA repair and genomic stability, was reported using a CNN on whole slide H&E images with an AUC of 0.77 (Wang et al., 2021). Using MIL, Cui et al. (2020) identified *IDH1/2* mutations in 1121 glioma cases (cropped region of interest images) with a good performance score (AUC 0.84). *IDH1/2* mutations are clinically highly relevant in glioma diagnosis and prognosis (Cui et al., 2020; Han et al., 2020). In colorectal cancer, several studies showed predictability for *APC, TP53, KRAS, SMAD4, PIK3CA, BRAF*, and MSI from histology images using various deep learning models (Bilal et al., 2021; Jang et al., 2020; Kather et al., 2020; Schrammen et al., 2022). Of note, a slide-level prediction of tumor detection and genetic aberrations was achieved with a single off-the-shelf neural network (SLAM—the slide-level assessment model) without manual annotations, improving on previous methods (Schrammen et al., 2022). In this study, after extensive validation with an external test set, an AUC of 0.900 and 0.821 was achieved for MSI and *BRAF* mutation in colorectal cancer. In thyroid carcinoma and melanoma, where *BRAF* mutation is prevalent, several studies have demonstrated that deep learning models have predictive capability with AUC of 0.78−0.95 in thyroid cancer and AUC of 0.71 in melanoma (Anand et al., 2021; Kim et al., 2022; Tsou & Wu, 2019). Hong et al. (2021) applied a multiresolution deep learning model to predict endometrial cancers with *POLE* ultramutated, MSI-high, copy number-low (CNV-L) and copy number-high (CNV-H) molecular subtypes and achieved good performance with AUC ranging from 0.827 (MSI-high) to 0.958 (CNV-high). In high-grade serous ovarian carcinomas, *BRCA1* and *BRCA2* mutation prediction and MSI status could be achieved using TCGA and institutional datasets (Zeng et al., 2021). Even in very rare types, such as rhabdomyosarcoma, deep learning models could predict embryonal versus alveolar rhabdomyosarcoma with an AUC of 0.85 in an independent test set (Milewski et al., 2023). In a recent metaanalysis of 24 studies targeting common cancers (GI, GU, GYN, lung, and head and neck), the AUC across the organs ranged from 0.92 for *BRAF* in thyroid carcinoma to 0.79 for *EGFR* detection in lung cancer with average of all mutation at AUC 0.64 (Alam et al., 2023). These studies show great promise but do not equate performance levels of the current methods for mutations detection; however, it can serve as prescreening tools for specific mutations due to high sensitivity.

Tumor mutation burden which refers to the number of somatic mutations accumulated in the specific cancer type is a prognostic and a therapeutic marker and is also a surrogate marker for tumor immunogenicity due to high association with increased neoantigen expression at cell surface (Chan et al., 2019; Galuppini et al., 2019). This test is FDA approved for therapy with immune checkpoint inhibitors in solid tumors (Center for Drug Evaluation, Research. FDA approves pembrolizumab for adults and children with TMB-H solid tumors. Available from: https://www.fda.gov/drugs/drug-approvals-and-databases/fda-approves-pembrolizumab-adults-and-children-tmb-h-solid-tumors). Several studies have assessed Tumor Mutation Burden (TMB) (high vs low) in many tumor types (lung, colorectal, renal cell carcinoma) and found predictability ranging from 0.65 to 0.817 inferring high potential of deep learning models, however, needs further testing with additional external validation studies (Dammak et al., 2023; Huang et al., 2022; Sadhwani et al., 2021). MSI, an approved biomarker for immunotherapy, is also an important determinant of tumor immunogenicity; an approved biomarker for immunotherapy is associated with high TMB, high tumor infiltrating lymphocytes and high neoantigen expression (Chang et al., 2018; Nebot-Bral et al., 2019; Yamashita et al., 2018). Predicting MSI through deep learning has been reported by several studies and in larger datasets can achieve excellent performance with AUC of 0.95 (Cao et al., 2020; Echle et al., 2020; Kather, Pearson, et al., 2019; Lee et al., 2021; Yamashita et al., 2021). Homologous recombinant deficiency (HRD) is an indicator of DNA damage repair and is a therapeutic biomarker for targeting PARP (Poly (ADP-Ribose) polymerase) in breast, ovarian, pancreatic, and prostate cancers (Hoppe et al., 2018). In breast and gastric cancer studies, deep learning models can predict HRD with an AUC of 0.70−0.77 (Diao et al., 2021; Valieris et al., 2020). The potential for deep learning model to enter the clinical arena is promising, especially for MSI based on published data (Kacew et al., 2021). Table 6.1 summarizes genetic alterations detected by ML in various tumor types.

Gene Expression: With the advent of spatial transcriptomics gene expression, data can yield important information correlating expression with cellular compartments (tumor vs stroma) and tumor heterogeneity. A deep learning model was employed by He et al. (2020) to assess gene expression on H&E images in breast cancer, and the model was able to predict the expression of over 200 genes in concordance with spatial transcriptomics data and subsequent validation with an external dataset. Other pan-cancer studies have shown that gene expression data (high vs low) can be acquired with deep learning models (Diao et al., 2021; Fu et al., 2020; Schmauch et al., 2020).

Image-based prognosis and outcome prediction

The prognosis of various tumor types is based on established clinicopathological criteria which include morphological parameters such as histological grade, nodal metastases, vascular invasion, pathological staging, and ancillary tests such as biomarker assessment. These criteria are overall robust and reproducible, with some

Table 6.1 Genetic alterations identified by machine learning in various tumor types.

Tumor type	Genetic alteration	References
Breast cancer	*BRCA*	Wang et al. (2021)
Glioma	*IDH1/2*	Han et al. (2020)
Colorectal cancer	*APC, TP53, KRAS, SMAD4, PIK3CA, BRAF*, MSI	Kather et al. (2019), Schrammen et al. (2022), Jang et al. (2020), Bilal et al. (2021), Kather et al. (2020)
Thyroid	*BRAF*	Tsou and Wu, (2019), Anand et al. (2021)
Melanoma	*BRAF*	Kim et al. (2022)
Endometrial carcinoma	*POLE*, MSI, CNV-low, CNV-High	Hong et al. (2021)
Serous ovarian carcinoma	*BRCA1, BRCA2*, MSI	Zeng et al. (2021)
Lung cancer	*STK11, EGFR, FAT1, SETBP1, KRAS and TP53*	Coudray et al. (2018), Alam et al. (2023)
Lung, colorectal, renal cell carcinomas	Tumor mutation burden	Dammak et al. (2023), Huang et al. (2022), Sadhwani et al. (2021)
Breast and gastric cancer	Homologous recombination deficiency	Diao et al. (2021), Valieris et al. (2020)

caveats associated with subjective estimations (e.g., mitoses count, % necrosis) for tumor grading and semiquantitative biomarker assessment. In colorectal cancer, several studies have employed deep learning modalities to extract prognostic information from whole slide H&E images. Kather, Krisam, et al. (2019) trained CNN to identify nine tissue classes (normal, tumor, and stroma) on colorectal cancer H&E images and used the best model to extract tissue characteristics in two large cohorts with independent validation and calculated a "a deep stroma score" from the nontumor components as an independent prognostic factor for overall survival (OS) in a multivariate Cox proportional hazard model (HR = 1.99). In a systematic review of deep learning-based characterization for prognostic or predictive biomarkers in gastrointestinal cancers, there were two more studies showing tumor to stroma ratio as a prognostic factor in colorectal carcinoma (Geessink et al., 2019; Kuntz et al., 2021; Zhao et al., 2020). In breast cancer, a machine-learning model (C-Path) applied to microscopic images of two cohorts was able to create a prognostic score which was strongly associated with OS and independent of clinical, pathological, and molecular features. In this model, stromal features were significantly associated with survival (Beck et al., 2011). In multiple publications, histology-based (e.g., mesothelioma, nonsmall cell lung cancer, embryonal rhabdomyosarcoma) and immune-markers-based (e.g., oral cavity squamous cell carcinoma, melanoma) deep learning models have shown OS and disease-specific recurrence predictability (Courtiol et al., 2019; Kulkarni et al., 2020; Milewski et al., 2023; Shaban et al., 2019; Wang et al., 2017).

Deep learning-based detection of genomic alterations, prognosis, and outcome predictions on H&E images is a rapidly growing field; however, currently, the field is not optimal for clinical implementation. Factors to consider include robust quality control measures: preanalytic (image preparation, color, quality), analytic (case selection, model approach for the task validation, external datasets), and postanalytic (outcome data, prospective trials) (Echle et al., 2021, 2022; Schömig-Markiefka et al., 2021). With technological advancement and performance improvements of the models, "explainability and interpretability" of the so-called "black boxes," and generation of "heat maps" as "class activation maps," these approaches can assist in better understanding and aid in new discoveries for morphology-molecular-clinical correlations (Holzinger et al., 2019; Wulczyn et al., 2021).

Summary

Digital pathology has been steadily growing and evolving in the last 10 years and has provided an opportunity to explore histological images in ways pathologists could not have imagined 15 years ago. From the simple task of storing images which can be retrieved in a matter of seconds which increases not only workflow efficiency but augments a multitude of pathology functions such as education, efficient tumor board presentations, immediate patient care from any remote site, case sharing with colleagues and consultants across the world, quality assurance, and research. The traditional practice of pathology via the microscope will likely continue for the near future. However, the move toward digital pathology is growing and is expected to have a similar impact as genomics, especially with the advancement of computational tools. The disruptive influence of genomics in science and current clinical care was due to the innovative technology of next-generation sequencing, allowing for faster, low-cost, and high-throughput sequencing whose innovation and scale continue to expand and grow. Similarly in this digital era, advances in technology and ML aka AI will drive digital pathology to parallel the genomic boom. Pathology practice has remained the same for 100 years driven by highly competent expert physicians rendering accurate diagnoses by highly skilled pattern recognition. However, the field has remained qualitative or semiquantitative, with subjective estimations of disease characteristics and disease-driving biomarkers. The advancing field of digital and computational pathology offers enormous benefits for efficiency, organization, telepathology, teaching, objective quantitation, and reproducibility. The approval of whole slide scanners for clinical use by Food and Drug Administration, with guidelines and recommendations for the digital pathology practice put forth by the CAP and DPA, has spurred pathology departments to take the first steps for adoption of the digital environment. The pandemic and postpandemic enforcement discretion to permit pathologists to review digital images remotely by CMS is a very positive step forward for wider adoption of pathology practices nationwide. In the next few years, it is to be expected that some of the repetitive tasks such as counting mitoses, searching for metastases in lymph nodes, estimating Ki-67 proliferative index, measuring tumor size on glass slides, and biomarker quantitation will be performed through a computer-assisted tool, freeing up pathologist's time for more innovative work such as integrated diagnostics and reporting. While these changes will have a profound impact in our 100-year-old discipline, the major innovation will come from ML and AI advancement. The pathologist's expertise is crucial in this area, and the hand-in-hand working relationship between pathologists and computational scientists is of paramount importance. Though clinical grade algorithms are not yet on the horizon, there is much potential to bring this into reality due to ever-improving deep learning models and generation of "explainable" neural networks, which will improve transparency and allow for future implementation into clinical practice.

References

Alam, M. R., Seo, K. J., Abdul-Ghafar, J., Yim, K., Lee, S. H., Jang, H.-J., Jung, C. K., & Chong, Y. (2023). Recent application of artificial intelligence on histopathologic image-based prediction of gene mutation in solid cancers. *Briefings in Bioinformatics, 24*(3). Available from https://doi.org/10.1093/bib/bbad151.

Allred, D. C., Harvey, J. M., Berardo, M., & Clark, G. M. (1998). Prognostic and predictive factors in breast cancer by immunohistochemical analysis. *Modern Pathology, 11*(2), 155–168.

Amin, M. B., Greene, F. L., Edge, S. B., Compton, C. C., Gershenwald, J. E., Brookland, R. K., Meyer, L., Gress, D. M., Byrd, D. R., & Winchester, D. P. (2017). The eighth edition AJCC cancer staging manual: Continuing to build a bridge from a population-based to a more "personalized" approach to cancer staging. *Cancer Journal for Clinicians, 67*(2), 93–99. Available from https://doi.org/10.3322/caac.21388, http://onlinelibrary.wiley.com/journal/10.3322/(ISSN)1542-4863.

Anand, D., Yashashwi, K., Kumar, N., Rane, S., Gann, P. H., & Sethi, A. (2021). Weakly supervised learning on unannotated H&E-stained slides predicts BRAF mutation in thyroid cancer with high accuracy. *Journal of Pathology, 255*(3), 232–242. Available from https://doi.org/10.1002/path.5773, http://onlinelibrary.wiley.com/journal/10.1002/(ISSN)1096-9896.

Aresta, G., Araújo, T., Kwok, S., Chennamsetty, S. S., Safwan, M., Alex, V., Marami, B., Prastawa, M., Chan, M., Donovan, M., Fernandez, G., Zeineh, J., Kohl, M., Walz, C., Ludwig, F., Braunewell, S., Baust, M., Vu, Q. D., To, M. N. N., ... Aguiar, P. (2019). BACH: Grand challenge on breast cancer histology images. *Medical Image Analysis, 56*, 122–139. Available from https://doi.org/10.1016/j.media.2019.05.010.

E. Arvaniti, K.S. Fricker, M. Moret, N. Rupp, T. Hermanns, C. Fankhauser, N. Wey, P.J. Wild, J.H. Rüschoff, M. Claassen, (2018). Automated Gleason grading of prostate cancer tissue microarrays via deep learning. Scientific Reports. 8 (1), http://www.nature.com/srep/index.html. doi: 10.1038/s41598-018-30535-1.

Ba, W., Wang, R., Yin, G., Song, Z., Zou, J., Zhong, C., Yang, J., Yu, G., Yang, H., Zhang, L., & Li, C. (2021). Diagnostic assessment of deep learning for melanocytic lesions using whole-slide pathological images. *Translational Oncology*, *14*(9)101161. Available from https://doi.org/10.1016/j.tranon.2021.101161.

Balar, A. V., Galsky, M. D., Rosenberg, J. E., Powles, T., Petrylak, D. P., Bellmunt, J., Loriot, Y., Necchi, A., Hoffman-Censits, J., Perez-Gracia, J. L., Dawson, N. A., van der Heijden, M. S., Dreicer, R., Srinivas, S., Retz, M. M., Joseph, R. W., Drakaki, A., Vaishampayan, U. N., Sridhar, S. S., ... Bajorin, D. F. (2017). Atezolizumab as first-line treatment in cisplatin-ineligible patients with locally advanced and metastatic urothelial carcinoma: a single-arm, multicentre, phase 2 trial. *The Lancet*, *389*(10064), 67–76. Available from https://doi.org/10.1016/S0140-6736(16)32455-2, http://www.journals.elsevier.com/the-lancet/.

Bashshur, R. L., Krupinski, E. A., Weinstein, R. S., Dunn, M. R., & Bashshur, N. (2017). The empirical foundations of telepathology: Evidence of feasibility and intermediate effects. *Telemedicine and e-Health*, *23*(3), 155–191. Available from https://doi.org/10.1089/tmj.2016.0278, http://www.liebertpub.com/publication.aspx?pub_id = 54.

Bauer, T. W., Schoenfield, L., Slaw, R. J., Yerian, L., Sun, Z., & Henricks, W. H. (2013). Validation of whole slide imaging for primary diagnosis in surgical pathology. *Archives of Pathology and Laboratory Medicine*, *137*(4), 518–524. Available from https://doi.org/10.5858/arpa.2011-0678-OA, http://www.archivesofpathology.org/doi/pdf/10.5858/arpa.2011-0678-OA.

Baxi, V., Lee, G., Duan, C., Pandya, D., Cohen, D. N., Edwards, R., Chang, H., Li, J., Elliott, H., Pokkalla, H., Glass, B., Agrawal, N., Lahiri, A., Wang, D., Khosla, A., Wapinski, I., Beck, A., & Montalto, M. (2022). Association of artificial intelligence-powered and manual quantification of programmed death-ligand 1 (PD-L1) expression with outcomes in patients treated with nivolumab ± ipilimumab. *Modern Pathology*, *35*(11), 1529–1539. Available from https://doi.org/10.1038/s41379-022-01119-2.

Beck, A. H., Sangoi, A. R., Leung, S., Marinelli, R. J., Nielsen, T. O., van de Vijver, M. J., West, R. B., van de Rijn, M., & Koller, D. (2011). Systematic analysis of breast cancer morphology uncovers stromal features associated with survival. *Science Translational Medicine*, *3*(108). Available from https://doi.org/10.1126/scitranslmed.3002564.

Betmouni, S. (2021). Diagnostic digital pathology implementation: Learning from the digital health experience. *Digital Health*, *7*. Available from https://doi.org/10.1177/20552076211020240, 205520762110202.

Bilal, M., Raza, S. E. A., Azam, A., Graham, S., Ilyas, M., Cree, I. A., Snead, D., Minhas, F., & Rajpoot, N. M. (2021). Development and validation of a weakly supervised deep learning framework to predict the status of molecular pathways and key mutations in colorectal cancer from routine histology images: A retrospective study. *The Lancet Digital Health*, *3*(12), e763. Available from https://doi.org/10.1016/S2589-7500(21)00180-1, https://www.sciencedirect.com/journal/the-lancet-digital-health.

Blumenthal, G. M., Kluetz, P. G., Schneider, J., Goldberg, K. B., McKee, A. E., & Pazdur, R. (2017). Oncology drug approvals: Evaluating endpoints and evidence in an era of breakthrough therapies. *The Oncologist*, *22*(7), 762–767. Available from https://doi.org/10.1634/theoncologist.2017-0152, http://theoncologist.alphamedpress.org/content/22/7/762.full.pdf.

Brinker, T. J., Hekler, A., Enk, A. H., Klode, J., Hauschild, A., Berking, C., Schilling, B., Haferkamp, S., Schadendorf, D., Fröhling, S., Utikal, J. S., von Kalle, C., Ludwig-Peitsch, W., Sirokay, J., Heinzerling, L., Albrecht, M., Baratella, K., Bischof, L., Chorti, E., ... Schrüfer, P. (2019). A convolutional neural network trained with dermoscopic images performed on par with 145 dermatologists in a clinical melanoma image classification task. *European Journal of Cancer*, *111*, 148–154. Available from https://doi.org/10.1016/j.ejca.2019.02.005, http://www.journals.elsevier.com/european-journal-of-cancer/.

Brinker, T. J., Schmitt, M., Krieghoff-Henning, E. I., Barnhill, R., Beltraminelli, H., Braun, S. A., Carr, R., Fernandez-Figueras, M. T., Ferrara, G., Fraitag, S., Gianotti, R., Llamas-Velasco, M., Müller, C. S. L., Perasole, A., Requena, L., Sangueza, O. P., Santonja, C., Starz, H., Vale, E., ... Kutzner, H. (2022). Diagnostic performance of artificial intelligence for histologic melanoma recognition compared to 18 international expert pathologists. *Journal of the American Academy of Dermatology*, *86*(3), 640–642. Available from https://doi.org/10.1016/j.jaad.2021.02.009, http://www.elsevier.com/inca/publications/store/6/2/3/3/7/0/index.htt.

Brunnström, H., Johansson, A., Westbom-Fremer, S., Backman, M., Djureinovic, D., Patthey, A., Isaksson-Mettävainio, M., Gulyas, M., & Micke, P. (2017). PD-L1 immunohistochemistry in clinical diagnostics of lung cancer: Inter-pathologist variability is higher than assay variability. *Modern Pathology*, *30*(10), 1411–1421. Available from https://doi.org/10.1038/modpathol.2017.59.

Brügmann, A., Eld, M., Lelkaitis, G., Nielsen, S., Grunkin, M., Hansen, J. D., Foged, N. T., & Vyberg, M. (2012). Digital image analysis of membrane connectivity is a robust measure of HER2 immunostains. *Breast Cancer Research and Treatment*, *132*(1), 41–49. Available from https://doi.org/10.1007/s10549-011-1514-2.

Buck, T. P., Dilorio, R., Havrilla, L., & O'Neill, D. G. (2014). Validation of a whole slide imaging system for primary diagnosis in surgical pathology: A community hospital experience. *Journal of Pathology Informatics*, *5*(1)43. Available from https://doi.org/10.4103/2153-3539.145731.

Bui, M. M., Riben, M. W., Allison, K. H., Chlipala, E., Colasacco, C., Kahn, A. G., Lacchetti, C., Madabhushi, A., Pantanowitz, L., Salama, M. E., Stewart, R. L., Thomas, N. E., Tomaszewski, J. E., & Hammond, M. E. (2019). Quantitative image analysis of human epidermal growth factor receptor 2 immunohistochemistry for breast cancer: Guideline from the college of American pathologists. *Archives of Pathology and Laboratory Medicine*, *143*(10), 1180–1195. Available from https://doi.org/10.5858/arpa.2018-0378-CP, https://www.archivesofpathology.org/doi/pdf/10.5858/arpa.2018-0378-CP.

Bulten, W., Kartasalo, K., Chen, P.-H. C., Ström, P., Pinckaers, H., Nagpal, K., Cai, Y., Steiner, D. F., van Boven, H., Vink, R., Hulsbergen-van de Kaa, C., van der Laak, J., Amin, M. B., Evans, A. J., van der Kwast, T., Allan, R., Humphrey, P. A., Grönberg, H., Samaratunga, H., ... Lee, H.-S. (2022). Artificial intelligence for diagnosis and Gleason grading of prostate cancer: The PANDA challenge. *Nature Medicine*, *28*(1), 154–163. Available from https://doi.org/10.1038/s41591-021-01620-2.

Bulten, W., Pinckaers, H., van Boven, H., Vink, R., de Bel, T., van Ginneken, B., van der Laak, J., Hulsbergen-van de Kaa, C., & Litjens, G. (2020). Automated deep-learning system for Gleason grading of prostate cancer using biopsies: A diagnostic study. *The Lancet Oncology*, *21*(2), 233–241. Available from https://doi.org/10.1016/s1470-2045(19)30739-9.

Buttner, R., Gosney, J. R., Skov, B. G., Adam, J., Motoi, N., Bloom, K. J., Dietel, M., Longshore, J. W., Lopez-Ríos, F., Penault-Llorca, F., Viale, G., Wotherspoon, A. C., Kerr, K. M., & Tsao, M. S. (2017). Programmed death-ligand 1 immunohistochemistry testing: A review of analytical assays and clinical implementation in non-small-cell lung cancer. *Journal of Clinical Oncology*, *35*(34), 3867–3876. Available from https://doi.org/10.1200/JCO.2017.74.7642, http://ascopubs.org/doi/pdf/10.1200/JCO.2017.74.7642.

Campanella, G., Hanna, M. G., Geneslaw, L., Miraflor, A., Werneck Krauss Silva, V., Busam, K. J., Brogi, E., Reuter, V. E., Klimstra, D. S., & Fuchs, T. J. (2019). Clinical-grade computational pathology using weakly supervised deep learning on whole slide images. *Nature Medicine*, *25*(8), 1301–1309. Available from https://doi.org/10.1038/s41591-019-0508-1, http://www.nature.com/nm/index.html.

Cao, R., Yang, F., Ma, S. C., Liu, L., Zhao, Y., Li, Y., Wu, D. H., Wang, T., Lu, W. J., Cai, W. J., Zhu, H. B., Guo, X. J., Lu, Y. W., Kuang, J. J., Huan, W. J., Tang, W. M., Huang, K., Huang, J., Yao, J., & Dong, Z. Y. (2020). Development and interpretation of a pathomics-based model for the prediction of microsatellite instability in colorectal cancer. *Theranostics*, *10*(24), 11080–11091. Available from https://doi.org/10.7150/thno.49864, http://www.thno.org/.

Chan, T. A., Yarchoan, M., Jaffee, E., Swanton, C., Quezada, S. A., Stenzinger, A., & Peters, S. (2019). Development of tumor mutation burden as an immunotherapy biomarker: Utility for the oncology clinic. *Annals of Oncology*, *30*(1), 44–56. Available from https://doi.org/10.1093/annonc/mdy495.

Chang, L., Chang, M., Chang, H. M., & Chang, F. (2018). Microsatellite instability: A predictive biomarker for cancer immunotherapy. *Applied Immunohistochemistry and Molecular Morphology*, *26*, 15–21, 2018.

Chang, S., Park, H. K., Choi, Y. L., & Jang, S. J. (2019). Interobserver reproducibility of PD-L1 biomarker in non-small cell lung cancer: A multi-institutional study by 27 pathologists. *Journal of Pathology and Translational Medicine*, *53*(6), 347–353. Available from https://doi.org/10.4132/jptm.2019.09.29, http://jpatholtm.org/about/about.php.

Chen, M., Zhang, B., Topatana, W., Cao, J., Zhu, H., Juengpanich, S., Mao, Q., Yu, H., & Cai, X. (2020). Classification and mutation prediction based on histopathology H&E images in liver cancer using deep learning. *NPJ Precision Oncology*, *4*(1). Available from https://doi.org/10.1038/s41698-020-0120-3.

Cheng, C. L., Azhar, R., Sng, S. H. A., Chua, Y. Q., Hwang, J. S. G., Chin, J. P. F., Seah, W. K., Loke, J. C. L., Ang, R. H. L., & Tan, P. H. (2016). Enabling digital pathology in the diagnostic setting: Navigating through the implementation journey in an academic medical centre. *Journal of Clinical Pathology*, *69*(9), 784–792. Available from https://doi.org/10.1136/jclinpath-2015-203600, http://jcp.bmj.com/.

Chuang, W. Y., Chang, S. H., Yu, W. H., Yang, C. K., Yeh, C. J., Ueng, S. H., Liu, Y. J., Chen, T. D., Chen, K. H., Hsieh, Y. Y., Hsia, Y., Wang, T. H., Hsueh, C., Kuo, C. F., & Yeh, C. Y. (2020). Successful identification of nasopharyngeal carcinoma in nasopharyngeal biopsies using deep learning. *Cancers*, *12*(2). Available from https://doi.org/10.3390/cancers12020507, https://www.mdpi.com/2072-6694/12/2/507/pdf.

Chuang, W. Y., Chen, C. C., Yu, W. H., Yeh, C. J., Chang, S. H., Ueng, S. H., Wang, T. H., Hsueh, C., Kuo, C. F., & Yeh, C. Y. (2021). Identification of nodal micrometastasis in colorectal cancer using deep learning on annotation-free whole-slide images. *Modern Pathology*, *34*(10), 1901–1911. Available from https://doi.org/10.1038/s41379-021-00838-2, http://www.nature.com/modpathol/index.html.

Cogswell, J., Inzunza, H. D., Wu, Q., Feder, J. N., Mintier, G., Novotny, J., & Cardona, D. M. (2017). An analytical comparison of Dako 28-8 PharmDx assay and an E1L3N laboratory-developed test in the immunohistochemical detection of programmed death-ligand 1. *Molecular Diagnosis & Therapy*, *21*(1), 85–93. Available from https://doi.org/10.1007/s40291-016-0237-9.

Coudray, N., Ocampo, P. S., Sakellaropoulos, T., Narula, N., Snuderl, M., Fenyö, D., Moreira, A. L., Razavian, N., & Tsirigos, A. (2018). Classification and mutation prediction from non–small cell lung cancer histopathology images using deep learning. *Nature Medicine*, *24*(10), 1559–1567. Available from https://doi.org/10.1038/s41591-018-0177-5, http://www.nature.com/nm/index.html.

Courtiol, P., Maussion, C., Moarii, M., Pronier, E., Pilcer, S., Sefta, M., Manceron, P., Toldo, S., Zaslavskiy, M., Le Stang, N., Girard, N., Elemento, O., Nicholson, A. G., Blay, J.-Y., Galateau-Sallé, F., Wainrib, G., & Clozel, T. (2019). Deep learning-based classification of mesothelioma improves prediction of patient outcome. *Nature Medicine*, *25*(10), 1519–1525. Available from https://doi.org/10.1038/s41591-019-0583-3.

Cui, D., Liu, Y., Liu, G., & Liu, L. (2020). A multiple-instance learning-based convolutional neural network model to detect the IDH1 mutation in the histopathology images of glioma tissues. *Journal of Computational Biology*, *27*(8), 1264–1272. Available from https://doi.org/10.1089/cmb.2019.0410, http://www.liebertonline.com/cmb.

Dammak, S., Cecchini, M. J., Breadner, D., & Ward, A. D. (2023). Using deep learning to predict tumor mutational burden from scans of H&E-stained multicenter slides of lung squamous cell carcinoma. *Journal of Medical Imaging*, *10*(1). Available from https://doi.org/10.1117/1.JMI.10.1.017502, http://medicalimaging.spiedigitallibrary.org/journal.aspx.

De Logu, F., Ugolini, F., Maio, V., Simi, S., Cossu, A., Massi, D., Nassini, R., & Laurino, M. (2020). Recognition of cutaneous melanoma on digitized histopathological slides via artificial intelligence algorithm. *Frontiers in Oncology*, *10*. Available from https://doi.org/10.3389/fonc.2020.01559.

Diao, J. A., Wang, J. K., Chui, W. F., Mountain, V., Gullapally, S. C., Srinivasan, R., Mitchell, R. N., Glass, B., Hoffman, S., Rao, S. K., Maheshwari, C., Lahiri, A., Prakash, A., McLoughlin, R., Kerner, J. K., Resnick, M. B., Montalto, M. C., Khosla, A., Wapinski, I. N., ... Taylor-Weiner, A. (2021). Human-interpretable image features derived from densely mapped cancer pathology slides predict diverse molecular phenotypes. *Nature Communications*, *12*(1). Available from https://doi.org/10.1038/s41467-021-21896-9, http://www.nature.com/ncomms/index.html.

Dick, V., Sinz, C., Mittlböck, M., Kittler, H., & Tschandl, P. (2019). Accuracy of computer-aided diagnosis of melanoma: A meta-analysis. *JAMA Dermatology*, *155*(11), 1291–1299,. Available from https://doi.org/10.1001/jamadermatol.2019.1375, http://archderm.jamanetwork.com/issues.aspx.

Dietz, R. L., Hartman, D. J., & Pantanowitz, L. (2020). Systematic review of the use of telepathology during intraoperative consultation. *American Journal of Clinical Pathology*, *153*(2), 198–209. Available from https://doi.org/10.1093/ajcp/aqz155, http://ajcp.oxfordjournals.org/content/by/year.

Dietz, R. L., Hartman, D. J., Zheng, L., Wiley, C., & Pantanowitz, L. (2018). Review of the use of telepathology for intraoperative consultation. *Expert Review of Medical Devices*, *15*(12), 883–890. Available from https://doi.org/10.1080/17434440.2018.1549987, http://www.tandfonline.com/loi/ierd20.

Dobson, L., Conway, C., Hanley, A., Johnson, A., Costello, S., O'Grady, A., Connolly, Y., Magee, H., O'Shea, D., Jeffers, M., & Kay, E. (2010). Image analysis as an adjunct to manual HER-2 immunohistochemical review: A diagnostic tool to standardize interpretation. *Histopathology*, *57*(1), 27–38. Available from https://doi.org/10.1111/j.1365-2559.2010.03577.x.

Dunn, B. E., Almagro, U. A., Choi, H., Sheth, N. K., Arnold, J. S., Recla, D. L., Krupinski, E. A., Graham, A. R., & Weinstein, R. S. (1997). Dynamic-robotic telepathology: Department of Veterans Affairs feasibility study. *Human Pathology*, *28*(1), 8–12. Available from http://www.elsevier.com/inca/publications/store/6/2/3/1/3/9/index.htt.

Echle, A., Ghaffari Laleh, N., Quirke, P., Grabsch, H. I., Muti, H. S., Saldanha, O. L., Brockmoeller, S. F., van den Brandt, P. A., Hutchins, G. G. A., Richman, S. D., Horisberger, K., Galata, C., Ebert, M. P., Eckardt, M., Boutros, M., Horst, D., Reissfelder, C., Alwers, E., Brinker, T. J., … Kather, J. N. (2022). Artificial intelligence for detection of microsatellite instability in colorectal cancer—A multicentric analysis of a pre-screening tool for clinical application. *ESMO Open*, *7*(2)100400. Available from https://doi.org/10.1016/j.esmoop.2022.100400.

Echle, A., Grabsch, H. I., Quirke, P., van den Brandt, P. A., West, N. P., Hutchins, G. G. A., Heij, L. R., Tan, X., Richman, S. D., Krause, J., Alwers, E., Jenniskens, J., Offermans, K., Gray, R., Brenner, H., Chang-Claude, J., Trautwein, C., Pearson, A. T., Boor, P., … Kather, J. N. (2020). Clinical-grade detection of microsatellite instability in colorectal tumors by deep learning. *Gastroenterology*, *159*(4), 1406. Available from https://doi.org/10.1053/j.gastro.2020.06.021, http://www.journals.elsevier.com/gastroenterology/.

Echle, A., Rindtorff, N. T., Brinker, T. J., Luedde, T., Pearson, A. T., & Kather, J. N. (2021). Deep learning in cancer pathology: A new generation of clinical biomarkers. *British Journal of Cancer*, *124*(4), 686–696. Available from https://doi.org/10.1038/s41416-020-01122-x, http://www.nature.com/bjc/index.html.

Ehteshami Bejnordi, B., Veta, M., Johannes van Diest, P., van Ginneken, B., Karssemeijer, N., Litjens, G., van der Laak, J. A. W. M., Hermsen, M., Manson, Q. F., Balkenhol, M., Geessink, O., Stathonikos, N., van Dijk, M. C. R. F., Bult, P., Beca, F., Beck, A. H., Wang, D., Khosla, A., Gargeya, R., … Venâncio, R. (2017). Diagnostic assessment of deep learning algorithms for detection of lymph node metastases in women with breast cancer. *JAMA: the Journal of the American Medical Association*, *318*(22), 2199. Available from https://doi.org/10.1001/jama.2017.14585.

Elmore, J. G., Longton, G. M., Carney, P. A., Geller, B. M., Onega, T., Tosteson, A. N. A., Nelson, H. D., Pepe, M. S., Allison, K. H., Schnitt, S. J., O'Malley, F. P., & Weaver, D. L. (2015). Diagnostic concordance among pathologists interpreting breast biopsy specimens. *JAMA - Journal of the American Medical Association*, *313*(11), 1122–1132. Available from https://doi.org/10.1001/jama.2015.1405, http://jama.jamanetwork.com/article.aspx?articleid = 2203798.

Emens, L. A., Molinero, L., Loi, S., Rugo, H. S., Schneeweiss, A., Diéras, V., Iwata, H., Barrios, C. H., Nechaeva, M., Nguyen-Duc, A., Chui, S. Y., Husain, A., Winer, E. P., Adams, S., & Schmid, P. (2021). Atezolizumab and nab-paclitaxel in advanced triple-negative breast cancer: Biomarker evaluation of the IMpassion130 study. *Journal of the National Cancer Institute*, *113*(8), 1005–1016. Available from https://doi.org/10.1093/jnci/djab004, http://jnci.oxfordjournals.org/.

Ertosun, M. G., & Rubin, D. L. (2015). Automated grading of gliomas using deep learning in digital pathology images: A modular approach with ensemble of convolutional neural networks. *AMIA. Annual Symposium Proceedings. AMIA Symposium.*, *2015*, 1899–1908.

Evans, A. J., Chetty, R., Clarke, B. A., Croul, S., Ghazarian, D. M., Kiehl, T. R., Perez Ordonez, B., Ilaalagan, S., & Asa, S. L. (2009). Primary frozen section diagnosis by robotic microscopy and virtual slide telepathology: The University Health Network experience. *Human Pathology*, *40*(8), 1070–1081. Available from https://doi.org/10.1016/j.humpath.2009.04.012.

Farahani, H., Boschman, J., Farnell, D., Darbandsari, A., Zhang, A., Ahmadvand, P., Jones, S. J. M., Huntsman, D., Köbel, M., Gilks, C. B., Singh, N., & Bashashati, A. (2022). Deep learning-based histotype diagnosis of ovarian carcinoma whole-slide pathology images. *Modern Pathology*, *35*(12), 1983–1990. Available from https://doi.org/10.1038/s41379-022-01146-z.

Feng, M., Deng, Y., Yang, L., Jing, Q., Zhang, Z., Xu, L., Wei, X., Zhou, Y., Wu, D., Xiang, F., Wang, Y., Bao, J., & Bu, H. (2020). Automated quantitative analysis of Ki-67 staining and HE images recognition and registration based on whole tissue sections in breast carcinoma. *Diagnostic Pathology*, *15*(1). Available from https://doi.org/10.1186/s13000-020-00957-5.

Fenstermaker, M., Tomlins, S. A., Singh, K., Wiens, J., & Morgan, T. M. (2020). Development and validation of a deep-learning model to assist with renal cell carcinoma histopathologic interpretation. *Urology*, *144*, 152–157. Available from https://doi.org/10.1016/j.urology.2020.05.094, http://www.elsevier.com/locate/urology.

Fitzgibbons, P. L., Page, D. L., Weaver, D., Thor, A. D., Allred, D. C., Clark, G. M., Ruby, S. G., O'Malley, F., Simpson, J. F., Connolly, J. L., Hayes, D. F., Edge, S. B., Lichter, A., & Schnitt, S. J. (2000). Prognostic factors in breast cancer. *Archives of Pathology and Laboratory Medicine*, *124*(7), 966–978.

Fu, Y., Jung, A. W., Torne, R. V., Gonzalez, S., Vöhringer, H., Shmatko, A., Yates, L. R., Jimenez-Linan, M., Moore, L., & Gerstung, M. (2020). Pan-cancer computational histopathology reveals mutations, tumor composition and prognosis. *Nature Cancer*, *1*(8), 800–810. Available from https://doi.org/10.1038/s43018-020-0085-8.

Galuppini, F., Dal Pozzo, C. A., Deckert, J., Loupakis, F., Fassan, M., & Baffa, R. (2019). Tumor mutation burden: From comprehensive mutational screening to the clinic. *Cancer Cell International*, *19*(1). Available from https://doi.org/10.1186/s12935-019-0929-4.

Geessink, O. G. F., Baidoshvili, A., Klaase, J. M., Ehteshami Bejnordi, B., Litjens, G. J. S., van Pelt, G. W., Mesker, W. E., Nagtegaal, I. D., Ciompi, F., & van der Laak, J. A. W. M. (2019). Computer aided quantification of intratumoral stroma yields an independent prognosticator in rectal cancer. *Cellular Oncology*, *42*(3), 331–341. Available from https://doi.org/10.1007/s13402-019-00429-z, http://link.springer.com/journal/volumesAndIssues/13402.

Gertych, A., Swiderska-Chadaj, Z., Ma, Z., Ing, N., Markiewicz, T., Cierniak, S., Salemi, H., Guzman, S., Walts, A. E., & Knudsen, B. S. (2019). Convolutional neural networks can accurately distinguish four histologic growth patterns of lung adenocarcinoma in digital slides. *Scientific Reports*, *9*(1). Available from https://doi.org/10.1038/s41598-018-37638-9.

Ghahremani, P., Marino, J., Dodds, R., & Nadeem, S. (2022). DeepLIIF, an online platform for quantification of clinical pathology slides. *Proceedings of the IEEE Computer Society Conference on Computer Vision and Pattern Recognition*, *2022*, 21399–21405.

Ginter, P. S., Idress, R., D'Alfonso, T. M., Fineberg, S., Jaffer, S., Sattar, A. K., Chagpar, A., Wilson, P., & Harigopal, M. (2021). Histologic grading of breast carcinoma: A multi-institution study of interobserver variation using virtual microscopy. *Modern Pathology*, *34*(4), 701–709. Available from https://doi.org/10.1038/s41379-020-00698-2, http://www.nature.com/modpathol/index.html.

Goacher, E., Randell, R., Williams, B., & Treanor, D. (2017). The diagnostic concordance of whole slide imaging and light microscopy: A systematic review. *Archives of Pathology & Laboratory Medicine*, *141*(1), 151–161. Available from https://doi.org/10.5858/arpa.2016-0025-ra.

Hammond, M. E. H., Hayes, D. F., Dowsett, M., Allred, D. C., Hagerty, K. L., Badve, S., Fitzgibbons, P. L., Francis, G., Goldstein, N. S., Hayes, M., Hicks, D. G., Lester, S., Love, R., Mangu, P. B., McShane, L., Miller, K., Osborne, C. K., Paik, S., Perlmutter, J., ... Wolff, A. C. (2010). American Society of Clinical Oncology/College of American Pathologists guideline recommendations for immunohistochemical testing of estrogen and progesterone receptors in breast cancer. *Journal of Clinical Oncology*, *28*(16), 2784–2795. Available from https://doi.org/10.1200/JCO.2009.25.6529, http://jco.ascopubs.org/content/28/16/2784.full.pdf + html.

Han, S., Liu, Y., Cai, S. J., Qian, M., Ding, J., Larion, M., Gilbert, M. R., & Yang, C. (2020). IDH mutation in glioma: Molecular mechanisms and potential therapeutic targets. *British Journal of Cancer*, *122*(11), 1580–1589. Available from https://doi.org/10.1038/s41416-020-0814-x, http://www.nature.com/bjc/index.html.

Hanna, M. G., Reuter, V. E., Ardon, O., Kim, D., Sirintrapun, S. J., Schüffler, P. J., Busam, K. J., Sauter, J. L., Brogi, E., Tan, L. K., Xu, B., Bale, T., Agaram, N. P., Tang, L. H., Ellenson, L. H., Philip, J., Corsale, L., Stamelos, E., Friedlander, M. A., ... Hameed, M. (2020). Validation of a digital pathology system including remote review during the COVID-19 pandemic. *Modern Pathology*, *33*(11), 2115–2127. Available from https://doi.org/10.1038/s41379-020-0601-5, http://www.nature.com/modpathol/index.html.

He, B., Bergenstråhle, L., Stenbeck, L., Abid, A., Andersson, A., Borg, Å., Maaskola, J., Lundeberg, J., & Zou, J. (2020). Integrating spatial gene expression and breast tumour morphology via deep learning. *Nature Biomedical Engineering*, *4*(8), 827–834. Available from https://doi.org/10.1038/s41551-020-0578-x.

Hekler, A., Utikal, J. S., Enk, A. H., Solass, W., Schmitt, M., Klode, J., Schadendorf, D., Sondermann, W., Franklin, C., Bestvater, F., Flaig, M. J., Krahl, D., von Kalle, C., Fröhling, S., & Brinker, T. J. (2019). Deep learning outperformed 11 pathologists in the classification of histopathological melanoma images. *European Journal of Cancer*, *118*, 91–96. Available from https://doi.org/10.1016/j.ejca.2019.06.012, http://www.journals.elsevier.com/european-journal-of-cancer/.

Helin, H. O., Tuominen, V. J., Ylinen, O., Helin, H. J., & Isola, J. (2016). Free digital image analysis software helps to resolve equivocal scores in HER2 immunohistochemistry. *Virchows Archiv*, *468*(2), 191–198. Available from https://doi.org/10.1007/s00428-015-1868-7.

Herbst, R. S., Baas, P., Kim, D. W., Felip, E., Pérez-Gracia, J. L., Han, J. Y., Molina, J., Kim, J. H., Arvis, C. D., Ahn, M. J., Majem, M., Fidler, M. J., De Castro, G., Garrido, M., Lubiniecki, G. M., Shentu, Y., Im, E., Dolled-Filhart, M., & Garon, E. B. (2016). Pembrolizumab versus docetaxel for previously treated, PD-L1-positive, advanced non-small-cell lung cancer (KEYNOTE-010): A randomised controlled trial. *The Lancet*, *387*(10027), 1540–1550. Available from https://doi.org/10.1016/S0140-6736(15)01281-7, http://www.journals.elsevier.com/the-lancet/.

Ho, D. J., Agaram, N. P., Jean, M. H., Suser, S. D., Chu, C., Vanderbilt, C. M., Meyers, P. A., Wexler, L. H., Healey, J. H., Fuchs, T. J., & Hameed, M. R. (2023). Deep learning-based objective and reproducible osteosarcoma chemotherapy response assessment and outcome prediction. *American Journal of Pathology*, *193*(3), 341–349. Available from https://doi.org/10.1016/j.ajpath.2022.12.004, http://ajp.amjpathol.org/.

Holten-Rossing, H., Møller Talman, M.-L., Kristensson, M., & Vainer, B. (2015). Optimizing HER2 assessment in breast cancer: Application of automated image analysis. *Breast Cancer Research and Treatment*, *152*(2), 367–375. Available from https://doi.org/10.1007/s10549-015-3475-3.

Holzinger, A., Langs, G., Denk, H., Zatloukal, K., & Müller, H. (2019). Causability and explainability of artificial intelligence in medicine. *WIREs Data Mining and Knowledge Discovery*, *9*(4). Available from https://doi.org/10.1002/widm.1312.

Hong, R., Liu, W., DeLair, D., Razavian, N., & Fenyö, D. (2021). Predicting endometrial cancer subtypes and molecular features from histopathology images using multi-resolution deep learning models. *Cell Reports Medicine*, *2*(9)100400. Available from https://doi.org/10.1016/j.xcrm.2021.100400.

Hoppe, M. M., Sundar, R., Tan, D. S. P., & Jeyasekharan, A. D. (2018). Biomarkers for homologous recombination deficiency in cancer. *Journal of the National Cancer Institute*, *110*(7), 704–713. Available from https://doi.org/10.1093/jnci/djy085, http://jnci.oxfordjournals.org/.

Huang, K., Lin, B., Liu, J., Liu, Y., Li, J., Tian, G., Yang, J., & Peng, H. (2022). Predicting colorectal cancer tumor mutational burden from histopathological images and clinical information using multi-modal deep learning. *Bioinformatics*, *38*(22), 5108–5115. Available from https://doi.org/10.1093/bioinformatics/btac641.

Huang, W., Randhawa, R., Jain, P., Iczkowski, K. A., Hu, R., Hubbard, S., Eickhoff, J., Basu, H., & Roy, R. (2021). Development and validation of an artificial intelligence-powered platform for prostate cancer grading and quantification. *JAMA Network Open*, *4*(11), e2132554. Available from https://doi.org/10.1001/jamanetworkopen.2021.32554.

Humphries, M. P., Hynes, S., Bingham, V., Cougot, D., James, J., Patel-Socha, F., Parkes, E. E., Blayney, J. K., O'Rorke, M. A., Irwin, G. W., McArt, D. G., Kennedy, R. D., Mullan, P. B., McQuaid, S., Salto-Tellez, M., & Buckley, N. E. (2018). Automated tumour recognition and digital pathology scoring unravels new role for PD-L1 in predicting good outcome in ER-/HER2 + breast cancer. *Journal of Oncology*, *2018*. Available from https://doi.org/10.1155/2018/2937012, http://www.hindawi.com/journals/jo/.

Ianni, J. D., Soans, R. E., Sankarapandian, S., Chamarthi, R. V., Ayyagari, D., Olsen, T. G., Bonham, M. J., Stavish, C. C., Motaparthi, K., Cockerell, C. J., Feeser, T. A., & Lee, J. B. (2020). Tailored for real-world: A whole slide image classification system validated on uncurated multi-site data emulating the prospective pathology workload. *Scientific Reports*, *10*(1), 3217. Available from https://doi.org/10.1038/s41598-020-59985-2.

Jang, H. J., Lee, A., Kang, J., Song, I. H., & Lee, S. H. (2020). Prediction of clinically actionable genetic alterations from colorectal cancer histopathology images using deep learning. *World Journal of Gastroenterology*, *26*(40), 6207–6223. Available from https://doi.org/10.3748/wjg.v26.i40.6207, https://www.wjgnet.com/1007-9327/full/v26/i40/6207.htm.

Jarkman, S., Karlberg, M., Pocevičiūtė, M., Bodén, A., Bándi, P., Litjens, G., Lundström, C., Treanor, D., & van der Laak, J. (2022). Generalization of deep learning in digital pathology: Experience in breast cancer metastasis detection. *Cancers*, *14*(21), 5424. Available from https://doi.org/10.3390/cancers14215424.

Kacew, A. J., Strohbehn, G. W., Saulsberry, L., Laiteerapong, N., Cipriani, N. A., Kather, J. N., & Pearson, A. T. (2021). Artificial intelligence can cut costs while maintaining accuracy in colorectal cancer genotyping. *Frontiers in Oncology*, *11*. Available from https://doi.org/10.3389/fonc.2021.630953, http://www.frontiersin.org/Oncology/about.

Kather, J. N., Heij, L. R., Grabsch, H. I., Loeffler, C., Echle, A., Muti, H. S., Krause, J., Niehues, J. M., Sommer, K. A. J., Bankhead, P., Kooreman, L. F. S., Schulte, J. J., Cipriani, N. A., Buelow, R. D., Boor, P., Ortiz-Brüchle, N., Hanby, A. M., Speirs, V., Kochanny, S., … Luedde, T. (2020). Pan-cancer image-based detection of clinically actionable genetic alterations. *Nature Cancer*, *1*(8), 789−799. Available from https://doi.org/10.1038/s43018-020-0087-6, https://www.nature.com/natcancer/.

Kather, J. N., Krisam, J., Charoentong, P., Luedde, T., Herpel, E., Weis, C. A., Gaiser, T., Marx, A., Valous, N. A., Ferber, D., Jansen, L., Reyes-Aldasoro, C. C., Zörnig, I., Jäger, D., Brenner, H., Chang-Claude, J., Hoffmeister, M., Halama, N., & Butte, A. J. (2019). Predicting survival from colorectal cancer histology slides using deep learning: A retrospective multicenter study. *PLoS Medicine*, *16*(1), e1002730. Available from https://doi.org/10.1371/journal.pmed.1002730.

Kather, J. N., Pearson, A. T., Halama, N., Jäger, D., Krause, J., Loosen, S. H., Marx, A., Boor, P., Tacke, F., Neumann, U. P., Grabsch, H. I., Yoshikawa, T., Brenner, H., Chang-Claude, J., Hoffmeister, M., Trautwein, C., & Luedde, T. (2019). Deep learning can predict microsatellite instability directly from histology in gastrointestinal cancer. *Nature Medicine*, *25*(7), 1054−1056. Available from https://doi.org/10.1038/s41591-019-0462-y, http://www.nature.com/nm/index.html.

Kim, R. H., Nomikou, S., Coudray, N., Jour, G., Dawood, Z., Hong, R., Esteva, E., Sakellaropoulos, T., Donnelly, D., Moran, U., Hatzimemos, A., Weber, J. S., Razavian, N., Aifantis, I., Fenyo, D., Snuderl, M., Shapiro, R., Berman, R. S., Osman, I., & Tsirigos, A. (2022). Deep learning and pathomics analyses reveal cell nuclei as important features for mutation prediction of BRAF-mutated melanomas. *Journal of Investigative Dermatology*, *142*(6), 1650. Available from https://doi.org/10.1016/j.jid.2021.09.034, http://www.nature.com/jid/index.html.

Kindler, C., Elfwing, S., Öhrvik, J., & Nikberg, M. (2023). A deep neural network-based decision support tool for the detection of lymph node metastases in colorectal cancer specimens. *Modern Pathology*, *36*(2)100015. Available from https://doi.org/10.1016/j.modpat.2022.100015.

Klimov, S., Miligy, I. M., Gertych, A., Jiang, Y., Toss, M. S., Rida, P., Ellis, I. O., Green, A., Krishnamurti, U., Rakha, E. A., & Aneja, R. (2019). A whole slide image-based machine learning approach to predict ductal carcinoma in situ (DCIS) recurrence risk. *Breast Cancer Research*, *21*(1). Available from https://doi.org/10.1186/s13058-019-1165-5, http://breast-cancer-research.com/.

Koomen, B. M., Voorham, Q. J. M., Epskamp-Kuijpers, C. C. H. J., van Dooijeweert, C., van Lindert, A. S. R., Deckers, I. A. G., & Willems, S. M. (2021). Considerable interlaboratory variation in PD-L1 positivity in a nationwide cohort of non-small cell lung cancer patients. *Lung Cancer*, *159*, 117−126. Available from https://doi.org/10.1016/j.lungcan.2021.07.012, http://www.elsevier.com/locate/lungcan.

Koopman, T., de Bock, G. H., Buikema, H. J., Smits, M. M., Louwen, M., Hage, M., Imholz, A. L. T., & van der Vegt, B. (2018). Digital image analysis of HER2 immunohistochemistry in gastric- and oesophageal adenocarcinoma: A validation study on biopsies and surgical specimens. *Histopathology*, *72*(2), 191–200. Available from https://doi.org/10.1111/his.13322, http://onlinelibrary.wiley.com/journal/10.1111/(ISSN)1365-2559.

Koopman, T., Buikema, H. J., Hollema, H., de Bock, G. H., & van der Vegt, B. (2019). What is the added value of digital image analysis of HER2 immunohistochemistry in breast cancer in clinical practice? A study with multiple platforms. *Histopathology*, *74* (6), 917–924. Available from https://doi.org/10.1111/his.13812, http://onlinelibrary.wiley.com/journal/10.1111/(ISSN)1365-2559.

Korbar, B., Olofson, A. M., Miraflor, A. P., Nicka, C. M., Suriawinata, M. A., Torresani, L., Suriawinata, A. A., & Hassanpour, S. (2017). Deep learning for classification of colorectal polyps on whole-slide images. *Journal of Pathology Informatics*, *8*(1)30. Available from https://doi.org/10.4103/jpi.jpi_34_17.

Kulkarni, P. M., Robinson, E. J., Sarin Pradhan, J., Gartrell-Corrado, R. D., Rohr, B. R., Trager, M. H., Geskin, L. J., Kluger, H. M., Wong, P. F., Acs, B., Rizk, E. M., Yang, C., Mondal, M., Moore, M. R., Osman, I., Phelps, R., Horst, B. A., Chen, Z. S., Ferringer, T., ... Saenger, Y. M. (2020). Deep learning based on standard H&E images of primary melanoma tumors identifies patients at risk for visceral recurrence and death. *Clinical Cancer Research*, *26*(5), 1126–1134. Available from https://doi.org/10.1158/1078-0432.ccr-19-1495.

Kuntz, S., Krieghoff-Henning, E., Kather, J. N., Jutzi, T., Höhn, J., Kiehl, L., Hekler, A., Alwers, E., von Kalle, C., Fröhling, S., Utikal, J. S., Brenner, H., Hoffmeister, M., & Brinker, T. J. (2021). Gastrointestinal cancer classification and prognostication from histology using deep learning: Systematic review. *European Journal of Cancer*, *155*, 200–215. Available from https://doi.org/10.1016/j.ejca.2021.07.012, http://www.journals.elsevier.com/european-journal-of-cancer/.

Larkin, J., Chiarion-Sileni, V., Gonzalez, R., Grob, J. J., Rutkowski, P., Lao, C. D., Cowey, C. L., Schadendorf, D., Wagstaff, J., Dummer, R., Ferrucci, P. F., Smylie, M., Hogg, D., Hill, A., Márquez-Rodas, I., Haanen, J., Guidoboni, M., Maio, M., Schöffski, P., ... Wolchok, J. D. (2019). Five-year survival with combined nivolumab and ipilimumab in advanced melanoma. *New England Journal of Medicine*, *381*(16), 1535–1546. Available from https://doi.org/10.1056/nejmoa1910836.

Lee, S. H., Song, I. H., & Jang, H. J. (2021). Feasibility of deep learning-based fully automated classification of microsatellite instability in tissue slides of colorectal cancer. *International Journal of Cancer*, *149*(3), 728–740. Available from https://doi.org/10.1002/ijc.33599, http://onlinelibrary.wiley.com/journal/10.1002/(ISSN)1097-0215.

Li, M., Jiang, Z., Shen, W., & Liu, H. (2022). Deep learning in bladder cancer imaging: A review. *Frontiers in Oncology*, *12*. Available from https://doi.org/10.3389/fonc.2022.930917.

Li, T., Xie, P., Liu, J., Chen, M., Zhao, S., Kang, W., Zuo, K., Li, F., & Khan, R. (2021). Automated diagnosis and localization of melanoma from skin histopathology slides using deep learning: A multicenter study. *Journal of Healthcare Engineering*, *2021*, 1–6. Available from https://doi.org/10.1155/2021/5972962.

Liang, W. Y., Hsu, C. Y., Lai, C. R., Ho, D. M. T., & Chiang, I. J. (2008). Low-cost telepathology system for intraoperative frozen-section consultation: Our experience and review of the literature. *Human Pathology*, *39*(1), 56–62. Available from https://doi.org/10.1016/j.humpath.2007.04.023.

Litjens, G., Bandi, P., Ehteshami Bejnordi, B., Geessink, O., Balkenhol, M., Bult, P., Halilovic, A., Hermsen, M., van de Loo, R., Vogels, R., Manson, Q. F., Stathonikos, N., Baidoshvili, A., van Diest, P., Wauters, C., van Dijk, M., & van der Laak, J.

(2018). 1399 H&E-stained sentinel lymph node sections of breast cancer patients: The CAMELYON dataset. *GigaScience*, *7*(6). Available from https://doi.org/10.1093/gigascience/giy065.

Liu, X., Faes, L., Kale, A. U., Wagner, S. K., Fu, D. J., Bruynseels, A., Mahendiran, T., Moraes, G., Shamdas, M., Kern, C., Ledsam, J. R., Schmid, M. K., Balaskas, K., Topol, E. J., Bachmann, L. M., Keane, P. A., & Denniston, A. K. (2019). A comparison of deep learning performance against health-care professionals in detecting diseases from medical imaging: A systematic review and meta-analysis. *The Lancet Digital Health*, *1*(6), e271. Available from https://doi.org/10.1016/S2589-7500(19)30123-2, https://www.sciencedirect.com/journal/the-lancet-digital-health.

Lodha, S., Saggar, S., Celebi, J. T., & Silvers, D. N. (2008). Discordance in the histopathologic diagnosis of difficult melanocytic neoplasms in the clinical setting. *Journal of Cutaneous Pathology*, *35*(4), 349–352. Available from https://doi.org/10.1111/j.1600-0560.2007.00970.x.

Longacre, T. A., Ennis, M., Quenneville, L. A., Bane, A. L., Bleiweiss, I. J., Carter, B. A., Catelano, E., Hendrickson, M. R., Hibshoosh, H., Layfield, L. J., Memeo, L., Wu, H., & O'Malley, F. P. (2006). , Interobserver agreement and reproducibility in classification of invasive breast carcinoma: An NCI breast cancer family registry study. *Modern Pathology*, *19*(2), 195–207. Available from https://doi.org/10.1038/modpathol.3800496.

Lucas, M., Jansen, I., Savci-Heijink, C. D., Meijer, S. L., de Boer, O. J., van Leeuwen, T. G., de Bruin, D. M., & Marquering, H. A. (2019). Deep learning for automatic Gleason pattern classification for grade group determination of prostate biopsies. *Virchows Archiv*, *475*(1), 77–83. Available from https://doi.org/10.1007/s00428-019-02577-x.

Lujan, G., Quigley, J. C., Hartman, D., Parwani, A., Roehmholdt, B., Van Meter, B., Ardon, O., Hanna, M. G., Kelly, D., Sowards, C., Montalto, M., Bui, M., Zarella, M. D., LaRosa, V., Slootweg, G., Retamero, J. A., Lloyd, M. C., Madory, J., & Bowman, D. (2021). Dissecting the business case for adoption and implementation of digital pathology: A white paper from the Digital Pathology Association. *Journal of Pathology Informatics*, *12*(1)17. Available from https://doi.org/10.4103/jpi.jpi_67_20.

Mahmood, T., Arsalan, M., Owais, M., Lee, M. B., & Park, K. R. (2020). Artificial intelligence-based mitosis detection in breast cancer histopathology images using faster R-CNN and deep CNNs. *Journal of Clinical Medicine*, *9*(3). Available from https://doi.org/10.3390/jcm9030749, https://www.mdpi.com/2077-0383/9/3/749/pdf.

Mercan, E., Mehta, S., Bartlett, J., Shapiro, L. G., Weaver, D. L., & Elmore, J. G. (2019). Assessment of machine learning of breast pathology structures for automated differentiation of breast cancer and high-risk proliferative lesions. *JAMA Network Open*, *2*(8), e198777. Available from https://doi.org/10.1001/jamanetworkopen.2019.8777.

Milewski, D., Jung, H., Brown, G. T., Liu, Y., Somerville, B., Lisle, C., Ladanyi, M., Rudzinski, E. R., Choo-Wosoba, H., Barkauskas, D. A., Lo, T., Hall, D., Linardic, C. M., Wei, J. S., Chou, H. C., Skapek, S. X., Venkatramani, R., Bode, P. K., Steinberg, S. M., … Khan, J. (2023). Predicting molecular subtype and survival of rhabdomyosarcoma patients using deep learning of H&E images: A report from the Children's Oncology Group. *Clinical Cancer Research*, *29*(2), 364–378. Available from https://doi.org/10.1158/1078-0432.CCR-22-1663.

Mok, T. S. K., Wu, Y. L., Kudaba, I., Kowalski, D. M., Cho, B. C., Turna, H. Z., Castro, G., Srimuninnimit, V., Laktionov, K. K., Bondarenko, I., Kubota, K., Lubiniecki, G. M., Zhang, J., Kush, D., Lopes, G., Gomez Aubin, G., Fein, L., Kaen, D.,

Kowalyszyn, R., ... Pham, X. D. (2019). Pembrolizumab versus chemotherapy for previously untreated, PD-L1-expressing, locally advanced or metastatic non-small-cell lung cancer (KEYNOTE-042): A randomised, open-label, controlled, phase 3 trial. *The Lancet*, *393*(10183), 1819–1830. Available from https://doi.org/10.1016/S0140-6736(18)32409-7, http://www.journals.elsevier.com/the-lancet/.

Mukhopadhyay, S., Feldman, M. D., Abels, E., Ashfaq, R., Beltaifa, S., Cacciabeve, N. G., Cathro, H. P., Cheng, L., Cooper, K., Dickey, G. E., Gill, R. M., Heaton, R. P., Kerstens, R., Lindberg, G. M., Malhotra, R. K., Mandell, J. W., Manlucu, E. D., Mills, A. M., Mills, S. E., ... Taylor, C. R. (2018). Whole slide imaging versus microscopy for primary diagnosis in surgical pathology: A multicenter blinded randomized noninferiority study of 1992 cases (pivotal study). *American Journal of Surgical Pathology*, *42*(1), 39–52. Available from https://doi.org/10.1097/PAS.0000000000000948, http://journals.lww.com/ajsp/pages/default.aspx.

Mun, Y., Paik, I., Shin, S. J., Kwak, T. Y., & Chang, H. (2021). Yet another automated Gleason grading system (YAAGGS) by weakly supervised deep learning. *NPJ Digital Medicine*, *4*(1). Available from https://doi.org/10.1038/s41746-021-00469-6, https://www.nature.com/npjdigitalmed/.

Nagpal, K., Foote, D., Liu, Y., Chen, P.-H. C., Wulczyn, E., Tan, F., Olson, N., Smith, J. L., Mohtashamian, A., Wren, J. H., Corrado, G. S., MacDonald, R., Peng, L. H., Amin, M. B., Evans, A. J., Sangoi, A. R., Mermel, C. H., Hipp, J. D., & Stumpe, M. C. (2019). Development and validation of a deep learning algorithm for improving Gleason scoring of prostate cancer. *NPJ Digital Medicine*, *2*(1). Available from https://doi.org/10.1038/s41746-019-0112-2.

Naik, N., Madani, A., Esteva, A., Keskar, N. S., Press, M. F., Ruderman, D., Agus, D. B., & Socher, R. (2020). Deep learning-enabled breast cancer hormonal receptor status determination from base-level H&E stains. *Nature Communications*, *11*(1). Available from https://doi.org/10.1038/s41467-020-19334-3.

Nebot-Bral, L., Coutzac, C., Kannouche, P. L., & Chaput, N. (2019). Why is immunotherapy effective (or not) in patients with MSI/MMRD tumors? *Bulletin du Cancer*, *106*(2), 105–113. Available from https://doi.org/10.1016/j.bulcan.2018.08.007, http://www.john-libbey-eurotext.fr/en/revues/bdc/index.htm.

Oberholzer, M., Fischer, H.-R., Christen, H., Gerber, S., Brühlmann, M., Mihatsch, M., Famos, M., Winkler, C., Fehr, P., Bächthold, L., & Kayser, K. (1993). Telepathology with an integrated services digital network—A new tool for image transfer in surgical pathology: A preliminary report. *Human Pathology*, *24*(10), 1078–1085. Available from https://doi.org/10.1016/0046-8177(93)90187-I.

Ozsahin, D., Emegano, I., Uzun, D., & Ozsahin, B. (2022). The systematic review of artificial intelligence applications in breast cancer diagnosis. *Diagnostics*, *13*(1), 2022.

Pantanowitz, L., Dickinson, K., Evans, A. J., Hassell, L. A., Henricks, W. H., Lennerz, J. K., Lowe, A., Parwani, A. V., Riben, M., Smith, C. D., Tuthill, J. M., Weinstein, R. S., Wilbur, D. C., Krupinski, E. A., & Bernard, J. (2014). American Telemedicine Association clinical guidelines for telepathology. *Journal of Pathology Informatics*, *5*(1)39. Available from https://doi.org/10.4103/2153-3539.143329.

Pantanowitz, L., Sinard, J. H., Henricks, W. H., Fatheree, L. A., Carter, A. B., Contis, L., Beckwith, B. A., Evans, A. J., Otis, C. N., Lal, A., & Parwani, A. V. (2013). Validating whole slide imaging for diagnostic purposes in pathology: Guideline from the College of American pathologists Pathology and Laboratory Quality Center.

Archives of Pathology and Laboratory Medicine, *137*(12), 1710–1722. Available from https://doi.org/10.5858/arpa.2013-0093-CP, http://www.archivesofpathology.org/doi/pdf/10.5858/arpa.2013-0093-CP.

Perincheri, S., Levi, A. W., Celli, R., Gershkovich, P., Rimm, D., Morrow, J. S., Rothrock, B., Raciti, P., Klimstra, D., & Sinard, J. (2021). An independent assessment of an artificial intelligence system for prostate cancer detection shows strong diagnostic accuracy. *Modern Pathology*, *34*(8), 1588–1595. Available from https://doi.org/10.1038/s41379-021-00794-x, http://www.nature.com/modpathol/index.html.

Pham, H. H. N., Futakuchi, M., Bychkov, A., Furukawa, T., Kuroda, K., & Fukuoka, J. (2019). Detection of lung cancer lymph node metastases from whole-slide histopathologic images using a two-step deep learning approach. *American Journal of Pathology*, *189*(12), 2428–2439. Available from https://doi.org/10.1016/j.ajpath.2019.08.014, http://ajp.amjpathol.org/.

Phillips, T., Simmons, P., Inzunza, H. D., Cogswell, J., Novotny, J., Taylor, C., & Zhang, X. (2015). Development of an automated PD-L1 immunohistochemistry (IHC) assay for non-small cell lung cancer. *Applied Immunohistochemistry and Molecular Morphology*, *23*(8), 541–549. Available from https://doi.org/10.1097/PAI.0000000000000256, http://www.appliedimmunohist.com.

Piepkorn, M. W., Longton, G. M., Reisch, L. M., Elder, D. E., Pepe, M. S., Kerr, K. F., Tosteson, A. N. A., Nelson, H. D., Knezevich, S., Radick, A., Shucard, H., Onega, T., Carney, P. A., Elmore, J. G., & Barnhill, R. L. (2019). Assessment of second-opinion strategies for diagnoses of cutaneous melanocytic lesions. *JAMA Network Open*, *2*(10). Available from https://doi.org/10.1001/jamanetworkopen.2019.12597, https://jamanetwork.com/journals/jamanetworkopen.

Pérez-Soler, R., Chachoua, A., Hammond, L. A., Rowinsky, E. K., Huberman, M., Karp, D., Rigas, J., Clark, G. M., Santabárbara, P., & Bonomi, P. (2004). Determinants of tumor response and survival with erlotinib in patients with non–small-cell lung cancer. *Journal of Clinical Oncology*, *22*(16), 3238–3247. Available from https://doi.org/10.1200/jco.2004.11.057.

Raciti, P., Sue, J., Ceballos, R., Godrich, R., Kunz, J. D., Kapur, S., Reuter, V., Grady, L., Kanan, C., Klimstra, D. S., & Fuchs, T. J. (2020). Novel artificial intelligence system increases the detection of prostate cancer in whole slide images of core needle biopsies. *Modern Pathology*, *33*(10), 2058–2066. Available from https://doi.org/10.1038/s41379-020-0551-y.

Rohr, J. M., Ginnebaugh, K., Tuthill, M., Pimentel, J., & Markin, R. (2023). Real-time telepathology is substantially equivalent to in-person intraoperative frozen section diagnosis. *Archives of Pathology & Laboratory Medicine*. Available from https://doi.org/10.5858/arpa.2022-0261-oa.

Sadhwani, A., Chang, H. W., Behrooz, A., Brown, T., Auvigne-Flament, I., Patel, H., Findlater, R., Velez, V., Tan, F., Tekiela, K., Wulczyn, E., Yi, E. S., Mermel, C. H., Hanks, D., Chen, P. H. C., Kulig, K., Batenchuk, C., Steiner, D. F., & Cimermancic, P. (2021). Comparative analysis of machine learning approaches to classify tumor mutation burden in lung adenocarcinoma using histopathology images. *Scientific Reports*, *11*(1). Available from https://doi.org/10.1038/s41598-021-95747-4, http://www.nature.com/srep/index.html.

Sato, S., Maki, S., Yamanaka, T., Hoshino, D., Ota, Y., Yoshioka, E., Kawachi, K., Washimi, K., Suzuki, M., Ohkubo, Y., Yokose, T., Yamashita, T., Ohtori, S., &

Miyagi, Y. (2021). Machine learning-based image analysis for accelerating the diagnosis of complicated preneoplastic and neoplastic ductal lesions in breast biopsy tissues. *Breast Cancer Research and Treatment*, *188*(3), 649–659. Available from https://doi.org/10.1007/s10549-021-06243-2.

Schmauch, B., Romagnoni, A., Pronier, E., Saillard, C., Maillé, P., Calderaro, J., Kamoun, A., Sefta, M., Toldo, S., Zaslavskiy, M., Clozel, T., Moarii, M., Courtiol, P., & Wainrib, G. (2020). A deep learning model to predict RNA-Seq expression of tumours from whole slide images. *Nature Communications*, *11*(1). Available from https://doi.org/10.1038/s41467-020-17678-4.

Schrammen, P. L., Ghaffari Laleh, N., Echle, A., Truhn, D., Schulz, V., Brinker, T. J., Brenner, H., Chang-Claude, J., Alwers, E., Brobeil, A., Kloor, M., Heij, L. R., Jäger, D., Trautwein, C., Grabsch, H. I., Quirke, P., West, N. P., Hoffmeister, M., & Kather, J. N. (2022). Weakly supervised annotation-free cancer detection and prediction of genotype in routine histopathology. *The Journal of Pathology*, *256*(1), 50–60. Available from https://doi.org/10.1002/path.5800.

Schömig-Markiefka, B., Pryalukhin, A., Hulla, W., Bychkov, A., Fukuoka, J., Madabhushi, A., Achter, V., Nieroda, L., Büttner, R., Quaas, A., & Tolkach, Y. (2021). Quality control stress test for deep learning-based diagnostic model in digital pathology. *Modern Pathology*, *34*(12), 2098–2108. Available from https://doi.org/10.1038/s41379-021-00859-x.

Shaban, M., Khurram, S. A., Fraz, M. M., Alsubaie, N., Masood, I., Mushtaq, S., Hassan, M., Loya, A., & Rajpoot, N. M. (2019). A novel digital score for abundance of tumour infiltrating lymphocytes predicts disease free survival in oral squamous cell carcinoma. *Scientific Reports*, *9*(1). Available from https://doi.org/10.1038/s41598-019-49710-z, http://www.nature.com/srep/index.html.

Sharma, H., Zerbe, N., Klempert, I., Hellwich, O., & Hufnagl, P. (2017). Deep convolutional neural networks for automatic classification of gastric carcinoma using whole slide images in digital histopathology. *Computerized Medical Imaging and Graphics*, *61*, 2–13. Available from https://doi.org/10.1016/j.compmedimag.2017.06.001.

Shimosato, Y., Yagi, Y., Yamagishi, K., Mukai, K., Hirohashi, S., Matsumoto, T., & Kodama, T. (1992). Experience and present status of telepathology in the National Cancer Center Hospital, Tokyo. *Zentralblatt für Pathologie*, *138*(6), 413–417.

Steiner, D. F., Macdonald, R., Liu, Y., Truszkowski, P., Hipp, J. D., Gammage, C., Thng, F., Peng, L., & Stumpe, M. C. (2018). Impact of deep learning assistance on the histopathologic review of lymph nodes for metastatic breast cancer. *American Journal of Surgical Pathology*, *42*(12), 1636–1646. Available from https://doi.org/10.1097/PAS.0000000000001151, http://journals.lww.com/ajsp/pages/default.aspx.

Ström, P., Kartasalo, K., Olsson, H., Solorzano, L., Delahunt, B., Berney, D. M., Bostwick, D. G., Evans, A. J., Grignon, D. J., Humphrey, P. A., Iczkowski, K. A., Kench, J. G., Kristiansen, G., van der Kwast, T. H., Leite, K. R. M., McKenney, J. K., Oxley, J., Pan, C.-C., Samaratunga, H., ... Eklund, M. (2020). Artificial intelligence for diagnosis and grading of prostate cancer in biopsies: A population-based, diagnostic study. *The Lancet Oncology*, *21*(2), 222–232. Available from https://doi.org/10.1016/s1470-2045(19)30738-7.

Taylor, C. R., Jadhav, A. P., Gholap, A., Kamble, G., Huang, J., Gown, A., Doshi, I., & Rimm, D. L. (2019). A Multi-institutional study to evaluate automated whole slide scoring of immunohistochemistry for assessment of programmed death-ligand 1 (PD-

L1) expression in non−small cell lung cancer. *Applied Immunohistochemistry & Molecular Morphology*, *27*(4), 263−269. Available from https://doi.org/10.1097/pai.0000000000000737.

Tsou, P., & Wu, C. J. (2019). Mapping driver mutations to histopathological subtypes in papillary thyroid carcinoma: Applying a deep convolutional neural network. *Journal of Clinical Medicine*, *8*(10). Available from https://doi.org/10.3390/jcm8101675, https://www.mdpi.com/2077-0383/8/10/1675/pdf.

Tuominen, V. J., Ruotoistenmäki, S., Viitanen, A., Jumppanen, M., & Isola, J. (2010). ImmunoRatio: A publicly available web application for quantitative image analysis of estrogen receptor (ER), progesterone receptor (PR), and Ki-67. *Breast Cancer Research*, *12*(4). Available from https://doi.org/10.1186/bcr2615, http://breast-cancer-research.com/content/12/4/R56.

van Dooijeweert, C., van Diest, P. J., & Ellis, I. O. (2022). Grading of invasive breast carcinoma: The way forward. *Virchows Archiv*, *480*(1), 33−43. Available from https://doi.org/10.1007/s00428-021-03141-2, https://www.springer.com/journal/428.

van Dooijeweert, C., van Diest, P. J., Willems, S. M., Kuijpers, C. C. H. J., van der Wall, E., Overbeek, L. I. H., & Deckers, I. A. G. (2020). Significant inter- and intralaboratory variation in grading of invasive breast cancer: A nationwide study of 33,043 patients in the Netherlands. *International Journal of Cancer*, *146*(3), 769−780. Available from https://doi.org/10.1002/ijc.32330, http://onlinelibrary.wiley.com/journal/10.1002/(ISSN)1097-0215.

Valieris, R., Amaro, L., Bueno, C. A., Osório, de T., Bueno, A. P., Rosales Mitrowsky, R. A., Carraro, D. M., Nunes, D. N., Dias-Neto, E., & da Silva, I. T. (2020). Deep learning predicts underlying features on pathology images with therapeutic relevance for breast and gastric cancer. *Cancers*, *12*(12), 3687. Available from https://doi.org/10.3390/cancers12123687.

Velmahos, C. S., Badgeley, M., & Lo, Y. C. (2021). Using deep learning to identify bladder cancers with FGFR-activating mutations from histology images. *Cancer Medicine*, *10*(14), 4805−4813. Available from https://doi.org/10.1002/cam4.4044, http://onlinelibrary.wiley.com/journal/10.1002/(ISSN)2045-7634.

Veta, M., van Diest, P. J., Willems, S. M., Wang, H., Madabhushi, A., Cruz-Roa, A., Gonzalez, F., Larsen, A. B. L., Vestergaard, J. S., Dahl, A. B., Cireşan, D. C., Schmidhuber, J., Giusti, A., Gambardella, L. M., Tek, F. B., Walter, T., Wang, C. W., Kondo, S., Matuszewski, B. J., . . . Pluim, J. P. W. (2015). Assessment of algorithms for mitosis detection in breast cancer histopathology images. *Medical Image Analysis*, *20*(1), 237−248. Available from https://doi.org/10.1016/j.media.2014.11.010, http://www.elsevier.com/inca/publications/store/6/2/0/9/8/3/index.htt.

Vitkovski, T., Bhuiya, T., & Esposito, M. (2015). Utility of telepathology as a consultation tool between an off-site surgical pathology suite and affiliated hospitals in the frozen section diagnosis of lung neoplasms. *Journal of Pathology Informatics*, *6*(1), 55. Available from https://doi.org/10.4103/2153-3539.168515.

Wang, X., Janowczyk, A., Zhou, Y., Thawani, R., Fu, P., Schalper, K., Velcheti, V., & Madabhushi, A. (2017). Prediction of recurrence in early stage non-small cell lung cancer using computer extracted nuclear features from digital H&E images. *Scientific Reports*, *7*(1). Available from https://doi.org/10.1038/s41598-017-13773-7.

Wang, X., Zou, C., Zhang, Y., Li, X., Wang, C., Ke, F., Chen, J., Wang, W., Wang, D., Xu, X., Xie, L., & Zhang, Y. (2021). Prediction of BRCA gene mutation in breast cancer based on deep learning and histopathology images. *Frontiers in Genetics*, *12*. Available from https://doi.org/10.3389/fgene.2021.661109.

Wang, Y., Acs, B., Robertson, S., Liu, B., Solorzano, L., Wählby, C., Hartman, J., & Rantalainen, M. (2022). Improved breast cancer histological grading using deep learning. *Annals of Oncology*, *33*(1), 89–98. Available from https://doi.org/10.1016/j.annonc.2021.09.007.

Wang, Y., Guan, Q., Lao, I., Wang, L., Wu, Y., Li, D., Ji, Q., Wang, Y., Zhu, Y., Lu, H., & Xiang, J. (2019). Using deep convolutional neural networks for multi-classification of thyroid tumor by histopathology: A large-scale pilot study. *Annals of Translational Medicine*, *7*(18), 468. Available from https://doi.org/10.21037/atm.2019.08.54, –468.

Weinstein, R. S. (1986). Prospects for telepathology. *Human Pathology*, *17*(5), 433–434. Available from https://doi.org/10.1016/S0046-8177(86)80028-4.

Weinstein, R. S., Bhattacharyya, A. K., Graham, A. R., & Davis, J. R. (1997). Telepathology: A ten-year progress report. *Human Pathology*, *28*(1), 1–7. Available from http://www.elsevier.com/inca/publications/store/6/2/3/1/3/9/index.htt.

Wellnitz, U., Binder, B., Fritz, P., Friedel, G., & Schwarzmann, P. (2000). Reliability of telepathology for frozen section service. *Analytical Cellular Pathology*, *21*(3-4), 213–222. Available from https://doi.org/10.1155/2000/904578.

Wolff, A. C., Hammond, M. E. H., Allison, K. H., Harvey, B. E., Mangu, P. B., Bartlett, J. M. S., Bilous, M., Ellis, I. O., Fitzgibbons, P., Hanna, W., Jenkins, R. B., Press, M. F., Spears, P. A., Vance, G. H., Viale, G., McShane, L. M., & Dowsett, M. (2018). Human epidermal growth factor receptor 2 testing in breast cancer: American Society of Clinical Oncology/College of American Pathologists clinical practice guideline focused update. *Journal of Clinical Oncology*, *36*(20), 2105–2122. Available from https://doi.org/10.1200/JCO.2018.77.8738, http://ascopubs.org/doi/pdf/10.1200/JCO.2018.77.8738.

Wulczyn, E., Steiner, D. F., Moran, M., Plass, M., Reihs, R., Tan, F., Flament-Auvigne, I., Brown, T., Regitnig, P., Chen, P.-H. C., Hegde, N., Sadhwani, A., MacDonald, R., Ayalew, B., Corrado, G. S., Peng, L. H., Tse, D., Müller, H., Xu, Z., . . . Mermel, C. H. (2021). Interpretable survival prediction for colorectal cancer using deep learning. *NPJ Digital Medicine*, *4*(1). Available from https://doi.org/10.1038/s41746-021-00427-2.

Yamashita, H., Nakayama, K., Ishikawa, M., Nakamura, K., Ishibashi, T., Sanuki, K., Ono, R., Sasamori, H., Minamoto, T., Iida, K., Sultana, R., Ishikawa, N., & Kyo, S. (2018). Microsatellite instability is a biomarker for immune checkpoint inhibitors in endometrial cancer. *Oncotarget*, *9*(5), 5652–5664. Available from https://doi.org/10.18632/oncotarget.23790.

Yamashita, R., Long, J., Longacre, T., Peng, L., Berry, G., Martin, B., Higgins, J., Rubin, D. L., & Shen, J. (2021). Deep learning model for the prediction of microsatellite instability in colorectal cancer: A diagnostic study. *The Lancet Oncology*, *22*(1), 132–141. Available from https://doi.org/10.1016/s1470-2045(20)30535-0.

Young, A. T., Xiong, M., Pfau, J., Keiser, M. J., & Wei, M. L. (2020). Artificial intelligence in dermatology: A primer. *Journal of Investigative Dermatology*, *140*(8), 1504–1512. Available from https://doi.org/10.1016/j.jid.2020.02.026, http://www.nature.com/jid/index.html.

Zeng, H., Chen, L., Zhang, M., Luo, Y., & Ma, X. (2021). Integration of histopathological images and multi-dimensional omics analyses predicts molecular features and prognosis in high-grade serous ovarian cancer. *Gynecologic Oncology*, *163*(1), 171–180. Available from https://doi.org/10.1016/j.ygyno.2021.07.015.

Zhang, R., Chen, H. J., Wei, B., Zhang, H. Y., Pang, Z. G., Zhu, H., Zhang, Z., Fu, J., & Bu, H. (2010). Reproducibility of the Nottingham modification of the Scarff-Bloom-

Richardson histological grading system and the complementary value of Ki-67 to this system. *Chinese Medical Journal*, *123*(15), 1976–1982. Available from https://doi.org/10.3760/cma.j.issn.0366-6999.2010.15.005, http://www.cmj.org/Periodical/PdfList.asp?id = LW201082363378008881.

Zhao, K., Li, Z., Yao, S., Wang, Y., Wu, X., Xu, Z., Wu, L., Huang, Y., Liang, C., & Liu, Z. (2020). Artificial intelligence quantified tumour-stroma ratio is an independent predictor for overall survival in resectable colorectal cancer. *EBioMedicine*, *61*. Available from https://doi.org/10.1016/j.ebiom.2020.103054, http://www.journals.elsevier.com/ebiomedicine.

Zhou, Z., & Li, M. (2022). Targeted therapies for cancer. *BMC Medicine*, *20*(1). Available from https://doi.org/10.1186/s12916-022-02287-3.

CHAPTER

Use cases in cytology

7

Brie Kezlarian-Sachs

Department of Pathology and Laboratory Medicine, Memorial Sloan Kettering Cancer Center, New York, NY, United States

Introduction

The digital pathology revolution has had a significant impact on the practice of modern cytopathology. Innovations such as whole slide imaging with z-stack technology and telecytology-assisted rapid onsite evaluation (teleROSE) highlight how cytopathology practice is uniquely positioned to benefit from these digital tools. However, cytopathology also has some unique challenges in regard to the deployment of digital pathology and artificial intelligence (AI) solutions. In this chapter, we will discuss specific examples of digital pathology solutions and their impact on modern cytopathology practice with an eye toward the future, including potential hurdles to implementation.

Whole slide imaging in cytopathology

Whole slide imaging in cytopathology shares many of the same benefits and potential pitfalls as surgical pathology. Like surgical pathology, the ability to digitize cytology slides opens myriad possibilities, such as digital slide archiving, remote sign-out, and deployment of whole slide imaging solutions for use as educational tools. They also share similar challenges, such as ensuring access to digital storage space and workflow considerations. However, there are additional nuances specific to the practice of cytopathology that are important to consider prior to clinical implementation. In this section, we will discuss those important considerations that are specific to cytopathology.

The first consideration has been the longest-standing and most widely discussed hurdle to the implementation of cytology slide digitization. Unlike surgical pathology slides, cytopathology specimens can be three-dimensional. If one scans a slide in only a single dimension, as is done with surgical pathology slides, there will be material above and below the focal plane that will not be ideally discernible. This may not necessarily preclude accurate diagnoses, particularly with

Digital Pathology. DOI: https://doi.org/10.1016/B978-0-443-13809-6.00009-9

monolayer preparations (Kim et al., 2022). However, if one is considering widespread implementation, particularly for primary diagnosis, having potentially diagnostic material that is not discernible is not ideal.

To address the three-dimensionality of cytology slides, many scanners now have the ability to scan in multiple planes or Z-stacks. Many systems allow the user to customize scanning with an adjustable number of levels and the ability to define the amount of space between levels. This allows the cytopathologist to scroll through multiple focus planes, similar to how one experiences three-dimensionality at the microscope. It is important to note that every additional z-stack level comes at the expense of both increased scan time and increased file size proportional to the number of levels selected (Capitanio et al., 2018). Where to draw the line between diagnostic sensitivity, workflow inefficiencies, and storage considerations is at the discretion of the individual laboratory. An internal validation study may be of use to discern ideal the number of and amount of space between z-stack levels for an individual practice.

Another mechanism by which scanner manufacturers may account for the three-dimensionality of cytology specimens is known as volumetric, all-in-focus, deep-focus, extended-focused, or extended-depth-of-field imaging (Attota, 2018; Capitanio et al., 2018; Liang et al., 2022; Matrecano et al., 2014). Simply put, this is a two-dimensional image that is presented at the optimized focus level for each area of the slide. The benefit of this methodology is that nonoverlapping cells that may be in different z-planes across a slide appear in focus on a single level, resulting in a file smaller than the z-stacked image. However, depending on how the image is constructed, overlapping cells may not be as clearly focused and discernible as they would be in a z-stacked image.

Another factor that hinders the large-scale implementation of whole slide imaging in cytopathology is that modern slide scanners often have difficulty scanning scantly cellular slides. Cytology specimens are frequently composed of scant single cells and small clusters that are not necessarily identifiable on a low-power prescan image. For example, a cerebrospinal fluid (CSF) specimen may have only rare cells across an entire slide. Many modern scanners will have difficulty identifying tissue to scan, often resulting in a tissue detection error and failure to scan the slide.

Not only may a scantly cellular specimen fail scanning due to a tissue detection error, but scantly cellular specimens also often demonstrate scanner focusing issues. This is easily understood when one considers the relative area of, for example, a scantly cellular CSF slide that is composed of blank space versus diagnostic tissue. Whether a scanner employs a "least squares" method or scans in continuous stripes, both rely on the random placement of focus points (Capitanio et al., 2018). If one randomly deploys focus points in a given area, the chance that a focus point lands on diagnostic tissue as opposed to blank space decreases as the cellularity of a specimen decreases. As a result, scanner failures due to out-of-focus tissue are a common issue standing in the way of full digitization of a cytopathology practice's workload.

Finally, many modern scanners will not scan the entire area of a slide. The amount of tissue missed varies between individual scanners but is often a very small proportion of the overall area of the slide involving the bottom or slides of the slide. This is not generally a concern for surgical pathology slides and certain cytology slides (such as monolayer and cytospin preparations) where the tissue is placed in the middle of the slide surrounded by ample blank space. In contrast, a smear slide will frequently have tissue that extends to all edges of the slide. If the scanner is incapable of reproducing the entirety of the tissue present on a slide, it is not advisable that the resulting image be used to render a primary diagnosis, as it would be impossible to know if the sole diagnostic cells are present in those unscanned areas. This issue must be resolved before the widespread implementation of digital cytology primary sign-out may proceed.

When evaluating whole slide image scanners, how a scanner negotiates specimen three-dimensionality, how well it images scantly cellular slides, and the proportion of the slide that is excluded from scanning should be considered in addition to the overall quality, time to scan, and file size considerations. Laboratories interested in purchasing a scanner may wish to compile a cohort of the same slides to use on each scanner to provide the ability for direct interscanner comparison. This cohort should include the entire range of stains and preparations with which one intends to use the scanner, thick smears, scantly cellular slides, and slides completely covered with diagnostic tissue to ensure these possible issues are thoroughly vetted.

Whole slide images for primary diagnosis

There are many benefits of rendering primary diagnoses on digital images as opposed to glass slides, including the capability for remote sign-out. Unfortunately, the implementation of primary diagnoses from whole slide images for cytopathology has lagged behind surgical pathology. To date, there have been no successful attempts to validate primary diagnosis of an entire cytopathology workload from whole slide images analogous to what has been achieved in surgical pathology (Hanna et al., 2020). This can likely be attributed, at least in part, to the difficulties digitizing cytology slides (Hanna & Pantanowitz, 2017). Despite this, some studies have been performed comparing primary diagnoses using glass slides versus digital representations of cytopathology specimens. Findings from 19 such studies are nicely compiled in a 2020 metaanalysis (Girolami et al., 2020) and are largely representative of the data available to date. The authors concluded the selected studies showed concordance rates comparable to those of surgical pathology (84.1% with the original diagnosis and 92.5% intra-observer concordance). Time to diagnosis was also notably increased for whole slide images. In time, increased comfort with digital images and implementation of improved strategies to deal with three-dimensionality may improve these statistics and lead to wider acceptance and subsequent implementation of primary diagnosis of cytology specimens from whole slide images.

AI applications for primary diagnosis of cytology images

With the improvement in the ability to scan cytology specimens and the subsequent increase in the availability of digitized cytology images comes the exciting possibility of the application of AI to the primary diagnosis of cytology specimens. As such, there have been numerous studies looking at the implementation of AI in primary diagnosis. Most studies have focused on cervical cytology (Bao et al., 2020a, 2020b; Bhatt et al., 2021; Cheng et al., 2021; Fang et al., 2022; Holmstrom et al., 2021; Landau & Pantanowitz, 2019; Li et al., 2021; Mohammed et al., 2021; Nambu et al., 2022; Sanyal et al., 2019; Tang et al., 2021; Wang et al., 2021, 2022) and urine specimens (Angeletti et al., 2005; Awan et al., 2021; Kaneko et al., 2022; Landau & Pantanowitz, 2019; Lebret et al., 2022; Lilli et al., 2021; Liu et al., 2022; Muralidaran et al., 2015; Ou et al., 2022; Sanghvi et al., 2019; Vaickus et al., 2019; Zhang et al., 2021), but studies have also investigated AI applications to thyroid (Elliott Range et al., 2020; Kezlarian & Lin, 2021; Landau & Pantanowitz, 2019; Wong et al., 2023), pancreas (Landau & Pantanowitz, 2019; Naito et al., 2021), breast (Landau & Pantanowitz, 2019; Saikia et al., 2019), pulmonary (Landau & Pantanowitz, 2019; Tao et al., 2022), cavity fluids (Landau & Pantanowitz, 2019; Xie et al., 2022), and CSF specimens (Landau & Pantanowitz, 2019; Yu et al., 2022). It should be noted that not all studies have utilized whole slide images as input. Studies that utilize image crops rather than whole slide images still validate the theory, even if they are not necessarily practical to implement clinically. While initial studies show promising results, more study needs to be done. At the time of this publication, the Hologic Genius Digital Diagnostics System for primary diagnosis in cervical cytology specimens is the only FDA approved AI or AI assisted system for cytopathology specimens (Hologic, 2024). This system is also CE marked for AI-assisted primary diagnosis in cervical specimens (Hologic, 2023) in Europe.

Practical considerations for the development and deployment of AI solutions for primary diagnosis

There are a few nuances particular to cytopathology that affect the design, training, and subsequent deployment of AI algorithms. The first is the relative difficulty of digitizing cytology specimens, as discussed above. Training and validating an AI algorithm requires many examples of a given diagnostic entity to learn all potential morphologic permutations. Therefore, a large volume of scanned material is essential to designing a clinically useful diagnostic tool. Given the paucity of digitized cytopathology slides in relation to surgical pathology slides, this places cytopathology at a relative disadvantage. However, with the more recent innovations as described above, this gap will narrow.

Compared to cytology specimens, surgical specimens are remarkably homogenous. Tissue sections are in one plane, cut at a consistent thickness, and practically all stained with hematoxylin and eosin—often with an automated stainer to ensure consistent staining from one slide to the next. Cytology specimens, by distinction, encompass a wide variety of preparations and stains. Further, due to the manual process by which some slides are prepared and stained (e.g., smears and touch preparations), there can be remarkable both intra-slide and inter-slide variability. While certainly not a contraindication to machine learning and AI applications, it is imperative that this variability is understood and mindfully taken into account. One must ensure that both the training and validation include all the stains and preparations on which one intends to ultimately apply the algorithm.

Furthermore, there can be tremendous variation in both preparation and stain quality from one laboratory to another. To train and ultimately deploy an algorithm that will perform acceptably in many different laboratory settings, it is essential to take this possible variation into account. It is therefore advisable to include specimens from more than one laboratory in the training and the validation of any algorithm that may be deployed in multiple laboratories.

One characteristic of cytopathology specimens that are uniquely advantageous to the application of AI is that many specimens (including cervical and voided urine specimens and, to some extent, thyroid fine needle aspiration specimens) have a limited range of expected diagnoses with relatively consistent morphologic appearance. For example, a low-grade squamous intraepithelial lesion looks remarkably similar from one specimen to another. Therefore, the training set to teach an AI algorithm to detect low-grade squamous intraepithelial lesions need not be as large or varied as, for example, all the morphologic permutations of lung carcinoma.

Further, these diagnoses are codified in strictly defined, widely accepted categories, as described in The Bethesda System for Reporting Cervical Cytology (Nayar & Wilbur, 2015), The Paris System for Reporting Urinary Cytology (Wojcik et al., 2022), and The Bethesda System for Reporting Thyroid Cytopathology (Ali & Cibas, 2018), respectively. By contrast, pancreas and salivary gland specimens, for example, while also having codified, relatively widely accepted diagnostic category schema (the WHO Reporting System for Pancreaticobiliary Cytopathology (IAC-IARC-WHO Joint Editorial Board, 2022) and The Milan System for Reporting Salivary Gland Cytopathology (Faquin et al., 2018), respectively), contain far more diagnostic entities the AI must consider. As a result, far more diagnostic entities (and all the morphologic permutations thereof) would need to be included in a teaching set to train a competent AI algorithm.

Digital imaging for education

Utilizing digital slides for teaching purposes has several advantages over physical slides, many of which are equally applicable to surgical and cytopathology. These include ease of transfer and sharing across distances, ease of storage and recall, and lack of degeneration or fading (Chiou, 2020; Rohr et al., 2022; Stergiou et al., 2009; Sura et al., 2022; Wilbur, 2016). Digitization may also allow for tracking the learner to glean information about the learning experience (Pantanowitz & Harrington, 2021). What is unique to cytopathology is that, with the exception of cell block slides, cytology slides are often irreplaceable. Therefore, digitizing slides for educational purposes, thereby limiting the opportunity for a slide to be damaged or broken, is of particular importance with cytopathology specimens. The digitization of particularly rare, educational, or informative glass slides ensures these cases will be available for future generations.

Telecytology

One of the most significant ways in which digital pathology is already transforming clinical practice is through the application of telecytology. At its very basic level, telecytology is the act of acquiring an image at one site and viewing it at another. Within that simplicity lie innumerable possible configurations and many exciting applications. After a brief overview of telecytology configurations, selected applications of telecytology in modern cytology practice will be discussed.

Telecytology configurations

While there is a significant amount of variability in telecytology setups, it is useful to conceptualize a telecytology platform as having three essential components: image capture, image transmission, and image review. There are many permutations of each of these separate components, each with its own benefits and drawbacks. The combination yields an abundance of possible configurations to suit a laboratory's specific needs.

The image capture component, very simply stated, seeks to convert the glass slide into a digital representation. Examples in the literature exist for microscope mounted cameras (Agarwal et al., 2016; Alsharif et al., 2010; Archondakis et al., 2020, 2021; Brooker et al., 2019; Buxbaum et al., 2012; Dudas et al., 2014; Heimann et al., 2012; Lin et al., 2019, 2020; McCarthy et al., 2015; McMahon et al., 2014; Monaco et al., 2020; Sahin et al., 2018; Sarode et al., 2021; Selvaggi, 2018; Sirintrapun, Rudomina, Mazzella, Feratovic, et al., 2017), including cellular telephones (Agarwal et al., 2016; Dudas et al., 2014; Sahin et al., 2018), robotic microscopes (Singh et al., 2002; Yao et al., 2018), and whole slide scanners (Canberk et al., 2020; Sirintrapun, Rudomina, Mazzella, et al., 2017; Yao et al., 2018) to serve as acquisition devices in telepathology configurations. There are several important factors one must consider when selecting a data acquisition device. The first is a staffing consideration. If there will be a skilled operator present at the slide acquisition site to identify and maneuver the slide to regions of interest, a simple camera-mounted microscope may suffice. Remote solutions, such as a whole slide imaging device or robotic microscope, will be required in clinical (or research) settings that do not have eligible staff available.

The second important consideration is image quality. While there are no official recommendations for the minimum resolution acceptable for telecytology applications, higher resolution is associated with higher satisfaction (McMahon et al., 2014). Each laboratory will have to decide what level of resolution is acceptable for the use case for which it will be deployed, assuming it performs adequately in validation.

The second part of the pathway is the mechanism by which one chooses to transmit images from the acquisition to the viewing site. Depending on the individual needs of the laboratory, this can be as simple as emailing or direct messaging (Sahin et al., 2018), saving images to a shared drive (Archondakis et al., 2020, 2021; Brooker et al., 2019), remote control of the on-site computer (Sirintrapun, Rudomina, Mazzella, et al., 2017), screen sharing via teleconferencing software (Agarwal et al., 2016; Dudas et al., 2014; Kouanda et al., 2023; Sirintrapun, Rudomina, Mazzella, et al., 2017) and internet broadcast (Buxbaum et al., 2012; Dudas et al., 2014; Lin et al., 2019, 2020; McCarthy et al., 2015; McMahon et al., 2014; Sarode et al., 2021; Selvaggi, 2018; Sirintrapun, Rudomina, Mazzella, Feratovic, et al., 2017; Xing et al., 2020). The most important decision one must make before choosing a telecytology transmission vehicle is whether a static or dynamic solution is needed. A static system describes a telecytology solution that transmits still, precaptured images, whereas a dynamic system will transmit live images. The decision whether to implement a static or dynamic system will in large part be determined by the intended use and expected case volume leveraged against budgetary and other constraints. A static system is in general more simple and less expensive to deploy. They are best reserved for applications such as case consultations (Archondakis et al., 2021) but may be considered for teleROSE if the expected volume is small (Asfahan et al., 2021). Dynamic systems are better suited for practices expecting a large volume of teleROSE cases.

Another important factor that cannot be overlooked when designing and deploying an image transmission solution is ensuring adequate security. This is particularly at issue concerning solutions involving teleconferencing software, such as WebEx, Zoom, or Microsoft Teams. If one relies on institutional firewall protection, additional consideration must be made for solutions that incorporate mobile technologies, such as cellular telephones, when that technology is not behind the institutional firewall.

The American Telemedicine Association recommends that telepathology systems should be validated as a single system rather than multiple components separately, that the types of specimens chosen for the validation should encompass that which one expects to encounter in the clinical setting, and that revalidation should occur in the event of "significant change in a component or use case," among other considerations (Pantanowitz et al., 2014). A review of this document is recommended before designing and deploying a telecytology solution for clinical use. Examples of cytology validations are present in the literature and vary not only in the number of slides chosen for validation (10–80) but also in the overall structure of the validation, with some employing a multiphase approach

(Agarwal et al., 2016; Goyal et al., 2012; McCarthy et al., 2015; Monaco & Pantanowitz, 2015; Trabzonlu et al., 2022).

Rapid onsite evaluation

Rapid onsite evaluation is the process by which a cytologic specimen is prepared and microscopically examined immediately after collection. This allows the opportunity for a proceduralist, such as an interventional bronchoscopist, intervention radiologist, or gastroenterologist, to incorporate real-time feedback regarding the collected sample. Specimens are generally either a fine needle aspiration biopsy (FNAB) or a touch preparation of a core biopsy. Not only can rapid onsite evaluation (ROSE) inform the proceduralist as to the adequacy of the specimen, but he/she/they can also be alerted to the need for ancillary testing, such as flow cytometry, molecular testing, or cell block material. The required additional material can subsequently be collected during the initial procedure, thereby potentially decreasing the need for a second procedure. ROSE has been shown to increase diagnostic adequacy rates, may decrease the number of needle passes per procedure, increase the amount of material available for genetic testing, and reduce the need for additional procedures (de Koster et al., 2016; Ecka & Sharma, 2013; Schmidt et al., 2013; Trisolini et al., 2015).

Despite the benefit ROSE presents to the hospital and patient care, reimbursement for ROSE is problematic. Reimbursement for ROSE interpretation falls under the billing codes 88172–3 for FNAs [*Cytopathology, evaluation of fine needle aspirate; immediate cytohistologic study to determine the adequacy of specimen(s) and 88333–4 [Pathology consultation during surgery; cytologic examination (e.g., touch prep, squash prep), initial site] for touch preparations]* (Lin et al., 2019). As these are both professional codes, a physician must provide the adequacy assessment for the procedure to be reimbursed.

This leaves the modern cytology practice with a conundrum. How does one provide the highest quality care when multiple concurrent procedures may require the pathologist's attention? Many have turned to telecytology solutions to resolve this conflict. Using a telecytology solution, a cytopathologist may remotely review multiple procedures occurring in different places from a single location. TeleROSE has been widely studied and has been shown to be both an effective and efficient mechanism by which to resolve this conflict (Agarwal et al., 2016; Alsharif et al., 2010; Heimann et al., 2012; Kouanda et al., 2023; Lin et al., 2019, 2020; Sarode et al., 2021; Sirintrapun, Rudomina, Mazzella, et al., 2017; Sirintrapun, Rudomina, Mazzella, Feratovic, et al., 2017; Yao et al., 2018).

AI applications in rapid onsite evaluation telecytology

As a matter of necessity, many have turned to converting slides into digital representations to provide large volumes of concurrent ROSE evaluations. Out of this necessity, a great opportunity arises. Can these digital representations be paired with AI applications to provide more value? While study in this area is limited, preliminary investigations into the use of AI for ROSE evaluations of fine needle aspiration biopsies of lung nodules (Ai et al., 2022), mediastinal lymph nodes (Asfahan et al., 2021), and pancreas (Lin et al., 2022; Zhang et al., 2022) have been undertaken and show promising results. Currently, there is no FDA-approved AI system for ROSE telecytology in the United States.

Practical considerations for AI applications for teleROSE mimic the issues discussed previously. The three-dimensional nature of specimens will need to be taken into account as it applies to visualization and nuclear segmentation. The interslide and intraslide variability is perhaps even more pronounced with intraoperative specimens, as they are by necessity manually prepared and (often) stained. The most important consideration that applies to AI applications of teleROSE specimens is that of time. The evaluation is performed intraoperatively, and the proceduralist must actively manage the patient on the table, often while the patient is under anesthesia, while awaiting the result. Therefore, the time taken to scan and evaluate the slide must be minimized into an acceptable timeframe.

Telecytology for primary diagnosis and expert consultation

Telecytology for primary diagnosis is not currently widely in use in the United States. However, one interesting description of the integration of telecytology into a modern cytopathology practice was reported by Bychkov et al. (2023). They report their 5-year experience incorporating telepathology into their practice in Kamogawa, Japan, including for the purpose of primary diagnosis (Bychkov et al., 2023). Their workflow includes prescreening followed by a dynamic sign-out session with the pathologist, thereby allowing sign-out to occur when the cytopathologist is not on site. While the volumes reported are relatively low (approximately 30 cases per week), this is nonetheless a unique and exciting example of the integration of digital technologies in modern cytopathology practice. In addition, there are several studies describing the application of telecytology for remote expert consultation in the literature (Archondakis et al., 2021; Mosquera-Zamudio et al., 2019; Yamashiro et al., 2019), reporting variable success. Far more commonly, remote telecytology solutions may be utilized for simple virtual "curbside" consults with colleagues or simply to share an interesting case.

In conclusion, there are exciting and innovative digital solutions that are already primed to integrate into the modern cytopathology laboratory, with more on the horizon. The cytopathologist of the future must not only become aware of the technologies that are already shaping the practice of cytopathology in the modern era but actively seek out new and innovative ways to integrate digital solutions in the best interest of patient care. Ultimately, the future of cytopathology is in our hands, and with the thoughtful and value-added integration of digital cytopathology solutions, the future is bright.

References

Agarwal, S., Zhao, L., Zhang, R., & Hassell, L. (2016). FaceTime validation study: Low-cost streaming video for cytology adequacy assessment. *Cancer Cytopathology, 124*(3), 213–220.

Ai, D., Hu, Q., Chao, Y.-C., Fu, C.-C., Wei, Y, Lv, L., Ye, Dexian, Li, C., Ye, M., Zhang, Y., Hong, Q., Hu, J., Xu, X., Zhang, L., Jiang, Q., Wang, X., Fang, Q., Wang, B., Hou, Y., & Zhang, X (2022). Artificial intelligence-based rapid on-site cytopathological evaluation for bronchoscopy examinations. *Intelligence-Based Medicine, 6.*

Ali, S. Z., & Cibas, E. S. (2018). *The Bethesda system for reporting thyroid cytopathology definitions, criteria, and explanatory notes* (2nd ed.). Cham: Springer International Publishing, Imprint: Springer; 2018.

Alsharif, M., Carlo-Demovich, J., Massey, C., Madory, J., Lewin, D., Medina, A., Recavarren, R., Houser, P., & Yang, J. (2010). Telecytopathology for immediate evaluation of fine-needle aspiration specimens. *Cancer Cytopathology, 118*(3), 119–126.

Angeletti, C., Harvey, N. R., Khomitch, V., Fischer, A. H., Levenson, R. M., & Rimm, D. L. (2005). Detection of malignancy in cytology specimens using spectral-spatial analysis. *Laboratory Investigation; A Journal of Technical Methods and Pathology, 85*(12), 1555–1564.

Archondakis, S., Roma, M., & Kaladelfou, E. (2021). The implementation of static telecytology for teleconsultation purposes during preoperative evaluation of thyroid fine-needle aspiration specimens. *Telemedicine Journal and e-Health: The Official Journal of the American Telemedicine Association, 27*(2), 207–212.

Archondakis, S., Roma, M., & Kaladelfou, E. (2020). The use of static telecytology for quality assessment purposes in the evaluation of cervical smears prepared by means of liquid-based cytology. *Telemedicine Journal and e-Health: The Official Journal of the American Telemedicine Association, 26*(12), 1522–1525.

Asfahan, S., Elhence, P., Dutt, N., Niwas Jalandra, R., & Chauhan, N. K. (2021). Digital-rapid on-site examination in endobronchial ultrasound-guided transbronchial needle aspiration (DEBUT): A proof of concept study for the application of artificial intelligence in the bronchoscopy suite. *The European Respiratory Journal: Official Journal of the European Society for Clinical Respiratory Physiology, 58*(4).

Attota, R. (2018). Through-focus or volumetric type of optical imaging methods: A review. *Journal of Biomedical Optics, 23*(7), 1–10.

Awan, R., Benes, K., Azam, A., Sonf, T., Shaban, M., Verril, C., Tsang, Y., Snead, D., Minhas, F., & Rajpoot, N. (2021). Deep learning based digital cell profiles for risk stratification of urine cytology images. *Cytometry. Part A: The Journal of the International Society for Analytical Cytology, 99*(7), 732–742.

Bao, H., Sun, X., Zhang, Y., Zhao, Y., Dong, Y., Luo, X., Zhou, D., You, Z., Wu, Y., Liu, Z., Zhang, Y., Liu, J., Fang, L., & Wang, L. (2020). Artificial intelligence-assisted cytology for detection of cervical intraepithelial neoplasia or invasive cancer: A multicenter, clinical-based, observational study. *Gynecologic Oncology, 159*(1), 171–178.

Bao, H., Sun, X., Zhang, Y., Pang, B., Li, H., Zhou, L., Wu, F., Cao, D., Wang, J., Turic, B., & Wang, L. (2020). The artificial intelligence-assisted cytology diagnostic system in large-scale cervical cancer screening: A population-based cohort study of 0.7 million women. *Cancer Medicine, 9*(18), 6896–6906.

Bhatt, A. R., Ganatra, A., & Kotecha, K. (2021). Cervical cancer detection in pap smear whole slide images using convNet with transfer learning and progressive resizing. *PeerJ Computer Science*, 7e348.

Brooker, A. J., Krimer, P. M., Meichner, K., & Garner, B. C. (2019). Impact of photographer experience and number of images on telecytology accuracy. *Veterinary Clinical Pathology/American Society for Veterinary Clinical Pathology*, *48*(3), 419–424.

Buxbaum, J. L., Eloubeidi, M. A., Lane, C. J., Varadarajulu, S., Linder, A., Crowe, A. E., Jhala, D., Jhala, N. C., Crowe, D. R., & Eltoum, I. A. (2012). Dynamic telecytology compares favorably to rapid onsite evaluation of endoscopic ultrasound fine needle aspirates. *Digestive Diseases and Sciences*, *57*(12), 3092–3097.

Bychkov, A., Yoshikawa, A., Munkhdelger, J., Hori, T., & Fukuoka, J. (2023). Integrating cytology into routine digital pathology workflow: A 5-year journey. *Virchows Archiv: an International Journal of Pathology*.

Canberk, S., Behzatoglu, K., Caliskan, C. K., Gelmez, S., Kayhan, K. C., Aydemir, S. F., Akbas, M., Yildiz, I., Veiga, R., Alrefae, N., Ince, U., & Schmitt, F. C. (2020). The role of telecytology in the primary diagnosis of thyroid fine-needle aspiration specimens. *Acta Cytologica*, *64*(4), 323–331.

Capitanio, A., Dina, R. E., & Treanor, D. (2018). Digital cytology: A short review of technical and methodological approaches and applications. *Cytopathology: Official Journal of the British Society for Clinical Cytology*, *29*(4), 317–325.

Cheng, S., Liu, S., Yu, J., Rao, G., Xiao, Y., Han, W., Zhu, W., Lv, X., Li, N., Cai, J., Wang, Z., Feng, X., Yang, F., Geng, X., Ma, J., Li, X., Wei, Z., Zhang, X., Quan, T., Zeng, S., Chen, L., Hu, J., & Liu, X. (2021). Robust whole slide image analysis for cervical cancer screening using deep learning. *Nature Communications*, *12*(1), 5639.

Chiou, P. Z. (2020). Learning cytology in times of pandemic: an educational institutional experience with remote teaching. *Journal of the American Society of Cytopathology*, *9*(6), 579–585.

de Koster, E. J., Kist, J. W., Vriens, M. R., Borel Rinkes, I. H., Valk, G. D., & de Keizer, B. (2016). Thyroid ultrasound-guided fine-needle aspiration: The positive influence of on-site adequacy assessment and number of needle passes on diagnostic cytology rate. *Acta Cytologica*, *60*(1), 39–45.

Dudas, R., VandenBussche, C., Baras, A., Ali, S. Z., & Olson, M. T. (2014). Inexpensive telecytology solutions that use the Raspberry Pi and the iPhone. *Journal of the American Society of Cytopathology*, *3*(1), 49–55.

Ecka, R. S., & Sharma, M. (2013). Rapid on-site evaluation of EUS-FNA by cytopathologist: An experience of a tertiary hospital. *Diagnostic Cytopathology*, *41*(12), 1075–1080.

Elliott Range, D. D., Dov, D., Kovalsky, S. Z., Henao, R., Carin, L., & Cohen, J. (2020). Application of a machine learning algorithm to predict malignancy in thyroid cytopathology. *Cancer Cytopathology*, *128*(4), 287–295.

Fang, S., Yang, J., Wang, M., Liu, C., & Liu, S. (2022). An improved image classification method for cervical precancerous lesions based on ShuffleNet. *Computational Intelligence and Neuroscience*, *20229675628*.

Faquin, W. C., Rossi, E. D., Baloch, Z., Barkan, G. A., Foschini, M. P., Kurtycz, D. F. I., Pusztaszeri, M., & Vielh, P. (2018). *The milan system for reporting salivary gland cytopathology* (1st ed). Cham: Springer International Publishing, Imprint: Springer; 2018.

Girolami, I., Pantanowitz, L., Marletta, S., Brunelli, M., Mescoli, C., Parisi, A., Barresi, V., Parwani, A., Neil, D., Scarpa, A., Rossi, E. D., & Eccher, A. (2020). Diagnostic concordance between whole slide imaging and conventional light microscopy in cytopathology: A systematic review. *Cancer Cytopathology*, *128*(1), 17–28.

Goyal, A., Jhala, N., & Gupta, P. (2012). TeleCyP (Telecytopathology): Real-time fine-needle aspiration interpretation. *Acta Cytologica*, *56*(6), 669–677.

Hanna, M. G., & Pantanowitz, L. (2017). Why is digital pathology in cytopathology lagging behind surgical pathology? *Cancer Cytopathology*, *125*(7), 519–520.

Hanna, M. G., Reuter, V. E., Ardon, O., Kim, D., Sirintrapun, S. J., Schuffler, P. J., Busam, K. J., Sauter, J. L., Brogi, E., Tan, L. K., Xu, B., Bale, T., Agaram, N. P., Tang, L. H., Ellenson, L. H., Philip, J., Corsale, L., Stamelos, E., Friedlander, M. A., Ntiamoah, P., Labasin, M., England, C., Klimstra, D. S., & Hameed, M. (2020). Validation of a digital pathology system including remote review during the COVID-19 pandemic. *Modern Pathology: An Official Journal of the United States and Canadian Academy of Pathology, Inc*, *33*(11), 2115–2127.

Heimann, A., Maini, G., Hwang, S., Shroyer, K. R., & Singh, M. (2012). Use of telecytology for the immediate assessment of CT guided and endoscopic FNA cytology: Diagnostic accuracy, advantages, and pitfalls. *Diagnostic Cytopathology*, *40*(7), 575–581.

Holmstrom, O., Linder, N., Kaingu, H., Mbuuko, N., Mbete, J., Kinyua, F., Tornquist, S., Muinde, M., Krogerus, L., Lundin, M., Diwan, V., & Lundin, J. (2021). Point-of-care digital cytology with artificial intelligence for cervical cancer screening in a resource-limited setting. *JAMA Netw Open*, *4*(3), e211740.

Hologic, I. (2023). Hologic announces CE mark of the genius digital diagnostics system for cervical cancer screening. 2020. (accessed 29.05.23).

Hologic, I. (2024). Hologic Announces First and Only FDA-Cleared Digital Cytology System – Genius™ Digital Diagnostics System. 2024. (accessed 4.10.24).

IAC-IARC-WHO Joint Editorial Board. (2022). (1st ed.). *WHO reporting system for pancreaticobiliary cytopathology*, (Vol 2). Lyon (France): International Agency for Research on Cancer.

Kaneko, M., Tsuji, K., Masuda, K., Ueno, K., Henmi, K., Nakagawa, S., Fujita, R., Suzuki, K., Inoue, Y., Teramukai, S., Konishi, E., Takamatsu, T., & Ukimura, O. (2022). Urine cell image recognition using a deep-learning model for an automated slide evaluation system. *BJU International*, *130*(2), 235–243.

Kezlarian, B., & Lin, O. (2021). Artificial intelligence in thyroid fine needle aspiration biopsies. *Acta Cytologica*, *65*(4), 324–329.

Kaneko, M., Tsuji, K., Masuda, K., Ueno, K., Henmi, K., Nakagawa, S., Fujita, R., Suzuki, K., Inoue, Y., Teramukai, S., Konishi, E., Takamatsu, T., & Ukimura, O. (2022). Evaluating the role of Z-stack to improve the morphologic evaluation of urine cytology whole slide images for high-grade urothelial carcinoma: Results and review of a pilot study. *Cancer Cytopathology*, *130*(8), 630–639.

Kouanda, A., McLean, R., Faggen, A., Demissie, E., Balassanian, R., Kamal, F., Avila, P., Arain, M., Dai, S. C., & Munroe, C. (2023). Telecytology versus in-room cytopathologist for EUS-guided FNA or fine-needle biopsy sampling of solid pancreatic lesions. *Gastrointestinal Endoscopy*, *97*(3), 466–471.

Landau, M. S., & Pantanowitz, L. (2019). Artificial intelligence in cytopathology: A review of the literature and overview of commercial landscape. *Journal of the American Society of Cytopathology*, *8*(4), 230–241.

Lebret, T., Pignot, G., Colombel, M., Guy, L., Rebillard, X., Savareux, L., Roumigue, M., Nivet, S., Coutade Saidi, M., Piaton, E., & Radulescu, C. (2022). Artificial intelligence to improve cytology performances in bladder carcinoma detection: Results of the VisioCyt test. *BJU International*, *129*(3), 356–363.

Li, X., Xu, Z., Shen, X., Zhou, Y., Xiao, B., & Li, T. Q. (2021). Detection of cervical cancer cells in whole slide images using deformable and global context aware faster RCNN-FPN. *Current Oncology (Toronto, Ontario)*, *28*(5), 3585–3601.

Liang, M., Bernadt, C., Wong, S. B. J., Choi, C., Cote, R., & Yang, C. (2022). All-in-focus fine needle aspiration biopsy imaging based on Fourier ptychographic microscopy. *Journal of Pathology Informatics*, *13*100119.

Lilli, L., Giarnieri, E., & Scardapane, S. (2021). A calibrated multiexit neural network for detecting urothelial cancer cells. *Computational and Mathematical Methods in Medicine*, *2021*5569458.

Lin, D. M., Tracht, J., Rosenblum, F., Kouba, E., Bahl, D., Patel, A., & Eltoum, I. E. (2020). Rapid on-site evaluation with telecytology significantly reduced unsatisfactory rates of thyroid fine-needle aspiration. *American Journal of Clinical Pathology*, *153*(3), 342–345.

Lin, O., Rudomina, D., Feratovic, R., & Sirintrapun, S. J. (2019). Rapid on-site evaluation using telecytology: A major cancer center experience. *Diagnostic Cytopathology*, *47* (1), 15–19.

Lin, R., Sheng, L. P., Han, C. Q., Guo, X. W., Wei, R. G., Ling, X., & Ding, Z. (2022). Application of artificial intelligence to digital-rapid on-site cytopathology evaluation during endoscopic ultrasound-guided fine needle aspiration: A proof-of-concept study. *Journal of Gastroenterology and Hepatology*.

Liu, Y., Jin, S., Shen, Q., Chang, L., Fang, S., Fan, Y., Peng, H., & Yu, W. (2022). A deep learning system to predict the histopathological results from urine cytopathological images. *Frontiers in Oncology*, *12*901586.

Matrecano, M., Paturzo, M., & Ferraro, P. (2014). Extended focus imaging in digital holographic microscopy: A review. *Optical Engineering*, *53*(11).

McCarthy, E. E., McMahon, R. Q., Das, K., & Stewart, J. I. I. I. (2015). Internal validation testing for new technologies: Bringing telecytopathology into the mainstream. *Diagnostic Cytopathology*, *43*(1), 3–7.

McMahon, R. Q., McCarthy, E. E., Hetzel, S. J., Das, K., & Stewart, J. I. I. I. (2014). Focus on technology: How important is resolution in telecytopathology? *Cancer Cytopathology*, *122*(7), 546–552.

Mohammed, M. A., Abdurahman, F., & Ayalew, Y. A. (2021). Single-cell conventional pap smear image classification using pre-trained deep neural network architectures. *BMC Biomedical Engineering*, *3*(1), 11.

Monaco, S. E., Han, M., Dietz, R., Xing, J., Cuda, J., & Pantanowitz, L. (2020). Assessing competency for remote telecytology rapid on-site evaluation using pre-recorded dynamic video streaming. *Cytopathology: Official Journal of the British Society for Clinical Cytology*, *31*(5), 411–418.

Monaco, S. E., & Pantanowitz, L. (2015). Telecytology value and validation: Developing a validation and competency tool for telecytology. *Diagnostic Cytopathology*, *43*(1), 1–2.

Mosquera-Zamudio, A., Hanna, M. G., Parra-Medina, R., Piedrahita, A. C., Rodriguez-Urrego, P. A., & Pantanowitz, L. (2019). Advantage of Z-stacking for teleconsultation between the USA and Colombia. *Diagnostic Cytopathology*, *47*(1), 35–40.

Muralidaran, C., Dey, P., Nijhawan, R., & Kakkar, N. (2015). Artificial neural network in diagnosis of urothelial cell carcinoma in urine cytology. *Diagnostic Cytopathology*, *43* (6), 443–449.

Naito, Y., Tsuneki, M., Fukushima, N., Koga, Y., Higashi, M., Notohara, K., Aishima, S., Ohike, N., Tajiri, T., Yamaguchi, H., Fukumura, Y., Kojima, M., Hirabayashi, K., Hamada, Y., Norose, T., Kai, K., Omori, Y., Sukeda, A., Noguchi, H., Uchino, K., Itakura, J., Okabe, Y., Yamada, Y., Akiba, J., Kanavati, F., Oda, Y., Furukawa, T., & Yano, H. (2021). A deep learning model to detect pancreatic ductal adenocarcinoma on endoscopic ultrasound-guided fine-needle biopsy. *Scientific Reports*, *11*(1), 8454.

Nambu, Y., Mariya, T., Shinkai, S., Umemoto, M., Asanuma, H., Sato, I., Hirohashi, Y., Torigoe, T., Fujino, Y., & Saito, T. (2022). A screening assistance system for cervical cytology of squamous cell atypia based on a two-step combined CNN algorithm with label smoothing. *Cancer Medicine*, *11*(2), 520–529.

Nayar, R., & Wilbur, D. C. (2015). *The bethesda system for reporting cervical cytology definitions, criteria, and explanatory notes* (3rd ed.). Cham: Springer International Publishing, Imprint: Springer; 2015.

Ou, Y. C., Tsao, T. Y., Chang, M. C., Lin, Y. S., Yang, W. L., Hang, J. F., Li, C. B., Lee, C. M., Yeh, C. H., & Liu, T. J. (2022). Evaluation of an artificial intelligence algorithm for assisting the Paris system in reporting urinary cytology: A pilot study. *Cancer Cytopathology*, *130*(11), 872–880.

Pantanowitz, L., Dickinson, K., Evans, A. J., Hassell, L. A., Henricks, W. H., Lennerz, J. K., Lowe, A., Parwani, A. V., Riben, M., Smith, C. D., Tuthill, J. M., Weinstein, R. S., Wilbur, D. C., Krupinski, E. A., & Bernard, J. (2014). American Telemedicine Association clinical guidelines for telepathology. *Journal of Pathology Informatics*, *5* (1)39.

Pantanowitz, L., & Harrington, S. (2021). Experience reviewing digital pap tests using a gallery of images. *Journal of Pathology Informatics*, *12*7.

Rohr, J. M., Mukherjee, M., Donnelly, A., Sprinkle, S., Martinez Duarte, E., & Yuil Valdes, A. (2022). Successful integration of thyroid cytopathology and surgical pathology education in an E-module format. *Journal of Pathology Informatics*, *13*100124.

Sahin, D., Hacisalihoglu, U. P., & Kirimlioglu, S. H. (2018). Telecytology: Is it possible with smartphone images? *Diagnostic Cytopathology*, *46*(1), 40–46.

Saikia, A. R., Bora, K., Mahanta, L. B., & Das, A. K. (2019). Comparative assessment of CNN architectures for classification of breast FNAC images. *Tissue & Cell*, *57*, 8–14.

Sanghvi, A. B., Allen, E. Z., Callenberg, K. M., & Pantanowitz, L. (2019). Performance of an artificial intelligence algorithm for reporting urine cytopathology. *Cancer Cytopathology*, *127*(10), 658–666.

Sanyal, P., Barui, S., Deb, P., & Sharma, H. C. (2019). Performance of a convolutional neural network in screening liquid based cervical cytology smears. *Journal of Cytology*, *36*(3), 146–151.

Sarode, V. R., Pena, V., Yan, S., Kirkpatrick, J., & Wanzer, D. (2021). Selective deployment of dynamic telecytology for rapid evaluation of cytology smears: Assessment of workflow processes and role of cytopathology fellows as on-site operators. *Journal of the American Society of Cytopathology*, *10*(6), 577–584.

Schmidt, R. L., Witt, B. L., Matynia, A. P., Barraza, G., Layfield, L. J., & Adler, D. G. (2013). Rapid on-site evaluation increases endoscopic ultrasound-guided fine-needle aspiration adequacy for pancreatic lesions. *Digestive Diseases and Sciences*, *58*(3), 872–882.

Selvaggi, S. M. (2018). On the job training: An educational program in ROSE of fine needle aspirates and telecytology for cytotechnologists. *Journal of the American Society of Cytopathology*, *7*(6), 306–310.

Singh, N., Akbar, N., Sowter, C., Lea, K. G., & Wells, C. A. (2002). Telepathology in a routine clinical environment: Implementation and accuracy of diagnosis by robotic microscopy in a one-stop breast clinic. *The Journal of Pathology*, *196*(3), 351–355.

Sirintrapun, S. J., Rudomina, D., Mazzella, A., Feratovic, R., Alago, W., Siegelbaum, R., & Lin, O. (2017). Robotic telecytology for remote cytologic evaluation without an on-site cytotechnologist or cytopathologist: A tale of implementation and review of constraints. *Journal of Pathology Informatics*, 832.

Sirintrapun, S. J., Rudomina, D., Mazzella, A., Feratovic, R., & Lin, O. (2017). Successful secure high-definition streaming telecytology for remote cytologic evaluation. *Journal of Pathology Informatics*, 833.

Stergiou, N., Georgoulakis, G., Margari, N., Aninos, D., Stamataki, M., Stergiou, E., Pouliakis, A., & Karakitsos, P. (2009). Using a web-based system for the continuous distance education in cytopathology. *International Journal of Medical Informatics*, *78* (12), 827–838.

Sura, G. H., Doan, J. V., & Thrall, M. J. (2022). Assessing the quality of cytopathology whole slide imaging for education from archived cases. *Journal of the American Society of Cytopathology*, *11*(5), 313–319.

Tang, H. P., Cai, D., Kong, Y. Q., Ye, H., Ma, Z. X., Lv, H. S., Tuo, L. R., Pan, Q. J., Liu, Z. H., & Han, X. (2021). Cervical cytology screening facilitated by an artificial intelligence microscope: A preliminary study. *Cancer Cytopathology*, *129*(9), 693–700.

Tao, Y., Cai, Y., Fu, H., Song, L., Xie, L., & Wang, K. (2022). Automated interpretation and analysis of bronchoalveolar lavage fluid. *International Journal of Medical Informatics*, *157*104638.

Tian, F., Liu, D., Wei, N., Fu, Q., Sun, L., Liu, W., Sui, X., Tian, K., Nemeth, G., Feng, J., Xu, J., Xiao, L., Han, J., Fu, J., Shi, Y., Yang, Y., Liu, J., Hu, C., Feng, B., Sun, Y., Wang, Y., Yu, G., Kong, D., Wang, M., Li, W., Chen, K., & Li, X. (2024). Prediction of tumor origin in cancers of unknown primary origin with cytology-based deep learning. *Nature Medicine*, *30*, 1309–1319. Available from https://doi.org/10.1038/s41591-024-02915-w.

Trabzonlu, L., Chatt, G., McIntire, P. J., Eshraghi, R., Lapadat, R., Atieh, M., Pambuccian, S. E., Wojcik, E. M., Mehrotra, S., & Barkan, G. A. (2022). Telecytology validation: Is there a recipe for everybody? *Journal of the American Society of Cytopathology*, *11*(4), 218–225.

Trisolini, R., Cancellieri, A., Tinelli, C., de Biase, D., Valentini, I., Casadei, G., Paioli, D., Ferrari, F., Gordini, G., Patelli, M., & Tallini, G. (2015). Randomized trial of endobronchial ultrasound-guided transbronchial needle aspiration with and without rapid on-site evaluation for lung cancer genotyping. *Chest*, *148*(6), 1430–1437.

Vaickus, L. J., Suriawinata, A. A., Wei, J. W., & Liu, X. (2019). Automating the Paris system for urine cytopathology—A hybrid deep-learning and morphometric approach. *Cancer Cytopathology*, *127*(2), 98–115.

Wang, C. W., Liou, Y. A., Lin, Y. J., Chang, C. C., Chu, P. H., Lee, Y. C., Wang, C. H., & Chao, T. K. (2021). Artificial intelligence-assisted fast screening cervical high grade squamous intraepithelial lesion and squamous cell carcinoma diagnosis and treatment planning. *Scientific Reports*, *11*(1)16244.

Wang, W., Tian, Y., Xu, Y., et al. (2022). 3cDe-Net: A cervical cancer cell detection network based on an improved backbone network and multiscale feature fusion. *BMC Medical Imaging*, *22*(1), 130.

Wilbur, D. C. (2016). Digital pathology and its role in cytology education. *Cytopathology: Official Journal of the British Society for Clinical Cytology*, *27*(5), 325–330.

Wojcik, E. M., Kurtycz, D. F. I., & Rosenthal, D. L. (2022). AG *The Paris system for reporting urinary cytology*. Cham: Springer International Publishing.

Wong, C. M., Kezlarian, B. E., & Lin, O. (2023). Current status of machine learning in thyroid cytopathology. *Journal of Pathology Informatics*, *14*100309.

Xie, X., Fu, C. C., Lv, L., Ye, Q., Yu, Y., Fang, Q., Zhang, L., Hou, L., & Wu, C. (2022). Deep convolutional neural network-based classification of cancer cells on cytological pleural effusion images. *Modern Pathology: An Official Journal of the United States and Canadian Academy of Pathology, Inc.*, *35*(5), 609–614.

Xing, J., Monaco, S. E., Cuda, J., & Pantanowitz, L. (2020). Telecytology rapid on-site evaluation: Diagnostic challenges, technical issues and lessons learned. *Cytopathology: Official Journal of the British Society for Clinical Cytology*, *31*(5), 402–410.

Yamashiro, K., Yoshimi, N., Itoh, T., Takino, H., Nakajima, M., Azuma, M., Taira, K., Makio, S., Shiina, S. I., Hata, S., Urabe, S., Fukuoka, J., & Mori, I. (2019). A small-scale experimental study of breast FNA consultation on the internet using Panoptiq. *Journal of the American Society of Cytopathology*, *8*(4), 175–181.

Yao, K., Shen, R., Parwani, A., & Li, Z. (2018). Comprehensive study of telecytology using robotic digital microscope and single Z-stack digital scan for fine-needle aspiration-rapid on-site evaluation. *Journal of Pathology Informatics*, *9*49.

Yu, W., Liu, Y., Zhao, Y., Huang, H., Liu, J., Yao, X., Li, J., Xie, Z., Jiang, L., Wu, H., Cao, X., Zhou, J., Guo, Y., Li, G., Ren, M. X., Quan, Y., Mu, T., Izquierdo, G. A., Zhang, G., Zhao, R., Zhao, D., Yan, J., Zhang, H., Lv, J., Yao, Q., Duan, Y., Zhou, H., Liu, T., He, Y., Bian, T., Dai, W., Huai, J., Wang, X., He, Q., Gao, Y., Ren, W., Niu, G., & Zhao, G. (2022). Deep learning-based classification of cancer cell in leptomeningeal metastasis on cytomorphologic features of cerebrospinal fluid. *Frontiers in Oncology*, *12*, 821594.

Zhang, Z., Fu, X., Liu, J., Huang, Z., Liu, N., Fang, F., & Rao, J. (2022). A deep learning-based segmentation system for rapid onsite cytologic pathology evaluation of pancreatic masses: A retrospective, multicenter, diagnostic study. *EBioMedicine*, *80*, 104022.

Zhang, Z., Fu, X., Liu, J., Huang, Z., Liu, N., Fang, F., & Rao, J. (2021). Developing a machine learning algorithm for identifying abnormal urothelial cells: A feasibility study. *Acta Cytologica*, *65*(4), 335–341.

CHAPTER

Regulatory aspects of digital pathology

8

S. Joseph Sirintrapun[1,2]

[1]Department of Pathology, Mass General Brigham, Massachusetts General Hospital, Brigham and Women's Hospital, Boston, MA, United States

[2]Harvard Medical School, Boston, MA, United States

Introduction

This chapter intends to guide essential information when implementing whole slide imaging (WSI) digital pathology on the current regulatory and best practices that maintain compliance and mitigate liability. The COVID-19 pandemic presented a tremendous challenge to laboratories, laboratory personnel, and pathologists on the front lines. This chapter heavily focuses on issues related to remote sign-out, which occurred during the COVID-19 public health emergency (PHE). Though comprehensive, this chapter cannot encompass all aspects with every contingency and permutation. Likewise, this chapter intends not to be prescriptive, and digital pathology considerations are ever-evolving. Now that the pandemic is endemic, there is an absolute need for institutions implementing primary WSI sign-out with remote WSI sign-out policies to exercise discretion and ensure these policies get enacted in the best interest of all stakeholders.

Food and Drug Administration regulations in digital pathology whole slide imaging devices

Digital pathology was employed for many decades. However, since the 2000s, WSI has witnessed significant technological advances. Likewise, WSI adoption into clinical practice was not commensurate to these technological advances, particularly in the United States. The lag in WSI adoption into clinical practice is mainly attributable to the clearance of WSI devices through regulatory authorities of the US Food and Drug Administration (FDA) (Abels & Pantanowitz, 2017).

When a medical device is labeled, promoted, and/or used in the United States in a manner that meets specific definitions according to the Federal Food, Drug, and Cosmetic (FD and C) Act (Title 21 Code of Federal Regulations part 201, [h]), the FDA regulates the medical device and in accordance to premarketing

Digital Pathology. DOI: https://doi.org/10.1016/B978-0-443-13809-6.00011-7

and postmarketing regulatory controls. Such a medical device and its intended use will drive the regulatory pathway. When devices are not promoted or sold but designed, developed, and validated for their specific intended use, they are not regulated by the FDA. Such nonregulated medical devices are laboratory-developed tests (LDTs) (Abels & Pantanowitz, 2017). As discussed in the final section of the chapter, the line between FDA regulation and LDTs is getting blurred.

The FDA has a three-tier medical device regulatory classification system (Abels & Pantanowitz, 2017):

- Class I: low to moderate risk
- Class II: moderate to high risk
- Class III: high risk

History of whole slide imaging and Food and Drug Administration regulation

In October 2009, the FDA publicly announced WSI devices fall under Class III. At the time, Class III meant approval of a premarket approval (PMA) application. A PMA is the FDA process of scientific and regulatory review to evaluate the safety and effectiveness of Class III medical devices. Likewise, a PMA is the most stringent type of device marketing application required by the FDA.

In the interim, some vendors devoted their efforts to obtaining regulatory approval for their WSI devices by the Conformité Européenne (European Conformity; CE marking) and in Canada (Têtu & Evans, 2014). The CE mark on a product indicates that the manufacturer or importer of that product affirms its compliance with the relevant EU legislation, and the product may get sold anywhere in the European Economic Area.

The successful use of WSI for routine diagnosis in general surgical pathology outside the United States provided evidence that this technology was safe and effective. In addition, several studies were published indicating that the risk of this technology was lower than originally perceived (Al-Janabi et al., 2012; Stathonikos et al., 2013; Thorstenson et al., 2014). These levels of evidence encouraged the use of de novo applications to the FDA for WSI devices rather than more burdensome PMAs. Following the first de novo classification, subsequent manufacturers can apply for a 510(k) clearance, which is the premarket submission made to FDA to demonstrate that the device to be marketed is as safe and effective, that is, substantially equivalent, to a legally marketed predicate device (Section 513(i)(1)(A) FD&C Act) ([last accessed on 4/24/2023]; https://www.fda.gov/medical-devices/premarket-submissions-selecting-and-preparing-correct-submission/premarket-notification-510k). This first authorized de novo WSI predicate device would pave the path to show substantial equivalence for other devices (Abels & Pantanowitz, 2017).

When applying to the FDA, manufacturers of a WSI system need to perform testing that includes both analytical testing and a clinical study to demonstrate that their device is safe and effective for its intended clinical use. By April 2016, the FDA released a guideline that contained recommendations regarding the technical performance assessment data that should be submitted for regulatory evaluation of a WSI system (FDA Technical Performance Assessment of Digital Pathology Whole Slide Imaging Devices, 2023).

Finally, in April 2017, through a successful de novo premarket review application as a Class III medical device, the Philips IntelliSite Pathology Solution (PIPS, Philips Medical Systems Nederland B.V.) gets the first FDA authorization as a Class II device to interpret WSIs created, viewed, and managed by PIPS, of surgical pathology slides prepared from formalin-fixed paraffin-embedded (FFPE) tissue ([last accessed on 4/24/2023]; https://www.fda.gov/drugs/resources-information-approved-drugs/intellisite-pathology-solution-pips-philips-medical-systems; [last accessed on 4/24/2023]; https://www.philips.com/a-w/about/news/archive/standard/news/press/2017/20170413-philips-receives-fda-clearance-to-market-philips-intellisite-pathology-solution-for-primary-diagnostic-use-in-the-us.html). Subsequent WSI systems, the Aperio AT2 DX system in May 2019 ([last accessed on 4/24/2023]; https://www.accessdata.fda.gov/cdrh_docs/pdf19/K190332.pdf; [last accessed on 4/24/2023]; https://www.fdanews.com/articles/191523-fda-clears-leica-biosystems-digital-pathology-system; [last accessed on 4/24/2023]; https://www.leicabiosystems.com/us/news-events/leica-biosystems-receives-fda-510k-clearance-to-market-a-digital-pathology-system-for); and the NanoZoomer S360MD Slide scanner system in September 2022 ([last accessed on 4/24/2023]; https://www.accessdata.fda.gov/cdrh_docs/pdf21/K213883.pdf; [last accessed on 4/24/2023]; https://www.hamamatsu.com/jp/en/news/products-and-technologies/2022/20221004000000.html; [last accessed on 4/24/2023]; https://www.fdanews.com/articles/209681-hamamatsu-photonics-nanozoomer-scanner-cleared) gets FDA authorization as Class II devices, cleared by the 510(k) pathway. Currently, these three WSI systems are the only ones that are FDA-approved for primary diagnosis of surgical pathology FFPE specimens with WSI.

Food and Drug Administration perspective on whole slide imaging components

FDA considers WSI systems comprising of two basic subsystems—(1) the image acquisition component, often referred to as the scanner, and (2) the workstation, or the image viewing software, computer, and monitor/display. With the initial Class II approvals, the FDA considered this entire WSI system a closed unit.

The term pixel pipeline is used to illustrate which components comprise a WSI system: those that acquire the image, store the image, transform the image, present the image, and display the image.

For FDA validation of a WSI system, there is no decoupling of the pixel pipeline with the entire WSI system validated holistically. Therefore if the pixel pipeline is "disrupted" by using, for example, a different monitor than the one approved by the FDA, then the system is no longer in compliance. Therefore when a monitor that a pathologist uses in a remote location like the home is different from the one approved by the FDA, the overall system is technically not FDA-cleared.

As discussed in the next section, the COVID-19 pandemic with declaration of the PHE accelerated the decoupling in device-type components of the "pixel pipeline" of prior Class II WSI systems, largely intended to enable remote work. In March 2020, the Sectra Digital Pathology Module got 510(k) clearance as a digital pathology image viewing and management software as a device-type component ([last accessed on 4/24/2023]; https://www.accessdata.fda.gov/cdrh_docs/pdf19/K193054.pdf; [last accessed on 4/24/2023]; https://medical.sectra.com/news-press-releases/news-item/B0A97D8EC803FF01/; [last accessed on 4/24/2023]; https://www.medtechdive.com/news/fda-clears-sectras-digital-pathology-module-for-primary-diagnostics-as-us/575346/; [last accessed on 4/24/2023]; https://www.accessdata.fda.gov/scripts/cdrh/cfdocs/cfPMN/pmn.cfm?start_search = 1&productcode = LLZ&knumber = &applicant = SECTRA%2DIMTEC%20AB). In July 2020, PAIGE also got 510(k) clearance as a digital pathology image viewing and management software as a device-type component ([last accessed on 4/24/2023]; https://www.fda.gov/news-events/press-announcements/fda-authorizes-software-can-help-identify-prostate-cancer; [last accessed on 4/24/2023]; https://www.accessdata.fda.gov/cdrh_docs/pdf20/DEN200080.pdf; [last accessed on 4/24/2023]; https://paige.ai/paige-receives-first-ever-fda-approval-for-ai-product-in-digital-pathology/).

Though manufacturers of the device-type components of the "pixel pipeline" had to use the 510(k) pathway in general for FDA clearance, during the PHE, FDA recognized the urgency for greater access to remote digital pathology devices may help facilitate the remote reviewing and reporting of pathology slides to facilitate continuity of patient care and reduce healthcare personnel contact and risk of exposure to SARS-CoV-2.

FDA issued a guidance document on enforcement policy for remote digital pathology devices during the PHE (FDA-2020-D-1138) stating that the FDA will not object to:

- Modifications to the FDA-cleared indications, functionality, hardware, and/or software of device-type components of the "pixel pipeline" to provide for use in a remote setting
- Marketing of new device-type components of the "pixel pipeline" intended for use in remote settings and not currently 510(k) cleared for any use and where such devices do not create undue risk ([last accessed on 4/26/2023]; https://www.fda.gov/regulatory-information/search-fda-guidance-documents/enforcement-policy-remote-digital-pathology-devices-during-coronavirus-disease-2019-covid-19-public).

FDA-2020-D-1138 intended to remain in effect during the PHE. However, on December 23, 2021, draft guidance on a transition plan for medical devices was released for preparation after the end of the PHE, basically stating that once the official end of the PHE is declared, FDA-2020-D-1138 will remain in effect for another 180 days after that date ([last accessed on 4/26/2023]; https://www.fda.gov/media/155038/download). With May 11, 2023, as the official end of the PHE, November 7, 2023, marks the 180-day end of the PHE transition period for device-type components of the "pixel pipeline." The ramifications of this loss in the flexibility in FDA-cleared indications, functionality, hardware, and/or software of device-type components of the "pixel pipeline" to provide for use in a remote setting are unclear.

Background and history of Clinical Laboratory Improvement Amendment and remote whole slide imaging sign-out

A brief history of Clinical Laboratory Improvement Amendment

The Clinical Laboratory Improvement Amendment (CLIA) began in the late 1960s with problems with Pap smears from error-prone cytology laboratories with overworked personnel. Through legislative passage in December 1967, CLIA'67 became the first laboratory regulations born. However, CLIA' primarily covered independent and hospital laboratories at the time, creating new gaps that gained recognition in the following decades.

Beginning in 1987, press pieces and Congressional hearings in 1988 focused on the faulty quality of laboratory testing, particularly those outside of independent and hospital laboratories that are not subject to federal or state regulations. Again, public interests questioned the accuracy of Pap smears and the validity of cholesterol screening.

By October 1988, a second amendment, CLIA'88, got introduced, allowing the Secretary of Health to develop comprehensive, quality standards for all laboratory testing to ensure patient test results' accuracy, reliability, and timeliness, regardless of where the test got performed. In addition, CLIA'88 oversees all facility types, not just the independent and hospital laboratories. Now CLIA'88 covers physician offices, nursing homes, and even health fairs.

CLIA'88 now requires certification of all laboratories performing tests and those participating in the Medicare/Medicaid programs to comply with CLIA-88 requirements. CLIA'88 clarified the definition of a laboratory as any facility which performs laboratory testing on specimens derived from humans to provide information for the diagnosis, prevention, treatment of disease, impairment of, or assessment of health. In addition, the final published provisions of CLIA'88 in February 1992 addressed the complexity of the test method; thus the more complicated the test, the more stringent the requirements. The three categories of test complexity are: waived complexity, moderate complexity, including the subcategory of provider-performed microscopy, and high complexity ([last accessed on 4/25/2023]; https://www.nmhealth.org/publication/view/general/2222/).

Twenty years post-CLIA'88 in 2008, the impact of CLIA'88 spans:

- >200,000 laboratories
- 333,000 individual laboratorians
- 8 billion laboratory tests (annually)
- Healthcare providers
- IVD manufacturing industry
- Healthcare consumers (patients, families, communities) ([last accessed on 4/25/2023]; https://www.cdc.gov/cliac/docs/addenda/cliac0208/Addendum-A.pdf)

One stipulation of CLIA'88, which posed downstream ramifications years after its passage, was the requirement for remote facilities and locations performing remote sign-out to have separate CLIA licenses.

Defining remote

Remote sign-out is the electronic verification of laboratory test results by qualified personnel under CLIA from a location that is NOT considered part of the central laboratory (e.g., a pathologist's home or a hotel room). In the context of anatomic pathology, remote sign-out includes the remote review and diagnostic interpretation of cases by digital pathology (e.g., whole slide image or WSI) or by glass slides and a microscope.

A remote facility is NOT within the central laboratory's CLIA address. Instead, it includes offices where images or data files get remotely reviewed and interpreted frequently. These locations can consist of the pathologist's home or hotel room.

Defining what constitutes a primary diagnosis in digital pathology

Reviewing WSIs in place of glass slides followed by electronic verification of a rendered pathology report (e.g., review of digital images without using a microscope and final reporting of the diagnoses) defines a primary diagnosis. Primary diagnosis is the original diagnosis used for patient care. The FDA does not consider secondary consultations, rapid on-site evaluation (ROSE), or frozen sections read by telepathology under the purview of primary diagnosis. In addition, the analysis of special stains, immunohistochemical stains, and other ancillary studies that aid in making a primary diagnosis is not considered primary diagnoses on their own.

Remote sign-out and CLIA'88

Permitted under CLIA'88 is the rendering of a primary diagnosis or interpretation of laboratory data from a remote site that already has its own CLIA license. Also permitted under CLIA'88 is when the final diagnosis is not changed from what was originally transcribed at the CLIA-licensed primary site. Such an example is a case that gets dictated by a pathologist at a CLIA-licensed primary site and a diagnosis rendered but not verified (i.e., not electronically signed-out). Later, from a remote site, the same pathologist securely logs into the anatomic pathology (AP) lab information system (LIS), reviews the report, and electronically signs it out. Or similarly, the pathologist remotely corrects typographical errors in the report or adds results of special stains.

NOT permitted under CLIA'88 is remote sign-out at a non-CLIA-licensed remote facility location. Likewise, the home office of a pathologist is NOT an extension of the CLIA-licensed laboratory in which they work. Therefore regular sign-out using glass slides or WSI from home means you must apply for a CLIA certificate (nonwaived) for your home.

Another situation when reading cases while traveling or sporadically/infrequently from home, one can ask for a temporary site exemption. Exemptions for remote sign-out were possible with a CMS Form 116 temporary certificate ([last accessed on 4/26/2023]; https://www.cms.gov/Medicare/CMS-Forms/CMS-Forms/CMS-Forms-Items/CMS012169) intended for sign-out. Interestingly, the quantification for what defines sporadic/infrequent is not stipulated but, in general, agreed upon as at most 10% of cases. Thus in a situation where one must review 100% of cases from home, CMS Form 116 is not applicable.

These regulatory hurdles in obtaining separate CLIA licenses or CMS Form 116 exemptions were well intended, keeping the low-quality laboratories from sprouting up everywhere with minimal oversight or quality standardization. However, CLIA'88 did not account for the emergence of primary diagnosis through digital pathology with the potential for remote sign-out locations outside the primary laboratory.

The COVID-19 pandemic introduced urgencies in overcoming challenges to deliver pathology services. Through grassroots advocacy efforts of the organizations like the Association of Pathology Informatics (API) that sparked the charge, along with the American Society of Clinical Pathology (ASCP) and the College of American Pathologists (CAP), these grassroots advocacy efforts persuaded Congress to action for more flexibility from CMS for enabling remote sign-out (Centers for Medicare, & Medicaid Services Memorandum, Clinical Laboratory Improvement Amendments CLIA Laboratory Guidance During COVID-19 Public Health Emergency, 2020; [last accessed on 4/25/2023]; https://www.ascp.org/content/news-archive/news-detail/2020/03/27/ascp-lab-community-secure-regulatory-flexibility-to-allow-remote-pathology-services#; [last accessed on 4/25/2023]; https://ascpcdn.s3.amazonaws.com/static/epolicy/2020/House_Sign_on_letter_ + CMS_Pathologist_Rules_032320.pdf; [last accessed on 4/25/2023]; https://www.voicebrook.com/blog/pathology-news-roundup-march-24-2020; [last accessed on 4/25/2023]; https://www.ascp.org/content/news-archive/news-detail/2020/03/27/ascp-lab-community-secure-regulatory-flexibility-to-allow-remote-pathology-services#; [LAST accessed on 4/25/2023]; https://www.pathologyinformatics.org/assets/annual_report/API%20Annual%20Report%202020-2021%20Sirintrapun.pdf; [last accessed on 4/25/2023]; https://www.cap.org/advocacy/latest-news-and-practice-data/

march-26-2020; [last accessed on 4/25/2023]; https://documents.cap.org/documents/cap-hhs-coronavirus-laboratories-regulations.pdf).

Eventually recognizing the urgency created by the current COVID-19 pandemic, the Centers for Medicare and Medicaid Services (CMS) exercised enforcement discretion to ensure pathologists may review slides and sign-out remotely. On March 26, 2020, CMS temporarily waived the requirement for remote locations to have separate CLIA licenses, provided that the designated primary site or laboratory has a CLIA certificate ([last accessed on 4/25/2023]; https://www.cms.gov/files/document/qso-20-21-clia.pdf-0).

Ramifications and requirements for the enforcement discretion "Waiver"

Required is a CLIA license under CLIA'88 at any site that performs testing, including where primary diagnoses and interpretations get rendered without the flexibility of remote sign-out at a remote non-CLIA-licensed facility. This requirement endangered an aging pathologist workforce susceptible to SARS-CoV-2 through exposure at busy hospitals and primary CLIA-licensed sites with a concentration of patients, specimens, and surfaces harboring SARS-CoV-2. This enforcement discretion allowed for remote sign-out activities provided that the designated primary site has a CLIA certificate during the PHE, the consequence of which optimized the safety of an aged demographic of pathologists. Likewise, the waiver fulfilled its intent to preserve the continuity of patient care through uninterrupted diagnostic pathology and laboratory services by getting cases to the appropriate pathologists digitally, ensuring deliverance of the most accurate diagnostic, prognostic, and predictive information to guide quality patient care.

With NO separate CLIA certificate for a home office before the waiver, glass slide and WSI remote sign-out could get performed on an INFREQUENT basis under a CMS Form 116 exemption. Now permissible under the waiver during the PHE is no limit on the number, percentage, or type of cases that get reviewed and signed out remotely through glass slides or WSI with certain conditions in Table 8.1 and under the umbrella of a primary laboratory's CLIA license.

As discussed in the next section, essential conditions for this waiver using digital pathology systems before using it for remote sign-out are proper validations with written procedures and documentation at non-CLIA-licensed locations affiliated with a primary site that CLIA licenses.

For glass slide remote sign-out, the home setup includes a microscope at home to allow remote review and sign-out of cases. The cases get delivered to their home as glass slides and accompanying laboratory requisitions. The pathologist also has a mechanism for sharing cases with colleagues for consultation and quality assurance.

For WSI remote sign-out, the home setup includes a digital pathology workstation to allow remote review and reporting of cases. The cases are accessible

Table 8.1 Conditions during the PHE, where cases can get reviewed and signed out remotely through glass slides or WSI under the umbrella of a primary laboratory's CLIA license.

The remote location has a designated primary site or primary laboratory with an active and valid CLIA certificate.
The specific process, assay, and/or application have been validated to be performed remotely with appropriate validation documentation.
A procedure has been written and approved by the Medical Director of the designated primary site or primary laboratory with the active and valid CLIA certificate.
Personnel signing out remotely have access to all procedures and policies required to perform their remote sign-out duties.
The remote location can access systems securely under the Health Insurance Portability and Accountability Act (HIPAA) Final Security Rule and related HIPAA regulations.
Personnel performing sign-out at such remote locations can securely view or store any protected health information (PHI) (e.g., slides, instrument printouts, documents) such that accidental or intentional unauthorized disclosure of PHI is extremely difficult or impossible.

through an electronic worklist in the anatomic pathology laboratory information systems (APLIS). The digital pathology system also facilitates sharing cases with colleagues for consultation and quality assurance. In addition, the pathologist has a mechanism for deferring cases to glass slide review when necessary.

For both glass slide and WSI remote sign-out, one needs secure (i.e., HIPAA-compliant) remote access to the APLIS and electronic patient records and where one can order and interpret any/all ancillary testing procedures required to generate complete diagnoses. CMS also requires that the final pathology report lists the physical address of the CLIA-licensed facility. This requirement introduces several permutations when signing a case out from home. Pathology reports that include work performed remotely, in part or whole, need to have the address of the remote site (e.g., someone's house) or a code traceable to the address. Labs also need to keep a list of staff working remotely or partially remotely. Likewise, if one has a CLIA license for their home, that home address must appear on the report. For sign-out at a remote location that does not have a CLIA license, under this waiver, the physical address of the designated primary CLIA-licensed facility and not the pathologist's remote non-CLIA-licensed site needs to appear in the report.

The ramifications for this enforcement discretion were sweeping, even going beyond digital and glass slide review to encompass clinical pathology images and data (e.g., electropherograms, gel images, fluorescence in situ hybridization, molecular results, flow cytometry dot plots, blood bank, and Human leukocyte antigen (HLA) data). As a result, the waiver created remote sign-out opportunities, not just for practicing anatomic pathologists but clinical pathologists and molecular pathologists.

Activities for clinical and molecular pathologists performed remotely include:

- Serum and urine protein electrophoresis.
- Hemoglobin electrophoresis.
- Next-generation sequencing.
- Genotyping assays.

- Results of polymerase chain reaction, Sanger sequencing, and other molecular assays.
- Karyotyping (banded chromosome analysis).
- Chromogenic in situ hybridization and fluorescence in situ hybridization.
- HLA results.
- Blood bank interpretations (e.g., antibody reactions, transfusion reaction workups).
- Microbiology and coagulation assay interpretations.

Moreover, other types of personnel, such as cytotechnologists, toxicologists, and cytogeneticists, could review slides/data/images remotely at a temporary location during this time without obtaining a separate CLIA number.

Continued enforcement discretion "Waiver"

The waiver began on March 26, 2020 with the temporary intent to persist until the declared end of the PHE for COVID-19, expected on May 11, 2023. Afterward, the granted flexibilities during the waiver revert to the pre-PHE statutes with the need for separate CLIA licenses and CMS Form 116 exemptions for remote facilities.

Recognizing this lack of specification for what happens to the enforcement discretion after the PHE, grassroots efforts by organizations, notably API, pushed for continued enforcement discretion past the end of the PHE to the CLIAC Advisory Committee on April 13, 2022 ([last accessed on 4/25/2023]; https://www.cms.gov/files/document/laboratories-cms-flexibilities-fight-covid-19.pdf; [last accessed on 4/25/2023]; https://www.mlo-online.com/management/regulatory/article/21282180/cms-continues-to-exercise-pathologists-and-other-laboratory-staff-enforcement-discretion; [last accessed on 4/25/2023]; https://www.cdc.gov/cliac/docs/april-2022/pc5_association-for-pathology-informatics.pdf; [last accessed on 4/25/2023]; https://www.cdc.gov/cliac/docs/april-2022/pc3_j-sirintrapun.pdf). In turn, the CLIAC Advisory Committee recommended continued enforcement discretion past the end of the PHE to CMS. As of February 24, 2023, CMS clarified that enforcement discretion (for pathologists) is *not contingent on PHE authority.*

Though a specific date for the lifting of the waiver is yet defined, CMS will continue to exercise enforcement discretion that allows *pathologists* to examine "digital images and laboratory data" at remote locations presumably past the declared end of the PHE for COVID-19, expected on May 11, 2023. An updated memo from CMS on May 12, 2023, further clarifies the post-PHE status of the temporary exercise of enforcement discretion and other flexibilities. CMS expanding the wording to "permit pathologists *and other laboratory personnel* to review digital laboratory data, digital results and digital images ('digital materials') remotely" ([last accessed on 5/19/2023]; https://www.cms.gov/files/document/qso-23-15-clia.pdf).

Interestingly, though the waiver applied to both digital and glass slide reviews during the PHE, past the end of the PHE, the specific wording of "digital images and laboratory data" explicitly omits remote glass review, which was previously allowed during the PHE. Likewise, there is hope that WSIs are under the definition of "digital images."

Accreditation

With accreditation currently, the CAP has WSI checklist requirements. Interestingly, the CAP Laboratory General Checklist houses these WSI checklist requirements instead of the CAP Anatomic Pathology checklist (College of American Pathologists. CAP Laboratory Accreditation Checklists, 2023). The likely reasons, from the checklist authors' perspective, stem from WSI systems having a broader application for diagnostic purposes (primary and consultation) across the laboratory than just anatomic pathology. Likewise, these CAP WSI checklist requirements do NOT apply to informal reviews without formal reporting or educational and research use (College of American Pathologists. CAP Laboratory Accreditation Checklists, 2023). Initially developed in 2013, these WSI checklist requirements, housed in the CAP Laboratory General Checklist, emerged after the early publications on WSI validation (College of American Pathologists. CAP Laboratory Accreditation Checklists, 2023; Pantanowitz et al., 2013; [last accessed on 4/30/2023]; Validation of Digital Pathology In a Healthcare Environment: A Whitepaper from the Digital Pathology Association, 2011 https://digitalpathologyassociation.org/_data/cms_files/files/DPA-Healthcare-White-Paper-FINAL_v1.0.pdf), with no recent updates since.

CAP inspectors will look for the sampling records (documentation) for training and system validation

Whole slide imaging user training

Required for compliance are records showing that all users of the WSI system got trained. User training is one of the critical aspects of ensuring the accuracy and reliability of WSI technology in clinical settings. All users of the WSI system, including individuals responsible for slide scanning and digital slide quality assessment and pathologists, must undergo proper training. The training procedure should include role-specific training, as defined in the approved laboratory procedures. Retraining may be required when significant system changes emerge. Evidence of compliance are the records of those with WSI training included in personnel files (College of American Pathologists. CAP Laboratory Accreditation Checklists, 2023). Training methods are not explicitly specified and likely deemed outside of the scope of the CAP Laboratory General Checklist and considered separate from validation/verification.

Whole slide imaging system validation/verification

To validate or verify WSI systems used for clinical diagnostic purposes, the laboratory must perform its studies, including approval for use by the laboratory director (or designee who meets CAP director qualifications) before the technology is used for the intended diagnostic purposes(s). Variables between institutions can affect performance, making manufacturer device validation (i.e., verification) alone insufficient (College of American Pathologists. CAP Laboratory Accreditation Checklists, 2023). Therefore the FDA's approval or clearance of WSI systems does not prevent the need for individual laboratories to verify the performance of these systems for the specific intended diagnostic purposes. For example, simple guidelines for cytology screening devices (which were FDA-approved) are insufficient. Therefore all pathology laboratories must carry out their validation or verification studies. The specific components of validation or verification studies are left to the laboratory's discretion.

Validation/verification of WSI systems ensure that a pathologist using this technique to view digitized glass slides can consistently make the same clinical interpretation as they would from viewing the glass slides using a traditional bright field microscope. Therefore the validation/verification process should address both technical and interpretative components. In addition, because WSI usage by laboratories in the clinical diagnostic setting is still relatively early, proper WSI system validation/verification is vital to ensure the maintenance of the quality, accuracy, and reliability of the technology of the WSI system.

Validation is not verification—clarification of confusing terminologies (CAP perspective)

Validation—"Did I do the right thing?". The process of testing a procedure or instrument to assess its performance for intended purposes and determine whether it is acceptable is called validation. It is necessary to prove that it performs as expected and achieves the intended result. For example, vendors of FDA-approved WSI systems underwent rigorous validation to obtain FDA clearance for intended purposes. WSI systems that are not FDA-approved or FDA-approved WSI systems but modified, including different intended purposes, are deemed for off-label use and get considered LDTs, thus requiring validation.

Verification—"Did I do the thing right?". In contrast, verification generally happens when a clinical lab wants to implement an FDA-approved WSI system. Verification is checking the manufacturer's claims for performance specifications before use and verifying the reproducibility of a product/test in the lab before patient testing (directed by the manufacturer's user manual).

General guiding principles for the whole slide imaging validation/verification process

The CAP Laboratory General Checklist outlines general guiding principles for the WSI validation/verification process. There are also 2021 updates to these general guiding principles. While the recommendations are largely similar, the main difference between the original guideline and the 2021 update is the different, more rigorous processes used to develop the guidelines. In collaboration with the ASCP and the API and endorsed by the CAP WSI Validation Expert Panel, the 2021 updates evaluated literature published since the 2013 original guideline release. These 2021 updates come in the form of nine good practice statements (GPSs) and three recommendations (Evans et al., 2022; Guyatt et al., 2016; Tugwell & Knottnerus, 2015).

GPSs are statements with a "high level of certainty that the recommendation will do more good than harm (or the reverse), but where there is little direct evidence." GPSs are not evidence-based; however (Guyatt et al., 2016; Tugwell & Knottnerus, 2015), nine of the recommendation statements from the 2013 guideline were either reaffirmed and/or revised as GPSs. GPSs are listed in Table 8.2 (Evans et al., 2022).

Table 8.2 Good practice statements (GPS).

GPS 1	All pathology laboratories implementing WSI technology for clinical diagnostic purposes should carry out their own validation studies.
GPS 2	Validation should be appropriate for and applicable to the intended clinical use and clinical setting of the application in which WSI will be used • Validation should involve specimen preparation types relevant to intended use (e.g., formalin-fixed, paraffin-embedded tissue; frozen tissue; and immunohistochemical stains). • Separate validation should be performed for a new application that differs materially from the previously validated use.
GPS 3	The validation study should closely emulate the real-world clinical environment in which the technology will be used.
GPS 4	The validation study should encompass the entire WSI systemNot necessary to separately validate each individual component (e.g., computer hardware, monitor, network, and scanner) of the system or the individual steps of the digital imaging process.
GPS 5	**Laboratories should have procedures in place to address changes to the WSI system that could impact clinical results.**
GPS 6	Pathologists adequately trained to use the WSI system must be involved in the validation process.
GPS 7	**The validation process should confirm that all of the material present on a glass slide to be scanned is included in the digital image.**
GPS 8	Documentation should be maintained recording the method, measurements, and final approval of validation for the WSI system to be used in the anatomic pathology laboratory
GPS 9	**Pathologists should review cases/slides in a validation set in random order** • **Applies to both the review modality (i.e., glass slides or digital) and the order in which slides/cases are reviewed within each modality.**

GPSs in bold are not yet integrated into the CAP Laboratory General Checklist.

Since sampling records (documentation) for system validation is mandatory for accreditation, following the CAP Laboratory General Checklist fulfills GPS1. However, only some of the 2021 updates have been fully incorporated in the CAP Laboratory General Checklist. For instance, GPSs 5, 7, and 9 are not yet integrated in the CAP Laboratory General Checklist. GPS7 states to confirm that the digital image shows all of the material on the corresponding scanned glass slide. Accurate digital reproduction of scanned glass slides is a critical consideration for diagnostic use. One of the best ways to do this is to check each WSI's macro image (not just the thumbnail). This method, unfortunately, is through human quality check evaluation and is only scalable with future automation. GPS9 states that during the validation/verification process, digital and glass slides can be evaluated in random or nonrandom order (as to which is examined first and second). Some believe that digital slides should be viewed before glass slides, considered the gold standard for making diagnoses. However, the order of viewing virtual versus glass slides has yet to be shown to affect interpretation, according to Koch et al. (2009). The evidence indicates it can go either way.

Currently, as stated in the Laboratory General Checklist, five general guiding principles for the validation/verification process should:

1. Closely emulate the real-world clinical environment and involve specimen preparation types and clinical settings relevant to the intended clinical use(s) in which WSI will be employed (College of American Pathologists. CAP Laboratory Accreditation Checklists, 2023). This guiding principle is updated as GPS2 and GPS3 (Evans et al., 2022). The goal of validation should be conducted in a manner that mimics how WSI will be used in the specific lab's work environment and after "go live." Different outcomes of WSI and glass slides with different specimen preparation types should be expected. Validation/verification of WSI systems thus should involve specimen preparation types relevant to the intended use (e.g., FFPE tissue, frozen tissue, immunohistochemistry (IHC) stains, cytology slides, and hematology blood smears).

   If a new intended use for WSI is contemplated, and this new use differs materially from the previously validated use, a separate validation for the new use should be performed. For example, a validation study used to support the diagnostic use of digitized slides for routine surgical pathology may not necessarily apply to frozen section digitized slides (e.g., with tissue folds, more pale staining, more mounting medium, etc.) For example, if rapid digitization of glass slides is required for clinical use (e.g., frozen sections), then timely preparation and reading of WSI should be included in the validation process.
2. Encompass the entire WSI system, with reevaluation if a significant change is made to a previously validated or verified system (College of American Pathologists. CAP Laboratory Accreditation Checklists, 2023). This guiding principle is updated as GPS2 and 4 (Evans et al., 2022). It is therefore not necessary to validate each individual component (e.g., computer hardware,

monitor, network, scanner) of the system nor the individual steps of the digital imaging process separately. The WSI system is made up of different components, including the scanner, hardware, software, network, viewing monitor, and pathologist. All components are essential and should not be separated, including the technical system ("tool") and observer ("pathologist"). The parameters of each component may impact digital image quality and, therefore, interpretation. The imaging process involves several steps, including image acquisition, storage, sharing, and viewing.

Revalidate/reverify when significant changes to any component of a previously validated or verified WSI system may affect the interpretation of digital slides (e.g., new scanner, major hardware, or software upgrade). The validation/verification process should be repeated for significant changes, with these new changes incorporated into the WSI system to demonstrate that it can still be employed for the intended use. Minor changes can be managed through a facilities change management procedure that includes assessment and calibration with or without documentation. Table 8.3 illustrates actions for assessment, calibration, and validation/verification in several hypothetical scenarios (Evans et al., 2022).

3. Be carried out by a pathologist(s) adequately trained to use the system and assess intraobserver concordance between digital and glass slides (College of American Pathologists. CAP Laboratory Accreditation Checklists, 2023). This guiding principle is updated as GPS6 (Evans et al., 2022). The validation/verification process should include individuals who will be using the system to make diagnoses. Published validation studies show an average number of evaluators at eight individuals per study (3–26 persons). The validation/verification team may include other pathology staff (e.g., image technician, histotechnologist, PA), IT personnel, and/or consultants.
4. Assess intraobserver concordance between digital and glass slides (College of American Pathologists. CAP Laboratory Accreditation Checklists, 2023). Establishing diagnostic concordance between digital and glass slides for the same observer, also known as intraobserver variability. Baseline intra/interobserver variability exists even with glass slides, but the aim is to evaluate the technology, not agreement between pathologists. Therefore we recommend measuring intrapathologist diagnostic reproducibility, meaning whether the pathologist can reach the same diagnosis with both modalities. Interobserver variability, which compares a pathologist's diagnosis to other pathologists, experts, or consensus, should not be measured.
5. Keep records of complete validation or verification study with review or approval as evidence of compliance (College of American Pathologists. CAP Laboratory Accreditation Checklists, 2023). This guiding principle is updated GPS8 (Evans et al., 2022). Documentation for compliance of complete validation/verification studies with review or approval should be maintained, recording the method, measurements, and final approval of validation for the WSI system to be used in the clinical laboratory.

Table 8.3 Additional actions for changes to previously validated/verified WSI systems.

	Action	**Scenario**
Scanner	*Full validation/verification with documentation*	When replacing a *scanner with different make/model*
	Full validation/verification with documentation for a new use case	When using a *previously validated scanner* for a *new, materially different use case*
	Full validation/verification with documentation for at least one site with assessment of applications, network, and histology issues at additional sites to determine if full validation is required for each new scanner	When introducing *scanners of different make/model to multiple sites across a hospital network*
	Calibration with documentation of images for same slide scanned with each scanner (function check)	When replacing a *scanner of the same make/model* for *previously validated applications*
	Assess applications, network, and histology issues at each site to determine if full validation/ verification is required for each new scanner with documentation	When deploying *scanners of the same make/model* as *previously validated at multiple sites across a hospital network*
Software	*Calibration* of images between old and new software to determine if full validation/verification is required with documentation	When introducing *completely different viewing software* including software from same vendor where legacy software has been sunset
	Calibration with documentation of images between old and upgraded software (potential function check) OR *partial validation/verification* dependent upon the criticality and extent of the upgrade	When employing a *software upgrade from same vendor*
Display	*Calibrate with or without documentation* image quality between old and new displays	When *replacing/upgrading displays* at pathologist workstations
Remote locations/ change in network	*Full validation with documentation*	When allowing *primary diagnosis to be remotely performed outside the hospital* (i.e., pathologist's home)
	Calibrate with or without documentation archived images immediately before and after the network changes are made	When there is a *change in pathology or hospital network*

In addition to GPSs, the 2021 updates included three recommendations (Evans et al., 2022). Like some of the GPSs, these recommendations have yet to be fully incorporated into the CAP Laboratory General Checklist. However, unlike GPSs, recommendations are evidenced-based and, when considered with the five general guiding principles stated in the Laboratory General Checklist, have more specificity.

The three 2021 updated recommendations state:

1. The validation process for digital pathology systems should include a representative sample set of *at least 60 cases* for one application (Evans et al., 2022), such as H&E stained sections of fixed tissue, frozen sections, cytology, or hematology, that reflects the spectrum and complexity of specimen types and diagnoses likely to be encountered during routine operation. Whether an individual validation case includes all slides and parts associated with a specific accession number or representative parts and slides is left to the discretion of pathologists completing the validation. The reasoning behind 60 cases in the validation/verification process stems from studies in the systematic review by the 2021 CAP WSI Validation Expert Panel showing that going beyond 60 cases did not improve mean concordance (Evans et al., 2022).

   An additional 20 cases should be included for each additional application (Stathonikos et al., 2013), such as immunohistochemistry or special stains, if these applications are relevant to intended use and were not included in the 60 cases mentioned above. Adequate sample size must be used to allow pathologists to negotiate any technology learning curve. The validation process should involve specimen preparation types relevant to the intended use but not specific organ systems, diseases, microscopic changes, or diagnoses.
2. Laboratories should investigate and attempt to remedy the cause if intraobserver concordance is less than 95% (Evans et al., 2022). Recommending that intraobserver concordance between light microscopy and WSI be greater than 95%. Though ideally, 100% concordance is desired, this does not reflect the subjective nature of pathology as practiced with glass slides where inter- and intraobserver variability is an established reality. The weighted mean percent concordance across the 33 studies in the systematic review by the 2021 CAP WSI Validation Expert Panel was 95.2% (Evans et al., 2022).

   Likewise, all discordant diagnoses should get reviewed and reconciled, with a particular emphasis on major discordances affecting patient management. For nonrectifiable major discordances attributable to WSI, using WSI to review similar cases should be carefully re-considered or deferred to glass slide review.
3. A washout period of at least 2 weeks should occur between viewing digital and glass slides (Evans et al., 2022). The washout period is the time interval between viewing the same case/slide using a different (glass or digital) modality. This washout period is important, as pathologists may recall pathologic images for lengthy periods after reviewing a case. With long washout periods, a pathologist's experience and/or diagnostic criteria could change over time.

These WSI guidelines should get considered in the proper context, not a standard of care. The Pathology and Laboratory Quality Center for Evidence-based Guidelines of the CAP employs a standard disclaimer in each of its guidelines, including the guideline for WSI validation. Guidelines are intended to assist physicians and patients in clinical decision-making and to identify questions and settings for further research and are continually updated. They cannot account for individual variation, and it is the responsibility of treating physicians, relying on independent experience and knowledge, to determine the best course of action according to particular circumstances.

As with any clinical evidence-based guideline, following the GPSs and recommendations of the CAP Laboratory General Checklist or the updated 2021 guidelines are not enforced as mandatory, though it is only highly encouraged to adopt these recommendations and GPSs as guiding principles.

Guiding principles are also ever-evolving. Considering recent advancements in WSI usage and experience, there are likely revisiting and revisions to the CAP Laboratory General Checklist with the integration of the 2021 updates around the horizon.

In general, the FDA focuses its validation studies on assessing the impact of potential technical issues that may affect the quality of digital slides, such as image compression, file format, and storage capacity. Laboratories generally focus their validation/verification studies more wholistically on the users, WSI systems, workflows, and operations involved in making the diagnoses.

Because of the early state of WSI use, all WSI system users involved in the validation/verification process should get adequately trained (GPS6). As WSI use becomes more widespread and commonplace, the validation/verification process could involve a diverse group of pathologists with varying levels of experience (possibly those with minimal training) and expertise to ensure that a wide range of users can use the technology.

Future accreditation checklist requirements may include quality assurance, and maintenance procedures should be implemented to ensure that the digital slides remain accurate and reliable over time. Such potential requirements consist of routine system testing and monitoring, regular software updates, and hardware maintenance. Likewise, future validation/verification processes could include a comprehensive review of the user interface and workflow to ensure that the system is easy to use and integrates smoothly with existing laboratory processes. These future validation/verification processes could assess the impact of digital pathology on laboratory workflow and efficiency. These processes include evaluating the time and resources required to prepare and scan slides and the time necessary for pathologists to review and interpret digital images compared to glass slides. Such metrics are currently challenging to capture because of the lack of "off-the-shelf ready" vendor solutions and customizable business analytic solutions to track such metrics.

Additional considerations around remote primary sign-out through digital pathology

CAP accreditation forms the basis of the information presented in this section and discussed in other sections of this chapter. Suppose one's state has a separate Department of Health or one's laboratory is not CAP accredited. In that case, that laboratory should follow guidance from the accreditation body for the primary CLIA-licensed facility.

Risks of using digital pathology for remote sign-out

Signing out remotely, one must be aware of risks when using any new technology and workflow and try and prevent and/or troubleshoot these risks.

Preanalytical variables of glass slide production affect downstream scanning of the tissue present (e.g., air bubbles, obstructed slide label information, tissue present under coverslips, or overhanging the glass slide edge). Re-coverslipping a slide may be needed, or a new histology section taken with properly centered tissue.

Depending on the resolution, specific WSI diagnoses are challenging compared to glass slides (e.g., identification of specific microorganisms such as helicobacter, quantification of mitotic figures, and evaluation of polarized tissue). Therefore scanning at 40 × may be better regarding an ideal image resolution.

With WSI, there are artifacts (e.g., out of focus, tiling/striping, tissue obstruction due to ink, etc.). The laboratory should ensure appropriate measures to minimize these artifacts, and there should be lines of communication per standard operating procedures (SOPs) to the CLIA-licensed laboratory for remediation. Artifacts may require rescanning.

Such variables can cause delays with remediation of the issues to achieve successful scanning and to complete rendering of a patient diagnosis. Regardless in all instances, the pathologist should have low thresholds for rescans. Deferral always is an option when the digital scan is inadequate ([last accessed on 4/30/2023]; https://www.cap.org/covid-19/remote-sign-out-faqs).

Standard operating procedures to cover for remote sign-out procedures

The fundamental elements of laboratory procedures and policies are the same as any other laboratory SOPs or policies. These SOPs or policies would apply to light microscopic and digital sign-out. One also needs to refer to the requirements of their respective laboratory accreditation organization and/or applicable local, state, and federal laws. Likewise, a qualified Laboratory Medical Director needs to approve the SOP for the designated CLIA-licensed primary site or primary laboratory associated with the remote location.

With remote sign-out, there is recognition that the risk for logistic errors is higher during the implementation of new SOPs. One needs special care to ensure the correct diagnosis gets reported for the right patient and that all patient materials (physical and digital) are received and reviewed. This re-emphasizes a consideration for developing a chain of custody for all patient materials, establishing tracking/auditing mechanisms for physical and digital assets with protected health information (PHI) transported or accessed remotely from the primary CLIA-licensed laboratory with appropriate procedures for timely reconciliation of missing patient materials, both physical and digital. Likewise, there can be documentation somewhere on how and where the primary diagnosis gets rendered for auditing purposes; this does not have to be stated explicitly on the final report but could be part of the laboratory or anatomic pathology information system. Moreover, SOPs should ensure that communication channels between the pathologist and clinical providers get maintained in the remote setting.

For all procedures in which the remote sign-out procedure is an adjunct to the usual procedure followed at the primary CLIA-licensed facility, the most important things to address are any differences in process or policy that occur when a case is signed-out at the remote location. Therefore if this approach to procedure writing gets taken, then the adjunct remote sign-out procedure must reference the procedure for sign-out that occurs at the primary CLIA-licensed facility.

Table 8.4 provides suggestions for what to include in a remote sign-out SOP.

Virtual tracking/auditing will be crucial during remote sign-out operations in the future. For example, all physical assets, like requisitions and slides leaving the CLIA lab and transported to non-CLIA remote sites, should leverage tracking-enabled barcodes or other logistical systems. In addition, tracking/auditing WSIs

Table 8.4 Suggestions for standard operating procedure elements in a remote sign-out procedure.

Procedure to access necessary laboratory information systems, electronic health records, images, and other data from the remote location.
Procedures to request intradepartmental consultation with case images to a different pathologist for review.
Mechanisms for compliance with HIPAA to protect patient confidentiality and privacy, particularly in a home setting where there is potential presence of family members.
If applicable.
Procedures to transport physical assets with PHI (e.g., confirmatory glass slides, tissue blocks, documents) between the primary CLIA-licensed laboratory and the remote location.
If remote setting is different from usual (on-premises) process.
Procedures to contacting the laboratory for remediation or to place orders for additional stains, request repeat testing or request ancillary testing.
Procedures to sign-out cases or to issue addendums or correct reports.
Communication channels between pathologists and clinical providers and mechanisms of reporting of critical values.

along with data and digital information accessed from remote sign-out locations and who accessed, assets remediated, etc., all are essential future quality data considerations with remote sign-out. An exemplary laboratory would have a quality management plan with periodic data analytics, reviewed by the Medical Director, to ensure that remote sign-out does not compromise patient care.

HIPAA/Privacy-related considerations

For the practice of remote sign-out, the same security and privacy requirements of HIPAA apply to both remote and on-site locations. Best practices for security and privacy should be part of the SOPs established by the Medical Director of the primary CLIA site. HIPAA issues are even more of a concern at a remote location that may have less security than a hospital office. Pathologists must ensure that any data taken or viewed outside the CLIA-licensed facility for sign-out gets protected from accidental or intentional unauthorized disclosure. Required is for care and attention that patient information on a computer screen, slide labels, paper requisitions, etc. is not easily accessed or viewed by unauthorized individuals like family members.

Most organizations already have secure remote access protocols built into a hospital or lab infrastructure, so extra steps and more vigilance may need to be taken when working with patient information at a remote site. The CAP recommends that the remote sign-out location has access to laboratory information systems and other data with PHI using secure protocols already established by the healthcare organization. In addition, any physical objects containing PHI, such as paper requisition forms, must be secured (e.g., under lock and key) such that it would be extremely difficult or impossible for an unauthorized individual to view them.

Each site must provide reasonable and expected confidentiality and data security in data storage and transmission. For example, pathologists should refrain from saving patient information such as clinical data, files, or images on their personal computers, which nonauthorized individuals might access. The pathologist should also be sure that the network connection used to view data or access a desktop in the primary healthcare facility uses a secure method such as a virtual private network (VPN). The IT support team can help with this implementation. Procedures include message security, system and user authentication, activity logs, encryption, and access restrictions. For patient identification—as is the case for any laboratory analysis—processes, procedures, and training must be in place to ensure that patient identification linked to glass and digital slides is accurate, maintained, and secure ([last accessed on 4/30/2023]; https://www.cap.org/covid-19/remote-sign-out-faqs).

Role of hospital IT when implementing remote sign-out

The Laboratory Medical Director should work with the IT department to develop HIPAA-compliant policies and procedures for signing out cases from home or other remote locations. Also required are policies and procedures for transcription, billing, and other activities performed from home without adversely impacting patient care. These plans would consist of developing or utilizing already existing secure remote access protocols which are in place in most organizations (e.g., VPN, two-factor authentication). The IT department can help determine if a pathologist's home computer can view WSIs (e.g., are the graphic cards good enough to handle these large image files). Laboratory Medical Directors should gather a list of applications needed for remote sign-out, including access to the healthcare organization's electronic health record, barcode scanners, laboratory data files, laboratory images, digital pathology images, gross photographic images, laboratory policies/procedures, and each applicable laboratory information system. It should ensure all remote personnel have access, have successfully set up these applications, and receive training to use them. Proper documentation, user manuals, and instructions should get provided. Also, Laboratory Medical Directors should work with their IT department to develop or provide instructions for remote access for other aspects of laboratory activities that are safely performable from a remote location, such as transcription, billing, and quality reviews ([last accessed on 4/30/2023]; https://www.cap.org/covid-19/remote-sign-out-faqs).

Validation of digital pathology platforms prior to engaging in remote sign-out, particularly during an emergency

As discussed in the prior section, the CAP Laboratory Accreditation Program (LAP) has only two requirements: documentation of (1) validation has been performed and approved by the Laboratory Medical Director and (2) training of users of the WSI system. It is the requirement for the former to validate a WSI system for diagnostic use that is not removed for implementation during an emergency like the PHE.

Validation/verification is critical to ensure that the data, images, and other materials that a pathologist would use are adequate to render a correct diagnosis or interpretation. However, LAP criteria do not dictate how validation/verification is performed and do not require laboratories to follow the CAP guidelines to the letter. As such, laboratories can use discretion when designing a WSI validation protocol with the aim being to demonstrate a WSI system will perform as expected (i.e., equivalent to glass slides and a microscope) for a specific intended use. During an emergency, however, no guidelines are developed to cover the fast-tracked validation of WSI systems.

Suppose laboratories have previously validated a digital pathology system for intraoperative consultations (frozen sections), consultations, or primary diagnosis. In that case, it is unnecessary (even impractical) to perform full validation studies on every single pathologist's home office, especially during an emergency. Alternatively, these laboratories should carry out verification studies using fewer cases before engaging in live remote sign-out. The number of cases can be determined at the discretion of the Laboratory Medical Director. It is prudent for each pathologist to sign-out remotely to participate in the validation process.

Additional alterations to the CAP guidelines that occurred due to the urgency created by the COVID-19 outbreak and are likely applicable to the next emergency is the 2-week washout period which was impractical. One proposed alternative approach to validate/verify WSI systems urgently is to use cases signed out in the past 2–6 months. Pathologists who initially reported these cases based on glass slides can review the same cases by WSI, removing the need for validation-specific glass slide review and its associated washout period.

For other assays and activities which do not have specific published guidelines, the Laboratory Medical Director for the designated primary CLIA-licensed facility must determine the appropriate level of validation/verification to ensure that the activity can be performed safely and accurately from a remote non-CLIA-licensed location, particularly during an emergency [last accessed on 4/30/2023]; https://www.cap.org/covid-19/remote-sign-out-faqs; [last accessed on 4/30/2023];"Validating Whole Slide Imaging (WSI) for Diagnostic Purposes in Pathology Guideline" https://www.cap.org/protocols-and-guidelines/cap-guidelines/current-cap-guidelines/validating-whole-slide-imaging-for-diagnostic-purposes-in-pathology.

Validating/verifying monitors or laptops from remote locations, particularly if the digital pathology system is already validated on-premises

Fortunately, it is unnecessary to validate each component (e.g., computer hardware, monitor, network, scanner) of the system or the individual steps of the digital imaging process separately. Instead, as discussed in the prior section, a validation/verification study encompasses the entire clinical and reporting pipeline using all components of the digital pathology system as would be used clinically. However, a validated system in the hospital may differ from the remote site. For example, the viewing screen may be a laptop display rather than a medical grade monitor. In addition, the bandwidth at the remote location may be different.

To account for the variabilities between on-premises and remote, one should verify and document that the system's performance at the remote site is acceptable or the minimum required specifications for use. As mentioned above, pathologists should engage their IT personnel to verify that their home computer is adequate for reviewing WSIs (e.g., the graphic cards are sufficient to handle these large image files). Finally, the Laboratory Medical Director establishes the criteria for that verification ([last accessed on 4/30/2023]; https://www.cap.org/covid-19/remote-sign-out-faqs; [last accessed on 4/30/2023]; "Validating Whole Slide Imaging WSI) for Diagnostic Purposes in Pathology Guideline" https://www.cap.org/protocols-and-guidelines/cap-guidelines/current-cap-guidelines/validating-whole-slide-imaging-for-diagnostic-purposes-in-pathology.

A minimal required specifications for hardware and software for use in remote sign-out

There are no definitive established minimum-required specifications for hardware or software, though remote workstations may have lower specifications (e.g., monitor resolution, contrast, brightness; CPU, network bandwidth, graphic cards) than department-configured workstations (may use medical grade monitors, high network bandwidth). To evaluate performed, again required is an adequate validation of the entire digital pathology system. One should use discretion as to whether they have sufficient resources to perform remote sign-out. Likewise, one should refer to vendor-provided minimum specifications to be used as guidelines for remote hardware and software specifications ([last accessed on 4/30/2023]; https://www.cap.org/covid-19/remote-sign-out-faqs).

Role of the Food and Drug Administration on remote whole slide imaging sign-out

A common misconception is that there is a requirement for an FDA-cleared system to perform WSI primary diagnosis including remote sign-out. The reality is that one is free to use any microscope or validated digital pathology system to render a diagnosis. The key is to validate or verify the digital pathology system for its intended clinical use. Likewise, if you validate a non-FDA-cleared system, or verify an FDA-cleared system for a purpose other than its intended use, it would be considered a LDT. With remote sign-out, using properly validated systems at non-CLIA-licensed locations affiliated with a primary site that is licensed by CLIA will enable pathologists to continue to provide the most accurate diagnostic, prognostic, and predictive information to guide quality patient care.

Currently, as discussed early in the chapter, the FDA oversees how vendors market their products, such as their claims and the specific intended use cases for the scanner (primary diagnosis, consultations, frozen sections, etc.). The FDA does not regulate healthcare professionals or medical practice (e.g., pathologists or the practice of pathology.) However, potential changes to this paradigm will get discussed in the next and final section of this chapter.

Medicolegal considerations

Many healthcare organizations allow secured remote access to the healthcare organization's clinical information systems. One is generally covered by the healthcare organization's malpractice insurance when functioning in the capacity of the medical staff of one's organization. It is essential to emphasize the importance of having the Medical Director for the CLIA site establish a policy/SOP for remote sign-out to legitimize the remote sign-out practice from a legal perspective.

Regarding incorrect diagnoses, such errors are always subject to malpractice suits, irrespective of how cases gets reviewed. One should check with your malpractice carrier regarding coverage for remote sign-out and/or reading digital slides or data instead of glass slides or instrument printouts.

The CAP is conservative in following the existing accreditation guidelines for data retention and with the applicable national, federal, state (or provincial), and local laws and regulations. Where specific guidelines for digital/electronic data retention are not specified, use the same retention strategy as would be used for the physical object. For example, the current CAP requirement is to keep glass slides for 10 years after diagnosis gets rendered, so an equivalent approach would be to keep digital slides for the same time period (10 years). But you can save both glass and digital images for less than 10 years. For example, if a diagnosis gets rendered with a digital image, it is acceptable to keep only the glass slide for 10 years and discard the digital images. There is only one scenario where a digital image must get retained: when the glass slide used to produce the digital image no longer exists (e.g., it was destroyed when tissue was scraped off for molecular analysis, including identity testing). Image retention policies should get developed in consultation with your institution's risk management department.

In another example, retention of clinical pathology data must include the original data to support primary results generated and re-analysis for a minimum of 2 years per CLIA. In addition, cytology slide preparations must get retained for at least 5 years from the date of examination; histopathology slides must get retained for at least 10 years from the date of examination; pathology specimen blocks must get retained for at least 2 years from the date of examination; and remnants of tissue for pathology examination must get preserved until a diagnosis gets made on the specimen ([last accessed on 4/30/2023]; https://www.cap.org/covid-19/remote-sign-out-faqs; [Last accessed on 4/30/2023]; https://www.ecfr.gov/current/title-42/chapter-IV/subchapter-G/part-493/subpart-J/section-493.1105).

Diagnostic procedures remotely reportable and billable from a non-CLIA-licensed facility under the waiver

Primary diagnosis in surgical pathology remains the most significant anatomic pathology use case for WSI and is the mainstay for remote primary sign-out. In addition, primary diagnosis in cytology (Pap smears and non-GYN specimens) and hematopathology is possible remotely under the waiver provisions. However, one must acknowledge that interpreting WSI scans in cytology and hematopathology, notably, smears are potentially challenging. Most WSI scanners have a maximum magnification of 40× with no focusing capability unless z-stacking gets performed, making interpreting such cases difficult. Likewise, the 40× magnification and z-stacking exponentially increase the scan times and storage requirements, which lessens the feasibility of scanning hundreds of cytology and hematopathology slides for primary diagnosis. Therefore one must always exercise clinical judgment regarding the quality of the scans and whether these WSIs are adequate for primary diagnosis.

Remote WSI review of special stains, immunohistochemical stains, and other ancillary studies, not under the purview of primary diagnosis but rather aid in the rendering of primary diagnosis, are permissible for remote billable reporting. Likewise, WSI image analysis of quantitative immunohistochemical biomarkers representing more complex testing is permissible for remote billable reporting.

Other anatomic pathology use cases are secondary consultations, permissible for remote billable reporting not under the purview of primary diagnosis. These are ROSE and frozen section interpretations by non-WSI digital pathology technologies like live video microscopy and robotic microscopes. With frozen sections, sometimes small-footprint WSI systems are preferable over robotic live dynamic technologies.

No matter the diagnostic procedure required for remote review with remote billable reporting, the process and application (i.e., remote access to a workstation to view and render a diagnosis for an intraoperative frozen section consultation) gets validated for performing the procedure remotely and there is appropriate documentation. For frozen section validation, each case examined during validation should compare the diagnosis rendered at the time of the frozen section using a traditional microscope against the diagnosis rendered when the same frozen section slide gets reviewed digitally.

Billing

Currently, the availability of Current Procedural Terminology (CPT) codes for digital pathology is fluid. For example, there is no Category I reimbursable digital pathology billing codes for primary diagnosis or remote sign-out beyond the existing Category I codes for the primary service. However, Category I reimbursable digital pathology codes exist for applied computer-assisted quantitative analysis of IHC stains: 88361. Furthermore, these codes are the same for sign-out of these procedures, whether at a CLIA-licensed facility or under the temporary CMS waiver, at a remote location.

In 2022 the CAP has worked with the American Medical Association CPT Editorial Panel to establish 13 new digital pathology Category III add-on codes (Table 8.5) ([last accessed on 4/30/2023]; https://www.ama-assn.org/system/files/may-2022-cpt-summary-panel-actions.pdf; [last accessed on 4/30/2023]; https://www.cap.org/advocacy/payments-for-pathology-services/2023-digital-pathology-codes).

These new digital pathology Category III add-on codes get reported in addition to the appropriate Category I primary service for primary diagnosis or test interpretation code (e.g., 88305, 88307, 81479, 88187, 88360, etc.) when the digitization of glass slides is performed and reported in conjunction with the Category I code for the primary service.

Table 8.5 Use digital pathology CPTs (0751T–0763T) in addition to the appropriate Category I service codes (88302–88360).

Category III digital pathology CPT codes	Category I primary service CPT codes	Description of primary service
0751T	88302	Level II, surgical pathology
0752T	88304	Level III, surgical pathology
0753T	88305	Level IV, surgical pathology
0754T	88307	Level V, surgical pathology
0755T	88309	Level VI, surgical pathology
0756T	88312	Special stain, group I, for microorganisms
0757T	88313	Special stain, group II, for other
0758T	88314	Histochemical stain on frozen tissue block
0759T	88319	Special stain, group III, for enzyme constituents
0760T	88342	IHC or immunocytochemistry (ICC), initial single antibody stain
0761T	88341	IHC or ICC, ex. additional single antibody stain
0762T	88344	IHC or ICC, ex. multiplex antibody stain
0763T	88360	Quantitative tumor IHC, ex. single antibody, manual

Table 8.6 When not to use the digital pathology CPTs.

Solely for archival purposes
Solely for educational purposes
Solely for developing a database for training or validation of AI algorithms
Solely for clinical conference presentations
For secondary diagnosis (i.e., 88321, 88323, 88325)
For frozen sections only (i.e., 88331, 88332)
For cytopathology

These new Category III codes are temporary for emerging technologies, services, and procedures. These new codes aim to capture and report additional clinical staff work and service requirements associated with digitizing glass microscope slides for primary diagnosis. Digitization of glass microscope slides enables remote examination by the pathologist and/or in conjunction with using artificial intelligence (AI) algorithms (Table 8.6).

As a result of CAP advocacy, the new codes will help pathologists, pathology practices, and laboratories providing digital pathology digitization procedures report these services. Therefore reporting these new Category III add-on digital pathology codes is vital to demonstrate the widespread use of the digital pathology digitization procedures to ensure that the codes get considered for future Category I status CPT codes with established payment.

The new digital pathology codes have been in effect since January 1, 2023, and demonstrated use will encourage the promotion of these Category III codes toward Category I codes.

Legislation potentially affecting digital pathology regulation

As defined by the FDA, LDTs are a type of in vitro diagnostic test designed, manufactured, and used within a single laboratory. While the uses of an LDT are often the same as the uses of FDA-cleared or approved in vitro diagnostic tests, many labs, especially academic labs, offer their tests as LDTs ([last accessed on 4/30/2023]; https://www.fda.gov/medical-devices/in-vitro-diagnostics/laboratory-developed-tests). Primary WSI sign-out is considered high-complexity testing with higher risk and has been implemented as an LDT (Hanna et al., 2020). In 2010 the FDA announced its intent to reconsider its policy of enforcement discretion for LDTs while strengthening its position of oversight of high-complexity high-risk testing ([last accessed on 4/30/2023]; https://www.fda.gov/medical-devices/in-vitro-diagnostics/laboratory-developed-tests).

The **V**erifying **A**ccurate **L**eading edge **I**VCT, and **D**evelopment (VALID) Act is an ACT that reflects this reconsideration of policy enforcement discretion for LDTs. This Act would give the FDA regulatory oversight of LDTs currently overseen by the CMS. It would also create a new category—in vitro clinical test.

What has happened so far with VALID?

Industry served as a critical driving partner with the FDA with VALID. During 2021 and 2022, the FDA and industry partners collaborated on VALID, tying it to various other legislation, all unsuccessfully. In 2022 VALID was cut twice from legislative packages, primarily due to concern by academic medical centers about VALID hampering their ability to deliver and develop new tests.

The FDA remains undiscouraged. Andi Fristedt, the FDA Deputy Commissioner for Policy, Legislation, and International Affairs, said, in a January 13, 2023 webinar hosted by the Alliance for a Stronger FDA, that “priorities that we didn’t see advance will remain important priorities for the agency” (Hanna et al., 2020).

Food and Drug Administration moving forward with rulemaking on diagnostic testing regulation

The overhauling of the FDA's oversight of in vitro diagnostics will see significant movement in 2023. VALID Act is one of **FDA's top legislative priorities** for Pandemic and All Hazards Preparedness Act reauthorization. Though the leading Senate champion Richard Burr from North Carolina retired in 2023, Representatives Larry Bucshon and Diana DeGette reintroduced VALID in late March 2023.

Despite the reinvigoration, the future of VALID still needs to be determined. There is intent for current versions of VALID to overhaul the most recent prior versions of the bill to help developers react more quickly and allow approval of one representative test from diagnostic categories versus every test within usual categories. Academic medical centers are still willing to **work on a policy agreement** and having more time to work on policy is a benefit, says Heather Pierce, Senior Director for Science Policy and Regulatory Counsel for the Association of American Medical Colleges.

The current VALID version resembles the version that came from the Senate Health, Education, Labor, and Pensions Committee in 2022, with some changes based on omnibus discussions. Interestingly, at the time of this publication, provisions have not included a carve-out for academic medical centers ([last accessed on 4/30/2023]; https://friendsofcancerresearch.org/news/diagnostics-world-news-current-perspectives-on-the-valid-act/).

An editorial on VALID and the feature of Food and Drug Administration overreach

Several well-intentioned patient advocacy groups and industry stakeholders have joined the FDA in championing VALID. Fears from labs, notably academic, however, traditionally considered good testing stewards, fear the unintended consequences of sweeping changes imposed through VALID. Compounding is November 7, 2023, which marks the end of the PHE transition period with possible rollbacks of flexibilities allowed by the FDA during the PHE under FDA-2020-D-1138.

Considering both VALID and the end of FDA-2020-D-1138 with loss in the flexibility in FDA-cleared indications, functionality, hardware, and/or software of device-type components of the "pixel pipeline" to provide for use in a remote setting, possible unanswered questions are:

- Given the historically technical focus of the FDA, will there be requirements in assessing the impact of potential technical issues that may affect the quality of digital slides, such as image compression, file format, storage capacity, and scanning errors, metrics usually captured by vendors already submitting to the FDA, but challenging for individual laboratories to capture due to no available vendor ready solutions?
- How does one's lab implement quality assurance and maintenance procedures to ensure the digital slides remain accurate and reliable

over time; trends which the FDA voiced focus on in future evaluations?
- What are the impacts on advancements in emerging digital pathology technologies and decoupling of components of the WSI pixel pipeline like monitors, software, and hardware, given the rollbacks of FDA-2020-D-1138 after November 7, 2023?
- What are impacts on remote WSI sign-out with components different from those of FDA-cleared pixel pipelines?
- Will labs get impractically constrained (time, cost, effort) into specific cost-prohibitive legacy monitors, software, and hardware; despite cheaper, more evolved, and better performing hardware and software?
- What is the road to clearance and deployment for individual labs that develop their own disruptive technologies that work effectively for their uses?
- What does VALID mean to regulation of pathology AI, which is heavily dependent on upstream digital pathology technologies?
- Is the FDA regulating the practice of medicine through VALID, which was previously not under the FDA's purview?

Concluding thoughts on legislation regulating laboratory-developed tests

LDTs, under a CMS regulatory framework, will remain controversial. Likewise, regulation of high-complexity testing is a top priority for the well-intentioned patient advocacy groups, FDA, and industry champions. Whether VALID is passed or not, legislative rules will come, and it is vital to continually monitor, remain aware, and stay actively informed of the legislative climate. No matter where one stands with FDA oversight over LDTs, consequences will affect medical practice.

References

Abels, E., & Pantanowitz, L. (2017). Current state of the regulatory trajectory for whole slide imaging devices in the USA. *Journal of Pathology Informatics*, *8*(23). Available from https://doi.org/10.4103/jpi.jpi_11_17.

Al-Janabi, S., Huisman, A., Vink, A., Leguit, R. J., Offerhaus, G. J., Ten Kate, F. J., van Dijk, M. R., & van Diest, P. J. (2012). Whole slide images for primary diagnostics in dermatopathology: A feasibility study. *Journal of Clinical Pathology*, *65*(2), 152–158. Available from https://doi.org/10.1136/jclinpath-2011-200277.

Centers for Medicare & Medicaid Services Memorandum, Clinical Laboratory Improvement Amendments (CLIA) Laboratory Guidance During COVID-19 Public Health Emergency [last accessed on 4.25.23], March 26, 2020. Available from https://www.cms.gov/medicare-provider-enrollment-and-certificationsurveycertificationgeninfopolicy-and-memos-states-and/clinical-laboratory-improvement-amendments-clia-laboratory-guidance-during-covid-19-public-health.

College of American Pathologists. CAP Laboratory Accreditation Checklists. Accessed April, 2023. https://www.cap.org/laboratory-improvement/accreditation/accreditation-checklists.

Evans, A. J., Brown, R. W., Bui, M. M., Chlipala, E. A., Lacchetti, C., Milner, D. A., Pantanowitz, L., Parwani, A. V., Reid, K., Riben, M. W., Reuter, V. E., Stephens, L., Stewart, R. L., & Thomas, N. E. (2022). Validating whole slide imaging systems for diagnostic purposes in pathology. *Archives of Pathology & Laboratory Medicine*, *146* (4), 440–450. Available from https://doi.org/10.5858/arpa.2020-0723-CP.

FDA Technical Performance Assessment of Digital Pathology Whole Slide Imaging Devices. [last accessed on 4/24/2023]. Available from https://www.fda.gov/regulatory-information/search-fda-guidance-documents/technical-performance-assessment-digital-pathology-whole-slide-imaging-devices.

Guyatt, G. H., Alonso-Coello, P., Schünemann, H. J., Djulbegovic, B., Nothacker, M., Lange, S., Murad, M. H., & Akl, E. A. (2016). Guideline panels should seldom make good practice statements: Guidance from the GRADE Working Group. *Journal of Clinical Epidemiology*, *80*, 3–7. Available from https://doi.org/10.1016/j.jclinepi.2016.07.006.

Hanna, M. G., Reuter, V. E., Ardon, O., Kim, D., Sirintrapun, S. J., Schüffler, P. J., Busam, K. J., Sauter, J. L., Brogi, E., Tan, L. K., Xu, B., Bale, T., Agaram, N. P., Tang, L. H., Ellenson, L. H., Philip, J., Corsale, L., Stamelos, E., Friedlander, M. A., ... Hameed, M. (2020). Validation of a digital pathology system including remote review during the COVID-19 pandemic. *Modern Pathology*, *33*(11), 2115–2127. Available from https://doi.org/10.1038/s41379-020-0601-5.

Koch, L. H., Lampros, J. N., Delong, L. K., Chen, S. C., Woosley, J. T., & Hood, A. F. (2009). Randomized comparison of virtual microscopy and traditional glass microscopy in diagnostic accuracy among dermatology and pathology residents. *Human Pathology*, *40*(5), 662–667. Available from https://doi.org/10.1016/j.humpath.2008.10.009.

Pantanowitz, L., Sinard, J. H., Henricks, W. H., Fatheree, L. A., Carter, A. B., Contis, L., Beckwith, B. A., Evans, A. J., Lal, A., & Parwani, A. V. (2013). College of American Pathologists Pathology and Laboratory Quality Center. Validating whole slide imaging for diagnostic purposes in pathology: Guideline from the College of American Pathologists Pathology and Laboratory Quality Center. *Archives of Pathology & Laboratory Medicine*, *137*(12), 1710–1722.

Stathonikos, N., Veta, M., Huisman, A., & van Diest, P. J. (2013). Going fully digital: Perspective of a Dutch academic pathology lab. *Journal of Pathology Informatics*, *4*, 15. Available from https://doi.org/10.4103/2153-3539.114206.

Têtu, B., & Evans, A. (2014). Canadian licensure for the use of digital pathology for routine diagnoses: One more step toward a new era of pathology practice without borders. *Archives of Pathology & Laboratory Medicine*, *138*(3), 302–304. Available from https://doi.org/10.5858/arpa.2013-0289-ED.

Thorstenson, S., Molin, J., & Lundström, C. (2014). Implementation of large-scale routine diagnostics using whole slide imaging in Sweden: Digital pathology experiences 2006-2013. *Journal of Pathology Informatics*, *5*(1), 14. Available from https://doi.org/10.4103/2153-3539.129452.

Tugwell, P., & Knottnerus, J. A. (2015). When does a good practice statement not justify an evidence based guideline? *Journal of Clinical Epidemiology*, *68*(5), 477–479. Available from https://doi.org/10.1016/j.jclinepi.2015.03.004.

CHAPTER

Standards in digital pathology

9

Rajesh C. Dash and William R. Jeck

Department of Pathology, Duke University Health System, Durham, NC, United States

Standards and interoperability in digital pathology

This chapter will review healthcare informatics standards in general and specific standards in support of digital pathology. It is important to recognize the definition of a standard and the types of standards within digital pathology specifically and in the field of clinical informatics more generally. Standards typically address a specific set of needs within an information domain, with digital pathology serving as an excellent example. A single standard may be designed to operate across multiple different domains of knowledge and processes, but this increases complexity and the cost of development of the standard. More typically, a standard will have a focused scope to solve a specific set of problems and will interact with other standards for addressing other areas. Within a healthcare organization, many standards interact together to holistically manage clinical data and associated processes and create infrastructure that is interoperable.

What is a standard and why is it important?

A standard represents a concept to normalize information in a way that reduces ambiguity, cost, and complexity of using that information. For example, a standard exists for describing the size of an automotive tire that benefits both consumers and tire manufacturers. It allows consumers to select the correct tire for their vehicle, recognizing that the standard stipulates (among other things) that the height, width, and diameter are documented in a specific order and with specific units of measure. The consumer does not have to be concerned about the proprietary idiosyncrasies of a given tire manufacturer and potential incompatibilities with their vehicle. The standard also allows tire manufacturers to create products that target a specific set of needs in the market and communicate to the market concisely and accurately. Standards help support vertical markets where multiple vendors can produce products that address a specific need in the marketplace, relying on other vendors to fulfill other needs in a way that complements

Digital Pathology. DOI: https://doi.org/10.1016/B978-0-443-13809-6.00012-9

their own offering to create a complete solution. Extending the tire analogy, standards for tire sizing help wheel and brake manufacturers create components that will be compatible (interoperable) with all manufacturers of tires for given sizes specified in the standard.

Standards in the healthcare setting take on some additional characteristics. According to the Healthcare Information and Management Systems Society (HIMSS), standards provide a common language and a common set of expectations that enable interoperability between systems and/or devices (Healthcare Information and Management Systems Society). They help systems seamlessly digest information about an individual and improve the overall coordination and delivery of healthcare. Standards support data sharing among various healthcare providers and patients regardless of application or market supplier. Data sharing is often referred to as interoperability. HIMSS defines interoperability as the ability of different information systems, devices, and applications (systems) to access, exchange, integrate, and cooperatively use data in a coordinated manner, within and across organizational boundaries, to provide timely and seamless portability of information and optimize the health of individuals and populations globally (Healthcare Information and Management Systems Society).

Who develops standards?

A standards development organization (SDO) is focused on developing, maintaining, and providing guidance on implementing standards that are intended to address the needs of a relatively wide base of stakeholders. To be recognized as a healthcare SDO, an organization is usually accredited by the American National Standards Institute (ANSI) or the International Organization for Standardization (ISO). There are over 40 different SDOs in the healthcare IT arena. Some entities create standards, such as Health Level Seven (HL7), Systematized Nomenclature of Medicine (SNOMED) International, and Digital Imaging and Communication in Medicine (DICOM). Others, like Integrating the Healthcare Enterprise (IHE), do not develop new standards but rather attempt to support real-world interoperability by describing workflow needs and challenges (i.e., "use cases") and how existing standards might be best leveraged or modified to support them. This creates a complementary effort that drives adoption of standards by providing guidance on how multiple standards can be used together to support interoperable health information exchange.

Types of standards in healthcare that support digital pathology

Standards can be categorized to better understand their role in health information exchange. The following categories are not exhaustive, and a single standard may span different categories, but this kind of schema can be helpful for thinking about the roles that different standards play.

1. Vocabulary/terminology/ontology standards. These standards address the ability to represent concepts in an unambiguous manner between a sender and receiver of information. Health information systems can communicate the meaning of transmitted data using structured vocabularies, terminologies, code sets, and classification systems. Some common vocabulary standards currently in use include:
   a. Current procedural terminology (CPT) is used for clinical billing of procedures; developed and maintained by the American Medical Association. The nomenclature is used across the country to report medical, surgical, radiology, laboratory, anesthesiology, genomic sequencing, evaluation, and management (E/M) services under public and private health insurance programs (American Medical Association, CPT).
   b. The international statistical classification of diseases and related health problems (ICD) is a medical classification list by the World Health Organization (WHO). It contains codes for diseases, signs and symptoms, abnormal findings, complaints, social circumstances, and external causes of injury or diseases (World Health Organization, International).
   c. Logical observation identifier names and codes (LOINC) is a universal code system for identifying health measurements, observations, and documents. These codes represent the "question" for a test or measurement. LOINC codes can be grouped into laboratory and clinical tests, measurement, and observations (Regenstrief Institute, Inc., Logical observation identifiers names and codes).
   d. Systematized nomenclature of medicine-clinical terms (SNOMED CT) represents a comprehensive clinical ontology that spans much of medicine. It enables the consistent, processable representation of clinical content in electronic health records (EHRs) and related systems (SNOMED International, Systematized).
   e. The unified code for units of measure (UCUM) is a code system to represent units of measure used in international science, engineering, and business to facilitate unambiguous electronic communication of quantities together with their units (Regenstrief Institute, Inc., Unified code for units of measure).
2. Messaging and transport standards. Transport standards define how a message is transmitted over local or wide-area network infrastructure. Messaging standards convey the way in which data is packaged so that sender and receiver can properly access the contents of the message. These standards primarily define the structure and organization of electronic messages. When it is possible to predict in advance, these standards can also attempt to define common sets of data content for specific message fields to facilitate reuse and interoperability.

a. HL7's Version 2.x (V2): A widely implemented messaging standard that allows the exchange of clinical data between systems. It is designed to support a central patient care system as well as a more distributed environment where data resides in departmental systems (Health Level Seven International, HL7 version 2 product suite).
b. HL7 Version 3 Clinical Document Architecture (CDA): An XML-based document markup standard that specifies the structure and semantics of "clinical documents" for the purpose of exchange between healthcare providers and patients. It defines a clinical document as having the following six characteristics: persistence, stewardship, potential for authentication, context, wholeness, and human readability (Health Level Seven International HL7 version 3 product suite).
c. Digital Imaging and Communications in Medicine (DICOM): The standard for the communication and management of medical imaging information and related data. DICOM enables the transfer of medical images across systems and facilitates the development and expansion of picture archiving and communication systems (The National Electrical Manufacturers Association, Digital imaging and communications in medicine).
d. Fast Healthcare Interoperability Resources (FHIR): An HL7 standard for exchanging healthcare information electronically. The basic building blocks of FHIR are "resources," which describe exchangeable health data formats and elements. FHIR also provides standardization for application programming interfaces (APIs). FHIR provides a number of benefits and improvements as a healthcare standard built upon communication protocols that serve as the foundation for the Internet (Health Level Seven International, HL7 fast healthcare interoperability resources).

3. Privacy and security standards. Privacy standards aim to protect an individual's (or organization's) right to determine whether, what, when, by whom, and for what purpose their personal health information is collected, accessed, used, or disclosed. Security standards define a set of administrative, physical, and technical actions to protect the confidentiality, availability, and integrity of health information.
   a. In the U.S., the Health Insurance Portability and Accountability Act (HIPAA) outlines standards that safeguard the privacy and security of protected health information. The HIPAA privacy rule establishes national standards to protect individuals' medical records and other personal health information. The rule sets limits and conditions on the uses and disclosures of information without patient authorization. The rule also gives patients more control over their own health information, including the right to examine and obtain a copy of their records and to request corrections. The HIPAA security rule sets national standards for protecting the confidentiality, integrity, and availability of electronically protected health information. The rule addresses the technical and nontechnical safeguards that healthcare organizations must have in place to secure an individual's

electronic protected health information (United States Department of Health and Human Services, Summary of the HIPAA privacy rule).

   b. In Europe, the General Data Protection Regulation (GDPR) outlines similar privacy and security regulations for all processing and storage of data relating to people in the European Union (EU) (European Union, 2016).

4. Identifier standards. Entities use identifier standards to uniquely identify patients, providers, specimen containers, blood products, or other trackable actors or assets within the healthcare ecosystem.
   a. National Provider ID (NPI): A unique 10-digit number to uniquely identify a healthcare provider in the United States. These NPIs are included in a freely accessible NPI Registry (United States Department of Health and Human Services, National plan & provider enumeration system).
   b. Object ID (OID): A globally unique ISO identifier and a preferred scheme for unique identifiers in HL7 (Health Level Seven International, Introduction for the HL7 object identifier (OID) registry).
   c. Specimen label identifiers (AUTO02): describes a barcode labeling standard for specimen container identification for use by laboratory automation (Clinical Laboratory Standards Institute, AUTO02).

Standards for digital pathology in more detail

The primary digital imaging standard in pathology (and every other specialty in medicine) has been and continues to be Digital Imaging Communication of Medicine (The National Electrical Manufacturers Association, Digital imaging and communications in medicine). The DICOM working group focused on digital pathology today is entitled "Working Group 26" (WG26) (The National Electrical Manufacturers Association, DICOM WG-26: Pathology). WG26 scope is to support and develop the DICOM standard in support of digital pathology workflows. WG26 has produced three supplements to augment the DICOM base standard in support of digital pathology (DICOM supplements 122, 145, and 222). To understand the purpose and nature of these supplements, some basic DICOM terminology and concepts should be clarified.

DICOM provides sufficiently detailed technical information to enable imaging instruments from different vendors to work together on an electronic network (via a local intranet and/or Internet). At the time of its inception, both HL7 and the European Standard CEN/TC 251/PT3−033 were deemed insufficient in providing a complete specification for communication of imaging. DICOM provides a protocol for message exchange that is compatible with common Internet standards (TCP/IP) and aims to provide a complete specification for image exchange. It does so by providing two groups of "services," termed "composite" and "normalized." The term composites refers to data derived from more than one real-world entity, while normalized refers to data originating from a single real-world entity. DICOM combines a service with an information object to create service–object pairs (SOP), a class of data that abstracts "how" and "what" information is transmitted between two systems. DICOM uses what are called Information Object Definitions (IODs) to abstractly specify information about real-world physical objects. It does not represent a specific instance but is a descriptor for a class of objects that share the same properties. A DICOM service element (DIMSE) is a lower-level (more technical) description of the message exchange. DICOM SOP classes specify the set of IODs and DIMSE services that are useful for a given purpose (i.e., use case) (Bidgood et al., 1997).

As a predecessor to WG26, working group 13 (WG13) produced supplement 15 as the DICOM standard targeting visible light photographic image for gross specimens and visible light slide-coordinate microscopic image for slide-based microscopic imaging (The National Electrical Manufacturers Association, DICOM supplement 15 final text). That supplement indicates that the American Society for Gastrointestinal Endoscopy and the College of American Pathologists (CAP) were early participants in the development of the standard but were joined by many other specialty societies, including those from radiology, dental, ophthalmology, dermatology, surgery, urology, and pulmonology. The supplement to the DICOM standard was developed with support and in accordance with procedures stipulated by the National Electrical Manufacturers Association (NEMA). The final text from WG13 was published in 1999. This supplement includes the single frame image and modality IODs for visible light and corresponding SOP classes for media storage and networking. It does so by defining subclasses of a generic IOD for visible light for each of four modalities: endoscopy, microscopy, automated stage microscopy, and photography. The scope of the IODs is to support visible light images produced by endoscopes, microscopes, and photographic cameras.

Supplement 122 to the DICOM standard introduces a new mechanism of pathology specimen identification and revisions to composite IODs to use the new specimen identification mechanism (The National Electrical Manufacturers Association, DICOM supplement 122 final text). It specifies the use of specimen identification attributes in the modality worklist (the "to-do list" organization paradigm in DICOM) and procedure step classes to support imaging workflows needed in the pathology department, recognizing that multiple images may be created for a single specimen. Existing IODs previously defined by DICOM for light microscopy (supplement 15 described above) did not have a strong mechanism for describing the specimen being imaged or associating a particular specimen with a particular image. While the relationship between patient and image is straightforward in other imaging fields and accurately captured by DICOM objects, in the field of anatomic pathology there was a need for a new specimen module to formally define specimen attributes at the image level (i.e., a digital specimen with associated metadata). The specimen attributes allow the processing history of the specimen to be encoded. This is rarely implemented today, but it may become a more important consideration as machine learning models require normalization of image data to create more stable and generalizable solutions. This supplement also augmented the standard image query service so that an information system application can obtain a list of images created for each specimen. Previously, image lists could only be obtained by patient, accession, or study.

Supplement 145 to the DICOM standard introduces the capability to efficiently support extremely large two-dimensional images needed for digital pathology WSIs (The National Electrical Manufacturers Association, DICOM supplement 145 final text). It also adds support for image subregions as "tiles" and support for multiple images at varying resolutions. There is accommodation for a third dimension with support for multiple stacked two-dimensional images that are at different focal depths, or "*z*-planes." These few modifications transform the DICOM base standard into one ideally suited for managing very large digital pathology WSIs. There is also significant addition of metadata to support WSIs (Singh et al., 2011). The supplement adds 56 new data elements, 36 new coded concepts, seven revised structured data templates, and 18 new defined value sets to the DICOM standard. This metadata enhancement provides significant improvement in support of interoperability of DICOM image data. Describing a typical real-world example and making some quick calculations can be helpful in clarifying the benefits afforded and scope of data managed by this supplement.

A standard microscope glass slide measures 75 × 25 mm, with a portion at one end usually reserved for human and machine-readable identifying information. This effectively leaves approximately 50 × 25 mm for placement of a tissue specimen. The thickness of tissue typically sandwiched between a cover slip and a glass slide ranges from 5–10 microns (0.005–0.01 mm) for surgical pathology but will be thicker and more variable for cytopathology specimens (10–30 microns or more). Typical digital pathology whole slide imaging scanners capture digital images at roughly the optical equivalence of a 40 × objective of a microscope (to the human eye, 400 ×, with the typical 10 × objective in series). For a typical digital scanner, this may equate to a resolution of 0.25 micrometers per pixel (mpp). A 50 × 25 mm image captured at 0.25 mpp equates to an image containing 200,000 × 100,000 pixels. These pixels are generally broken down into a grid of tiles, where each tile represents a much smaller subregion of the entire large image. This is necessary for digital pathology as DICOM image IOD pixel coordinates are limited to 16-bit unsigned integers, which have a maximum value of 65,535. This value represents the maximum number of rows and columns in a 2-D image. Assuming only a single z-stack and a typical 24-bit color depth (8 bits for each color channel of red, green, and blue as defined in prior imaging standards) for visible light, a single base-level image could consume over 55 gigabytes (GB) of storage. This is further increased if multiple "zoom" levels are stored simultaneously to prevent the delay associated with creating the desired zoom level on the fly. However, there are some remedies provided by the supplement to address these concerns.

There are two main factors to consider given the very large image size and data requirements. The first is the ability to transfer data quickly for viewing on a display monitor. Network bandwidth in most healthcare facilities is limited to 1 Gbps (gigabits per second) or in some cases 10 Gbps. For a 1 Gbps network backbone, the transfer rate is a theoretical maximum of 1/8 of a GB per second (as there are 8 bits in a byte). For a 55 GB image, this would equate to a delay of over 7 minutes for a single image to be transferred over the network. Even for a very fast network backbone of 10 Gbps, a single image would take over 40 seconds to transfer. However, the entire image is not needed in nearly all circumstances. A 1080p (high definition) monitor displays a little over 2 million pixels, and a 4K display monitor displays a little over 8 million pixels. With three bytes per pixel typical (one for each color channel), that equates to a rough maximum of 24 megabytes (MB, millions of bytes) needed at any given time for display. With the tiling functionality introduced in Supplement 145, only those tiles needed for the current display subregion need to be transferred over the network, not the entire 55 GB of base image data. Furthermore, at selected digital magnifications less than the maximum, a precomputed lower resolution image can be sent (called "down sampling"). In general, accessing a previously stored-down sampled image improves display performance for the end user in certain situations. The supplement's ability to precompute and store multiple magnifications of a base image from a low-resolution "thumbnail" all the way up to the fully detailed base image is often referred to as a "pyramid" of image data (Fig. 9.1). This allows for sending only those parts of the image needed for display without wasting network bandwidth on parts of the image that are not being viewed. The ability to precompute various common magnification levels can serve to minimize use of computing bandwidth at the time of image display. DICOM can also organize and encode the tiled pixel matrix in full or sparsely. In the latter, the

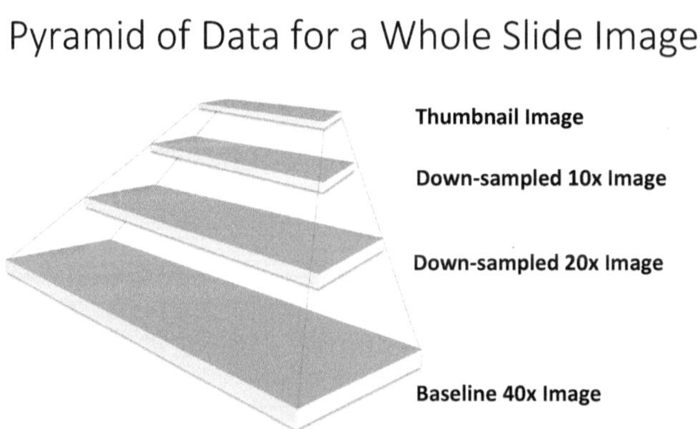

FIGURE 9.1

Pyramid of image data at various resolutions for a whole slide image modeled within DICOM.

Adapted from Supplement 145, Whole Slide Microscopic Image IOD and SOP Classes, Fig. 9.4.

encoding wastes little to no storage space on parts of an image that convey little value (e.g., if there is no tissue on certain parts of a glass slide). There is one final aspect to consider that assists with both network bandwidth management and data storage management: compression of the image. Lossless compression of the image can yield a three to five times reduction in size, whereas lossy compression can yield an order of magnitude better reduction. The most common compression used in digital pathology WSI applications is lossy compression provided by JPEG or JPEG2000, as these can result in up to a 50 × reduction in storage size with no "significant" loss of information detail (Herrmann et al., 2018; Singh et al., 2011). Data can remain compressed over the network and be uncompressed at the viewer, maximizing throughput of display in situations where compute power at a local workstation is sufficient to manage decompression of image data in real time. The concept of what represents "significant" loss of information detail has been studied extensively, with most studies recognizing that a pathologist remains able to render accurate diagnoses with even significant compression applied (Helin et al., 2018).

Four basic types of annotation capabilities that are part of the DICOM standard are described as particularly relevant for WSIs in Supplement 145 (The National Electrical Manufacturers Association, DICOM supplement 145 final text). Presentation states (described in additional supplements, such as 120, but also others) convey metadata such as display area, rotation, zoom, coloring (i.e., grayscale vs various color profiles), and basic region of interest (ROI) graphic indicators (i.e., rectangles, arrows, ellipses, crosshairs, dashed lines, etc.) with associated text labels. Presentation state annotations can apply to multiple tiles (e.g., all the focal *z*-planes at the same location). These annotations are generally for human use only. The textual information does not use a controlled or coded vocabulary and conveys no meaning, machine understanding, or relationship to other annotations.

Segmentation provides a mechanism for categorization of areas of an image typically rendered as an overlay. It usually represents an image derived from the original to indicate the presence of a specific feature. For example, a segmentation may identify all the nuclei in a WSI with a 1 bit per pixel compact representation. This is typically referred to as an "image mask." Alternatively, encoding can be one byte per pixel to allow for gradient visualization using an overlay. In conjunction with presentation state functionality, this can be used to generate a color overlay with variable opacity. For example, this can be used to provide a visualization of how likely a particular nucleus is to be cancerous.

Structured reporting captures observations such as measurements and analytics typically required for image analysis. It represents the standard for exchange of structured clinical data observations within images but does not convey how that data is to be presented. Structured reporting is defined within the main DICOM standard in parts 3.3 and 3.5 (where part 3 of the base DICOM standard conveys the IODs), with templates used to constrain reports for specific use cases defined in DICOM part 3.16. Unlike presentation states, a controlled vocabulary such as

SNOMED can be used as "triplet code," described later. For digital pathology, this provides a mechanism for attaching semantics (meaning) to annotations, such as an indication of metastatic tumor cells within an image of a lymph node, which can then later be queried and reliably identified.

Finally, real-world mapping conveys a mapping of pixel values into real-world defined units. For example, pixel values can be mapped to activity, concentration, or other physical measurements. IODs allow for conversion from pixel values to a real-world physical measurement through a lookup table or linear equation. In digital pathology, this functionality might be used to represent quantification of biomarker expression based on color values of pixels.

All the aforementioned annotation mechanisms are designed for manual creation and lack the ability to accommodate the numerous annotations that may be generated by a machine algorithm. This latter use case has become prevalent with more robust image analysis tools and machine learning algorithms being applied to whole slide images, resulting in thousands of machine-drawn polygons of different colors to label different cell types within a WSI. This limitation is addressed in Supplement 222.

Supplement 222 to the DICOM standard specifies a new IOD and storage SOP class for storing what is referred to as "Microscopy Bulk Simple Annotations" (MBSA) such as points, open polylines, closed polygons, and simple geometric shapes such as a rectangle and ellipse (The National Electrical Manufacturers Association, DICOM supplement 222 final text). These annotations can be grouped and sequenced but do not have further defining relationships, such as one group being embedded in another group as might be required for a Venn Diagram with associated Boolean logic. However, this latter capability is present elsewhere in the DICOM standard to support clinical radiotherapy planning. Those capabilities may be re-used, but this is out of scope at the present time for WSIs. This supplement does call out that MBSA are usually created by machine algorithms operating on high-resolution WSIs.

The DICOM standard was founded in the domain of radiology imaging. The data model in pathology is managed within laboratory information systems (LISs) and is strongly specimen-centric (Fig. 9.2). This is quite different from radiology and requires some adaptation of the original DICOM standard. Herrmann et al., in their 2018 paper on implementing the DICOM standard, demonstrated that DICOM can be a practical format for digital pathology and, with recent advancements, allows for efficient interoperable access to image data as well as associated metadata (Herrmann et al., 2018). A few concepts should be called out and explained in more detail. Data records are broken out into different information entities designated, for example, as "patient," "study," "series," and "instance or image." These are founded in radiology technology (Fig. 9.3), and some of the concepts (namely the 2nd and 3rd) do not apply well to pathology without some translation. A radiology "study" is best thought of as a pathology "case." This information entity has an accession number to uniquely identify the study or case as part of their metadata. In this situation, the accession number is equivalent

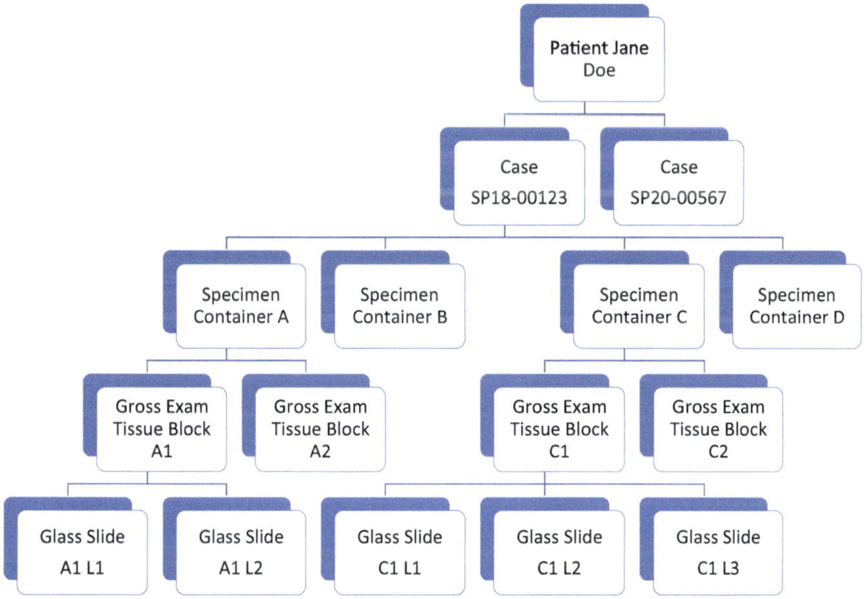

FIGURE 9.2

Anatomic pathology data model typically supported within laboratory information systems.

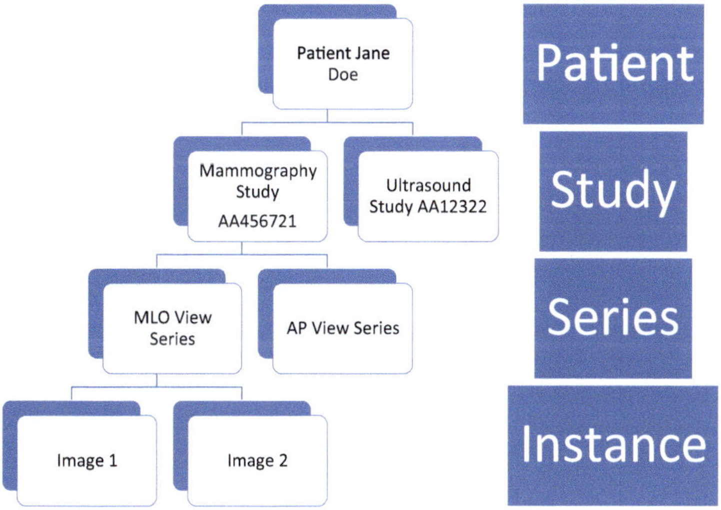

FIGURE 9.3

Digital imaging and communication in medicine data model developed originally for radiology imaging.

between radiology and pathology. A "series" is best thought of as digital representations of a single glass slide. Even a single digital WSI will contain multiple images comprised of tiles stored in a hierarchy of different magnifications. There may be multiple series associated with a study, as many glass slides may be associated with a pathology case. Additional series may be added to the study also, for example, if a re-scan is needed due to issues with the first scan. Each series will contain several images and associated metadata elements associated with a single WSI, such as the base image layer, downsampled lower magnification images, an image of the label, etc. Macroscopic images in pathology also need to be incorporated into the model. Fig. 9.4 represents a proposed adaptation of the data model to incorporate pathology data elements. A DICOM series is limited to data created by a single modality/piece of equipment in a single procedure step. Thus, other imaging data associated with a single WSI, such as annotations of a WSI, must be stored as a separate series. It should be noted that DICOM supports different series being encoded in the same electronic file or different electronic files. Supplement 122, introduced earlier, defines the information entities used to describe containers and specimens. The glass slide in the DICOM model is considered a "container" for a thin slice of tissue laid on top, the "specimen."

A final footnote on the DICOM standard is that network storage and retrieval has been extended in recent years to take advantage of internet-based communications with hypertext transfer protocol (http)-based resources and transactions. DICOM PS3.18, also known as DICOMweb, takes advantage of what are known as "RESTful" web services to support web-based imaging communication (The

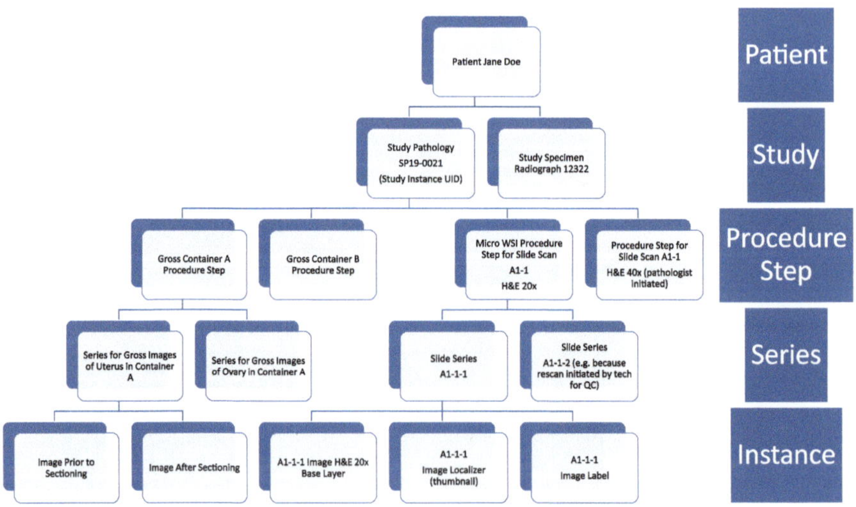

FIGURE 9.4

Proposed adaptation of the digital imaging and communication in medicine data model for digital pathology.

National Electrical Manufacturers Association, DICOM PS3.18 web services). The transactions are called QIDO-RS for querying, WADO-RS for retrieval, and STOW-RS for storage.

DICOM relies on a few other standards to support digital pathology interoperability. A key standard is the ontology called SNOMED CT (SNOMED International, Systematized). DICOM introduced a mechanism to express meaning (semantics) associated with image data through use of a subset of SNOMED terms, or microglossary. The purpose was to limit the domain of value sets to enhance interoperability and provide an efficient mechanism for embedding clinical knowledge into digital image communication. This microglossary is called the SNOMED-DICOM microglossary (SDM) (Bidgood, 1998). Coded information is conveyed in a DICOM message as an ordered set of data elements, each of which consists of three components: a code value, a coding scheme descriptor, and a code meaning. For each of these components, a triplet of three message elements is constructed, consisting of a data element tag, the data element length, and the value. Thus, nine total message components are generated for each coded entry. The DICOM triplet encoding is a flexible mechanism to support machine-interpretable representation of concepts using any coding scheme (vocabulary). The SDM database supports version control and contains four types of tables: a single context group summary table, multiple context group tables, a single template summary table, and many template tables. The context group summary table contains one entry for each context group in the SDM and describes the concept types. The multiple context group tables provide value-set domains for coded entry sequence data elements. An example provided by Bidgood in his paper describing the SDM references context group summary table entry of "visible light image type," individual context groups might include "endoscopy" and "bronchoscopy" (Bidgood, 1998). A template summary table might include "diagnoses on upper endoscopy," and the template tables themselves would define fields to be captured, such as "gross diagnosis" with a value set of "normal, ulceration, discoloration, submucosal mass, exophytic tumor" with appropriate SNOMED code values.

Enhancing interoperability

DICOM represents a strong foundation for digital pathology image standardization, with other standards such as SNOMED and HL7 also playing key roles. What is missing is how these standards should work together to solve real-world challenges. Integrating the Healthcare Enterprise (IHE, ihe.net) is an international initiative that brings together healthcare providers, scientists, academicians, vendors, and government agencies to solve real-world interoperability problems (Integrating the Healthcare Enterprise, Making healthcare interoperable). IHE takes a vendor-agnostic approach to describe problems or challenges, describing various systems as "actors" and the ways in which they interact as "transactions." IHE does not attempt to create a new standard per se but instead describes how best to leverage (or possibly modify) existing standards to support interoperability requirements. The primary output of IHE is system implementation guides, called IHE Integration Profiles. The Pathology and Laboratory Management (PALM) domain within IHE has published a plan to create a series of interoperability profiles in support of a set of predetermined digital pathology use cases (Dash et al., 2021; Integrating the Healthcare Enterprise, Pathology and laboratory medicine (PaLM)). Only one of the multiple profiles specified in the plan has been completed at the time of this writing.

As mentioned before, use cases represent a thoughtful approach to solving real-world challenges by first systematically describing the problems that need to be solved. Use cases support real-world interoperability by describing workflow needs and challenges. Solutions can then be devised by evaluating how existing standards might be best leveraged or augmented. The following use case summary represents high priority, high prevalence needs within the clinical domain of digital pathology as determined by a group of subject matter experts spanning the pathology and lab domain, clinical informatics, and technology vendor space. This list of use cases was originally published by the primary author of this chapter and is not exhaustive (Dash et al., 2021). Rather, it represents the current focus for interoperability and standards efforts.

1. Digital pathology: Primary diagnosis represents the evolution of surgical pathology from traditional microscope-based diagnostics to fully electronic digital image-based diagnostics. There are many aspects to consider relative to the analog to digital transformation beyond just the glass slide containing a thin slice of human tissue. For one, there are other images involved in the pathology workflow that precede the creation of the glass slide. The specimen radiograph and gross examination still images are digital assets that represent components of a holistic digital pathology workflow and are created in advance of the glass slide. As another example, the glass slide label must have a barcode, but this barcode will generally only encode a single identifier. This identifier is typically unique for a single glass slide in the organization that created it. This identifier will generally not be unique outside the organization. Using multiple identifiers on a physical asset is considered a best practice to allow for confirmation that both identifiers are accurate or to allow for failure of a single identifier to be readable. With a single barcode, there is not a second "machine-readable" identifier available to allow information systems to capture other important information, such as the patient's name, medical record number, etc. These examples illustrate the multitude of challenges that must be addressed as solutions are designed around this use case.
2. Imaging glass slides for historical review, secondary review, and consultation provides a significant value in the surgical pathology diagnostic

process. Historical review of digital images of prior pathology for a patient reflects a critical, high-yield use case by reducing the turnaround time to render a diagnosis. The usual lengthy wait time to retrieve physical glass slides from storage can be eliminated. Pathology consultation in traditional glass slide-based workflows usually require physical transport of slides and sometimes associated paperwork. In a digital pathology workflow, this manual effort is eliminated. However, new challenges arise, such as notification of the pathologist from whom a consultation is requested. This becomes more challenging as one considers that pathologist availability for consultation is generally not managed in today's information systems, particularly if the pathologist is not within the originating pathologist's organization.

3. Use of digital pathology for immunohistochemistry positive control slides can provide significant cost savings to an organization. This is achieved by creating digital copies of immunohistochemistry (IHC) positive control slides to preclude the need for creating multiple positive glass control slides for distribution to pathologists. There are some information system challenges that need to be addressed in this use case. Firstly, the positive control used in the same batch of patient test slides is generally not linked electronically. It is important for quality control for the pathologist to receive the control slide from the same batch (same run) in which the tissue slides for the patient originated from. This can potentially be solved by colocating the positive control onto the glass slide itself, but this adds a level of complexity to the cutting procedure and digital scanning process.
4. Digitization, sharing, and consulting gross examination images is another vitally important use case for larger specimens. Capturing and annotating gross exam images at various points in the gross dissection process is helpful in conveying complex findings and correlating gross and microscopic tissue relationships. This may be the only way to understand, for example, the true three-dimensional size and shape of an invasive tumor within a specimen.
5. Digitization of archived glass slides can help accelerate the digital transformation of a pathology department and support multiple use cases in a short amount of time. Some unique challenges must be addressed, including accurate association of the digital images to the corresponding pathology and patient data within electronic systems, lack of machine-readable barcode glass slide labels, re-labeling as needed, and repair of glass slides and/or coverslips that may have aging artifacts precluding digitization in some whole slide scanners. Conversion of legacy digital images can also represent a significant challenge. Unfortunately, many different image file formats exist today within laboratories due to the evolution of digital imaging standards and technologies over time. A number of different file formats span still images (gross and microscopic) as well as whole slide images (microscopic only). File formats for whole slide images (WSIs) today include those originally produced by Aperio (.svs, .tif), Hamamatsu (.vms,

⊡vmu, ⊡ndpi), Leica (.scn), MIRAX (.mrxs), Philips (.tiff), Sakura (. svslide), Trestle (.tif), Ventana (.bif, ⊡tif), and generic tiled TIFF (.tif).

6. Image analysis, computational pathology, and machine learning algorithms represent the largest potential return on investment for engaging in a digital transformation within pathology. Recent publications have demonstrated the capability of machine learning algorithms to fulfill complex diagnostic tasks, such as identification of lymph node metastases on digitized H&E images of lymph nodes. Some studies purport that machine algorithms already have the potential to exceed the capabilities of human pathologists, particularly with humans and machines working together. Some examples will be discussed in the concluding remarks.
7. Quality control, assurance, and error correction workflows represent critical efforts to ensure that the digitization and digital manipulation techniques employed do not negatively affect the diagnostic process. Challenges include data collection mechanisms to facilitate tracking quality metrics over time, including preanalytic factors such as tissue cutting and staining. It should be noted that the quality of the digitization process needs to be validated at the time of initial implementation, at the time of any significant changes to the system infrastructure or processes, and periodically to ensure maintained quality over time. Any automated algorithms implemented as part of a digital pathology workflow should also have quality assurance metrics incorporated to reflect the possibility of performance failure over time.
8. Digital pathology in support of clinical conferences represents a significant improvement in efficiency in support of this best practice in clinical care. In a traditional workflow, significant time is spent collecting and collating glass slides for consensus or multidisciplinary panel review of challenging patient care issues. Digital pathology allows for specific images to be flagged as most relevant for future review. Specific regions of interest disclosing key pathologic findings are tagged (e.g., positive margins, lymph node involvement, etc.).
9. Image coregistration functions involve the creation of a cross-image coordinate system of multiple images for the same tissue block to allow for useful image comparisons across "levels" of the same block. This is particularly useful for tissue in which findings are focal and scattered and comparisons across multiple slides are challenging. This often occurs in situations where immunohistochemistry (IHC) is performed across a number of tissue block sections to better characterize scattered tumor cells. An imaging system identifies landmarks within the image for correction of orientation, allowing the viewer to recognize the same cells across different IHC stains.
10. Support of intraoperative consultations or "frozen sections" represents the final use case documented in the IHE white paper. Intraoperative consultations allow surgeons to be guided by pathological evaluations of tissue in real time during a surgical procedure. Rapid imaging of frozen

tissue sections provides a mechanism for remote viewing by the pathologist on call as well as consultation from the primary pathologist to a subspecialty pathologist. Challenges in this use case revolve around the rapidity of image acquisition required by the pathologist for providing useful information back to the surgeon in "real time." In some circumstances, the digital image may be created in advance of the surgical pathology case within the lab information system, making downstream linkage of images to the appropriate pathology case challenging.

Digital pathology augments the traditional anatomic pathology workflow by creating digital images and associated data during one or more steps in the overall workflow. In IHE, the computer systems that perform work are described as "actors" having an abstract name based on their role, such as "order placer" and "order filler." In the IHE diagram, the "order placer" is usually an electronic health record (EHR), the "order filler" is usually a laboratory information system (LIS), an "acquisition manager" is typical image management software (IMS), an "acquisition modality" is typically a WSI scanner, an image manager/archive may be a picture archival and communication system (PACS) or a vendor neutral archive (VNA), an "evidence creator" may be a digital analysis algorithm, and an "image display" may be a WSI viewer. An implementor of this profile need not follow these typical designations, however. For example, a PACS system may request an LIS system to create another glass slide with a digital scan. In this case, the order placer is the PACS, and the order filler remains the LIS. As depicted in Fig. 9.5, the scope for proposed new profiles includes the following:

1. Digital pathology image acquisition (DPIA) was the first profile to be published and focuses on the role of the digital scanner in acquiring images (Integrating the Healthcare Enterprise, IHE PaLM digital pathology workflow–image acquisition (DPIA)). It describes transactions between the software that controls the scanner (acquisition manager actor) and the scanner itself (acquisition modality actor). It also describes how the newly scanned images are stored, displayed, and viewed.
2. Digital pathology ordering and workflow (DPOW) speaks to interactions between the PACS system and other actors to provide information necessary for image interpretation by a pathologist.
3. Digital Pathology Evidence Creation (DPEC) speaks to the creation of new data based on an existing image or across multiple images. The new data created may represent metadata associated with the image, new annotations, or an entirely new image derived from the original.

The IHE pathology study ordering process describes real-world tissue collection, subsequent pathology exam, and report of results back to the order placer. This process already exists as a well-established profile in IHE called the laboratory testing workflow (LTW) and speaks to pathology examination orders being placed by an order placer (usually the order entry system, or EHR), which are

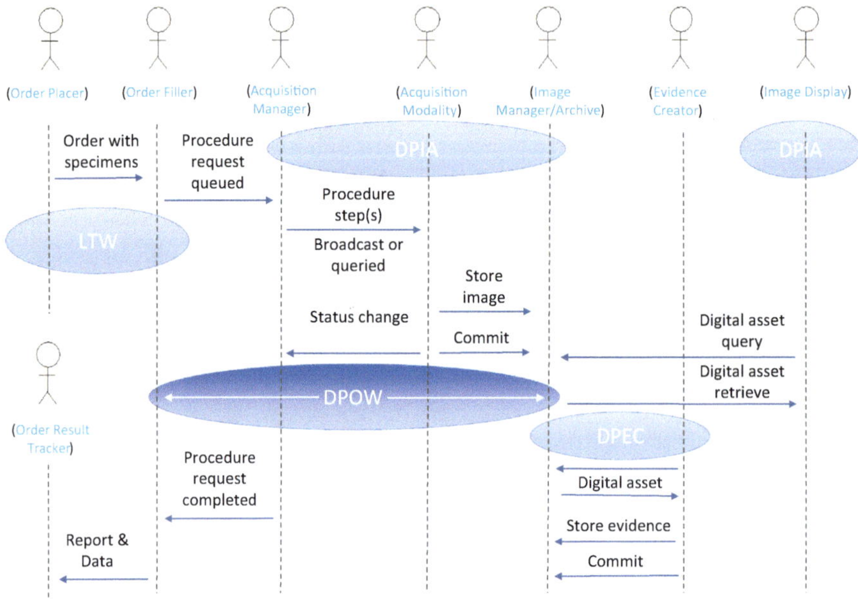

FIGURE 9.5

Integrating the healthcare enterprise digital pathology actor overview described by interoperability profiles including LTW (lab testing workflow), DPIA (digital pathology image acquisition), DPEC (digital pathology evidence creation), and DPOW (digital pathology ordering and workflow).

fulfilled by an order filler (usually the LIS) (Integrating the Healthcare Enterprise, 2017).

The acquisition, subsequent unique identification, storage, and display of a digital image is documented in the recently published digital pathology image acquisition (DPIA) profile (Integrating the Healthcare Enterprise, IHE PaLM digital pathology workflow–image acquisition (DPIA)). This profile includes the set of actors and transactions necessary for support of DICOM images with notification of availability of images to the relevant actor (software system). This profile recognizes that robust communication is needed between the scanner and LIS to populate useful metadata within a DICOM image (e.g., the stain[s] applied to the tissue on the glass slide that has been imaged).

Following image acquisition, any subsequent acquisition of digital information to support interpretation of an image by a pathologist is detailed in the digital pathology ordering and workflow (DPOW) integration profile currently under development. These transactions describe how data is communicated between the image manager/archive (PACS) and all other actors. A PACS system may need to retrieve an endoscopy report from the EHR, it may need to retrieve a gross examination report from the LIS, it may need to retrieve analytic data from an image analysis too, and it may need to return quantitative data back to the LIS for integration into a pathology report.

The manipulation of digital images by humans and machine algorithms, potentially resulting in the creation of new evidence or alteration of existing digital assets, will be documented in the digital pathology evidence creation (DPEC) workflow. The pathology laboratory may study both physical specimens and digital specimens. The production of observations (margin measurements, recognition of morphologic abnormality, histologic type, etc.) on digital specimens combines actors and transactions of existing IHE profiles with actors of future digital pathology specific profiles. Fig. 9.5 depicts how the new IHE digital pathology profiles work together to support interoperability of analytic data within a digital pathology workflow building upon the existing LTW profile.

The LIS needs to manage the "manufacturing process" of the digital image(s) and the final pathology diagnostic report, usually including structured and unstructured data. The LIS may break down the clinical order into a set of work orders similar to what is done in the clinical laboratory space. Work orders are in turn broken down into granular "work order steps" for instrumentation to follow for a given type of specimen and lab test to be performed. Middleware is uncommon in AP laboratories today, but it may be reasonable to assume that it will play a larger role as instrumentation and automation advances within AP laboratories. Specimen processing "events" tracked by a LIS are largely captured manually or using barcodes today, but technological advances should allow for greater automation and granularity in the future. It is important to recognize that many tasks fulfilled by the LIS today involve physical processing of the tissue (e.g., creating an unstained slide or then applying a stain to it), but in the future these may include processing of digital images (e.g., applying a mitotic count algorithm to a designated region of interest on a whole slide image). It is also possible that vended systems outside the LIS take on some of these roles (i.e., specialized "evidence creator" actors taking on greater roles rather than relying solely on "order filler" actor functionality).

In IHE, work order steps are the way to describe atomic actions performed by instrumentation actors (slide scanners, analyzers, etc.) on physical specimens or digital assets. Work order steps that acquire digital assets are defined as Imaging Work Order Steps (IWOS). Work order steps that produce analytical observations on physical or digital specimens are called Analytical Work Order Steps (AWOS). Work order steps representing a single operation to be performed on a specimen by an automated device (e.g., transferring a glass slide specimen container from a cover slipper to a whole slide scanner) are termed specimen work order steps (Integrating the Healthcare Enterprise, Laboratoryb device automation). Work order steps may be performed on both physical specimens and digital assets for patient diagnosis and quality control (QC) purposes.

Image acquisition workflows may be categorized into a few different paradigms with actors and transactions depicted in Fig. 9.6. An order- or step-driven workflow may also be described as a scheduled or queued workflow. An acquisition modality instrument (scanner) will typically query an acquisition manager

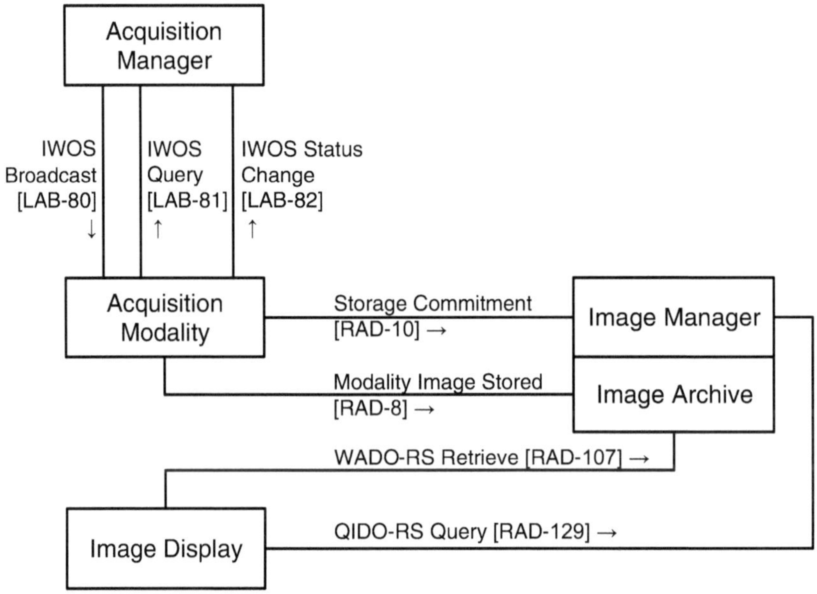

FIGURE 9.6

Integrating the healthcare enterprise digital pathology image acquisition interoperability profile transactions.

(the workstation that runs the scanner) for instructions on how best to acquire a digital image after recognizing the presence of a physical asset (glass slide). This is felt to likely represent the most common implementation in the near term. This will require a bidirectional interface between the acquisition manager and the acquisition modality, and it reflects a relatively standard workflow in radiology today (querying of an imaging modality worklist). Situations in which a physical asset unexpectedly appears within an imaging modality (i.e., a user places a glass slide unannounced into a scanner) may also be readily accommodated within this acquisition paradigm.

Another paradigm would be a unidirectional interface representing a broadcast message from an acquisition manager to one or more acquisition modalities. This represents a simpler paradigm for managing image acquisition. In this scenario, physical asset identifiers along with acquisition "protocols" will typically be broadcast to all acquisition modalities that are capable of acquiring a digital image from the physical asset. When the physical asset is recognized within any of the acquisition modalities, digital image acquisition can begin according to the previously broadcast parameters.

Finally, reconciliation workflows represent a category of exception handling to manage situations that fall outside the scope of the paradigms described above. Typically, these situations will require some level of manual intervention by a technologist. The technologist will have to guide the acquisition modality in situations where the instrumentation is not able to ascertain how to acquire an image due to a workflow or instrumentation fault of some kind (e.g., unreadable label).

It should be noted that an LIS may not use a traditional PACS in the sense of an image workflow and communication tool but rather leverage the LIS for workflow, image management software for digital asset organization, file format-specific image viewer(s) for display, and use an image archive for local, networked, database, or cloud storage. Vended software may choose to incorporate the roles of multiple actors, for example, a "Pathology PACS" might choose to serve as an image acquisition manager, image viewer, and image archive.

In IHE, the results fulfilling the order of tissue examination are messaged from the Order Filler (e.g., the LIS) to the actor tracking results for orders, abstractly called the Order Result Tracker. This is an application (usually the EHR) used by healthcare providers for reading the diagnostic information generated by pathologists. The "result" is represented by the preliminary or final pathology report, which may include digital assets too (such as a representative image of a tumor involving a lymph node or quantitative biomarker assessments). Management of digital assets and physical assets are part of the workflow and transactions of the IHE profiles slated to support digital pathology.

The data model for the interoperability profiles will leverage the HL7 specimen domain analysis model (SDAM), which has been aligned with what started as DICOM supplement 122 "Specimen Module and Revised Pathology SOP Classes" and is now part of DICOM Part 3: Information Object Definitions that includes a specimen module (C.7.6.22) (The National Electrical Manufacturers Association, DICOM PS3.3 2022d – Information object definitions. C.7.6.22 specimen module).

In summary, IHE attempts to create a consensus among all stakeholders, including multiple vendors, on how best to establish interoperability for the most common use cases, leveraging existing standards and best practices. With the recent publishing of the first digital pathology integration profile focused on image acquisition (DPIA), a solid foundation has been established upon which to build a robust digital pathology interoperable ecosystem.

Closing remarks

In this chapter, we have explored what a standard represents, why it is necessary, and the current state of standards in support of digital pathology. The intention is not to serve as a primer on how to implement these standards, but to understand the high-level goal of digital pathology standards and generally how they go about trying to reach those goals. Much work remains to be done, but this is to be expected as the field continues to rapidly evolve. Standards exist to enhance interoperability of content and information systems, but many standards need to work together to solve interoperability challenges. Some challenges still exist where standards development remains immature. A good example for digital pathology today is the current state of support for DICOM annotations. Many different IODs exist to support different flavors of annotations, but guidance on how best to implement a particular aspect of the standard to support a specific use case requires ongoing effort. The lack of guidance around the encoding of annotations leads to proprietary solutions and/or manual effort to allow different systems to work together. This is a barrier to transitioning between information systems or service providers, or when transitioning back and forth between a research context and a clinical context (i.e., using clinical annotations to aid research or translating research tools into clinical practice). DICOM files are not trivial to create due to the complexity of the specification. Certain important use cases, such as de-identification of image files, can be a daunting undertaking for those inexperienced with the format. Storing the image pyramid layers as different "series" and deciding if an image series should be stored as a single file or multiple files in a folder requires a broad understanding of the way in which the image data is to be used. There remains a lack of mature tools that interact with whole slide images and fully support the DICOM standard for digital pathology. Standards to evaluate and validate such tools are also lacking (Griffin & Treanor, 2017). However, progress continues to improve the state of the art, and productivity gains associated with digital pathology workflows continue to increase (Griffin & Treanor, 2017). In the research space, computer algorithms leveraging digital pathology infrastructure have been well-documented to outperform human performance, such as in breast cancer biomarker assessment (Stålhammar et al., 2016) and detection of lymph node metastases in breast cancer patients (Bejnordi et al., 2017). The ultimate promise of digital pathology is to raise the standard of patient care through the implementation of computer algorithms augmenting what humans alone are capable of (i.e., "augmented intelligence"). Standards in digital pathology represent the foundational efforts allowing this potential to be realized.

References

American Medical Association. CPT® overview and code approval. https://www.ama-assn.org/practice-management/cpt/cpt-overview-and-code-approval.

Bejnordi, B. E., Veta, M., van Diest, P. J., van Ginneken, B., Karssemeijer, N., Litjens, G., van der Laak JAWM., the CAMELYON16 Consortium., Hermsen, M., Manson, Q. F., Balkenhol, M., Geessink, O., Stathonikos, N., van Dijk, M. C., Bult, P., Beca, F., Beck, A. H., Wang, D., Khosla, A., & Venâncio, R. (2017). Diagnostic assessment of deep learning algorithms for detection of lymph node metastases in women with breast cancer. *JAMA: The Journal of the American Medical Association*, *318*(22), 2199–2210.

Bidgood, W. D., Jr. (1998). The SNOMED DICOM microglossary: Controlled terminology resource for data interchange in biomedical imaging. *Methods of Information in Medicine*, *37*(4-5), 404–414.

Bidgood, W. D., Jr, Horii, S. C., Prior, F. W., & Van Syckle, D. E. (1997). Understanding and using DICOM, the data interchange standard for biomedical imaging. *JAMIA*, *4*(3), 199–212.

Clinical Laboratory Standards Institute. AUTO02-A2 laboratory automation: Bar codes for specimen container identification. https://clsi.org/standards/products/automation-and-informatics/documents/auto02/.

Dash, R., Jones, N., Merrick, R., Haroske, G., Harrison, J., Haarselhorst, N., Wintell, M., Herrmann, M. D., & Macary, F. (2021). Integrating the healthcare enterprise (IHE) pathology and laboratory medicine (PALM) guideline for digital pathology interoperability. *Journal of Pathology Informatics*, *12*, 16.

European Union. Regulation (EU) 2016/679 of the European Parliament and of the Council of 27 April 2016 on the Protection of Natural Persons with Regard to the Processing of Personal Data and on the Free Movement of Such Data, and repealing Directive 95/46/EC (General Data Protection Regulation). https://eur-lex.europa.eu/legal-content/EN/TXT/PDF/?uri = CELEX:02016R0679-20160504.

Griffin, J., & Treanor, D. (2017). Digital pathology in clinical use: where are we now and what is holding us back? *Histopathology*, *70*(1), 134–145.

Health Level Seven International. HL7 version 2 product suite. https://www.hl7.org/implement/standards/product_brief.cfm?product_id = 185.

Health Level Seven International. HL7 version 3 product suite. https://www.hl7.org/implement/standards/product_brief.cfm?product_id = 186.

Health Level Seven International. HL7 fast healthcare interoperability resources. https://www.hl7.org/fhir/.

Health Level Seven International. Introduction for the HL7 object identifier (OID) registry. https://www.hl7.org/Oid/index.cfm.

Healthcare Information and Management Systems Society. Interoperability in healthcare. https://gkc.himss.org/resources/interoperability-healthcare.

Herrmann, M. D., Clunie, D. A., Fedorov, A., Doyle, S. W., Pieper, S., Klepeis, V., Le, L. P., Mutter, G. L., Milstone, D. S., Schultz, T. J., Kikinis, R., Kotecha, G. K., Hwang, D. H., Andriole, K. P., Iafrate, A. J., Brink, J. A., Boland, G. W., Dreyer, K. J., Michalski, M., Golden, J. A., Louis, D. N., & Lennerz, J. K. (2018). Implementing the DICOM Standard for Digital Pathology. *Journal of Pathology Informatics*, *9*, 37.

Helin, H., Tolonen, T., Ylinen, O., Tononen, P., Näpänkangas, J., & Isola, J. (2018). Optimized JPEG 2000 compression for efficient storage of histopathological whole-slide images. *Journal of Pathology Informatics*, *9*, 20.

Integrating the Healthcare Enterprise. Making healthcare interoperable. https://www.ihe.net/.

Integrating the Healthcare Enterprise. Pathology and laboratory medicine (PaLM). https://wiki.ihe.net/index.php/Pathology_and_Laboratory_Medicine_(PaLM).

Integrating the Healthcare Enterprise. IHE PaLM digital pathology workflow—image acquisition (DPIA). https://wiki.ihe.net/index.php/Digital_Pathology_Workflow_-_Image_Acquisition.

Integrating the Healthcare Enterprise. Laboratory device automation. https://wiki.ihe.net/index.php/Laboratory_Device_Automation.

Integrating the Healthcare Enterprise. (2017). *Laboratory testing workflow*. Available from https://wiki.ihe.net/index.php/Laboratory_Testing_Workflow.

Regenstrief Institute, Inc. Unified code for units of measure. https://ucum.org/.

Regenstrief Institute, Inc. Logical observation identifiers names and codes. https://loinc.org/about/.

Singh, R., Chubb, L., Pantanowitz, L., & Parwani, A. (2011). Standardization in digital pathology: Supplement 145 of the DICOM standards. *Journal of Pathology Informatics*, *2*, 23.

SNOMED International. Systematized nomenclature of medicine. https://www.snomed.org/about-us.

Stålhammar, G., Martinez, N. F., Liiper, M., Tobin, N. P., Mølholm, I., Kis, L., Rosin, G., Rantalainen, M., Pedersen, L., Bergh, J., Grunkin, M., & Hartman, J. (2016). Digital image analysis outperforms manual biomarker assessment in breast cancer. *Modern Pathology: An Official Journal of the United States and Canadian Academy of Pathology, Inc*, *29*(4), 318–329.

The National Electrical Manufacturers Association. Digital imaging and communications in medicine. https://www.dicomstandard.org/about-home.

The National Electrical Manufacturers Association. DICOM WG-26: Pathology. https://www.dicomstandard.org/activity/wgs/wg-26.

The National Electrical Manufacturers Association. DICOM supplement 15 final text. https://www.dicomstandard.org/News-dir/ftsup/docs/sups/sup15.pdf

The National Electrical Manufacturers Association. DICOM supplement 122 final text. https://dicom.nema.org/medical/dicom/Final/sup122_ft2.doc.

The National Electrical Manufacturers Association. DICOM supplement 145 final text. https://dicom.nema.org/dicom/dicomwsi/.

The National Electrical Manufacturers Association. DICOM supplement 222 final text. https://dicom.nema.org/medical/dicom/Final/sup222_ft_WSIAnnotations.pdf.

The National Electrical Manufacturers Association. DICOM PS3.18 web services. https://dicom.nema.org/medical/dicom/current/output/chtml/part18/ps3.18.html.

The National Electrical Manufacturers Association. DICOM PS3.3 2022d – Information object definitions. C.7.6.22 specimen module. https://dicom.nema.org/medical/dicom/2022d/output/chtml/part03/sect_C.7.6.22.html.

United States Department of Health and Human Services. Summary of the HIPAA privacy rule. https://www.hhs.gov/hipaa/for-professionals/privacy/laws-regulations/index.html.

United States Department of Health and Human Services. National plan & provider enumeration system. https://nppes.cms.hhs.gov/.

World Health Organization. International statistical classification of diseases and related health problems (ICD). https://www.who.int/standards/classifications/classification-of-diseases.

CHAPTER

The basics of machine learning and artificial intelligence

10

Stanley Cohen[1,2,3]

[1]Department of Pathology, Rutgers-NJMS, Newark, NJ, United States

[2]Perelman School of Medicine at the University of Pennsylvania, Philadelphia, PA, United States

[3]Kimmel School of Medicine, Jefferson University, Philadelphia, PA, United States

Introduction

In addition to its many other uses and advantages for our field, digital pathology is essential for the implementation of artificial intelligence (AI) by providing information as input into machine learning (ML) algorithms. Without digital pathology, there could be no AI for pathology. This chapter will provide a broad overview of AI and ML without a requirement for either computer programming knowledge or advanced mathematics. As such, it is presented in narrative form without specific references. However, even within these constraints, it is possible to explain the inner workings, intuitions, and strategies involved. A major focus will be on "deep learning," which has become synonymous with multilayer neural networks. Actual examples of the use of AI in pathology will be discussed in other chapters. The main purpose of this chapter is to explain AI approaches in general, as well as to provide a vocabulary useful in understanding the literature, as well as new methodologies as they unfold.

Unfortunately, AI has become an acronym-rich discipline. For this reason, the major acronyms are provided for each technical term or concept. While this may slow the exposition, it is necessary to know these terms in order to be able to understand the literature. To begin, it is necessary to define artificial intelligence (AI). As an example of the power of AI, I asked a modern example of AI (ChatGPT) to explain AI. The result was much better than the paragraph I originally wrote:

> *Artificial Intelligence (AI) refers to the ability of computer systems to perform tasks that normally require human intelligence, such as understanding natural language, recognizing images, and making decisions based on complex data. AI systems use algorithms, machine learning techniques, and statistical models to learn from data and improve their performance over time. Although*

Digital Pathology. DOI: https://doi.org/10.1016/B978-0-443-13809-6.00013-0

traditional computing algorithms make use of a set of rules that perform a predetermined task. Machine learning uses algorithms to learn from data without being explicitly programmed. Deep learning is a type of machine learning that uses artificial neural networks to learn from data. Deep learning models are particularly good at image and speech recognition.

"Shallow" learning

Prior to machine learning, it was necessary to provide a set of instructions or rules to the computer. For example, if we wanted to identify a lesion as benign or malignant, we would define a set of measurements to describe the features, or attributes, of these. This might include some quantitation of nuclear size and shape, spatial distribution, etc. Both identification of critical attributes and quantification of these are obviously difficult. Machine learning takes a different approach. To illustrate this, it is helpful to begin with a simple example. Suppose we are trying to distinguish between caterpillars and ladybugs. The minimal set of defining attributes is small; in this example, length and width will suffice. The first step is to create a dataset of known examples of these bugs. If we are using a rules-based approach, the program would calculate the average height and width for each class and then would compare the unknown to this information. However, instead of a statistical analysis, we can simply take advantage of the database to construct a plot with height on one axis and width on the other. We can do this without any knowledge of statistics. Simply plot each sample on an *XY* axis where *Y* represents height and *X* represents width (or vice versa). Now enter the unknown on that graph. The simplest approach might be to look for its nearest known neighbor and assume that it is a member of the neighbor's class. If the neighbor is a ladybug, then the unknown is a ladybug. However, there are a number of reasons why this test might fail, especially if we are in a dense region of the plot. However, if we compare the unknown to several of its closed neighbors (e.g., K of these), then the results are more reliable. This is known as the KNN algorithm. This is an example of a geometric machine learning model and is one of the earliest forms of machine learning.

Another method might be to see if it is possible to draw a line of the graph such that one class is on one side and the other class on the other. The location of the unknown relative to that line tells you which class it belongs to. The basic feature here is that we have drawn a line (one-dimensional) to define a boundary between two regions of a plane. There are two problems. First, there may be many boundary lines possible. Second, in real life, there will always be overlapping cases that don't fall on even the best boundary line we can draw. Third, not all data is linear. For example, if the data falls into concentric circular regions, like the cross section of a jelly donut, we can't separate the classes by a single line. This one is easy; by moving one class along a different axis (the *Z* axis), we can now draw a plane between them. More generally, we can find a hyperplane separating multidimensional data by adding an additional dimension. The machine learning algorithm that solves all these problems is known as a Support Vector Machine (SVM). These issues are illustrated in Fig. 10.1.

KNN and SVM are examples of geometric learning. There are also probabilistic learning strategies, in which we try to find a probability of the class based on its attributes. It is easy to find the probability of each attribute based on the class; we can simply find it from our dataset. From this, we can work backwards and

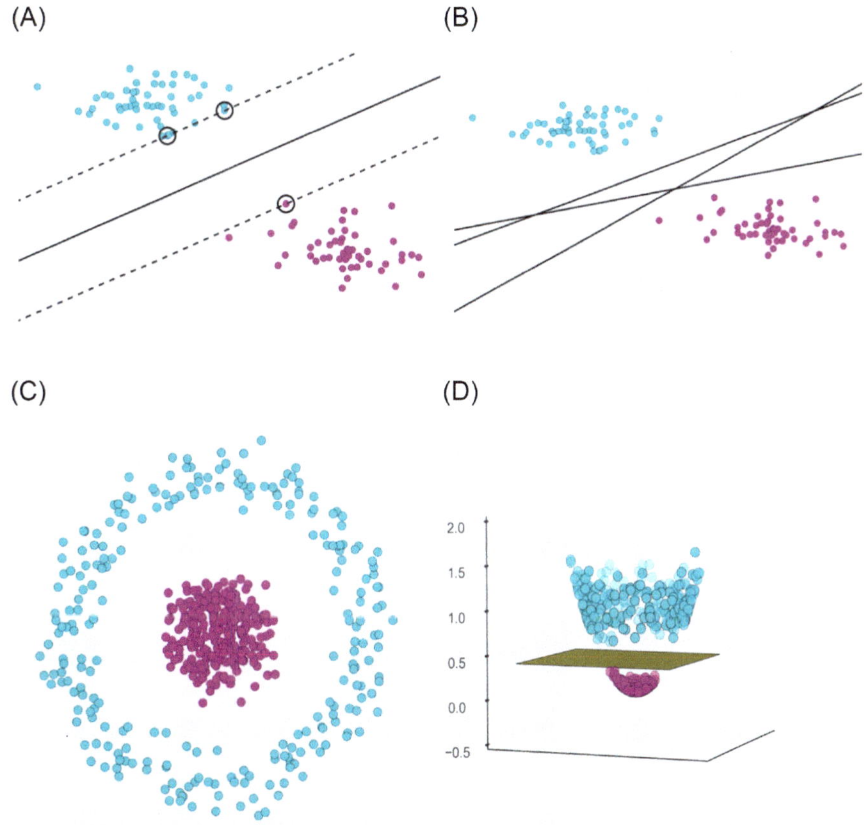

FIGURE 10.1

(A) Possible linear separators for the data clusters shown. (B) Support vectors are used to construct the line that best separates them. (C) The clusters in this example form a pattern not separable by a single line. (D) Adding a dimension to one of the clusters allows them to be separated by a plane. In general, for any N-dimensional space, a linear construct such as this is called a hyperplane.

derive the probability of the class from the probability of the attributes. This is known as Bayesian analysis, and the algorithm is known as Naïve Bayes (naïve because it requires a number of assumptions about the data). In addition to geometric and probabilistic learning, there are also stratification models, such as decision trees, that attempt to partition the data to arrive at a conclusion, sort of like the game of twenty questions on steroids. This is usually done as a random forest, which involves multiple decision trees with different subsets of the data and/or of the branches.

In all these examples, the first step is to create a dataset, which typically is a matrix in which each row is an attribute or feature, except for the last row, which defines the class of the sample. The columns are all the examples in the dataset. The collection of features for each sample is known as its feature vector. The bigger the dataset, the better the results. In effect, the algorithm improves its performance based on experience, which is pretty much the definition of learning.

The good news about these three kinds of models is that they can be used for classes defined by many thousands of attributes, which can't be visualized as each sample is a point in a multidimensional space array. The bad news is that it's really hard to define thousands of attributes. The trick here is to let the computer itself find the attributes. This is where the power of deep learning comes into play.

Deep learning

After approximately 1.9 million years of organic brain evolution, it turns out that the brain itself represents a pretty good model for computation. In the brain, a neuron collects information from many other neurons in the form of quantitative signals. It integrates that information, and if it exceeds a threshold value, it passes it on to a new set of neurons. In other words, the brain processes data in layers, in which each layer integrates information from the previous layer. An artificial neural network (ANN) mimics this architecture, but in a very simplistic way.

The ANN has an input layer and an output layer separated by a number of layers that do sequential processing. These are known as hidden layers because, in practice, we do not need to access the intermediate information stored in them. The architecture of a simple neural network (NN) is shown in Fig. 10.2.

The essential features are:

1. Each neuron in one layer connects to all the neurons in the next layer.
2. Neurons in a layer do not connect to other neurons in that layer (this restriction is an oversimplification of the way the brain works).
3. Each input neuron (node) gets only one input directly from the outside; those of the hidden layers and the output layer get many. The connections have numeric values (weights), and each node sums its incoming weights and

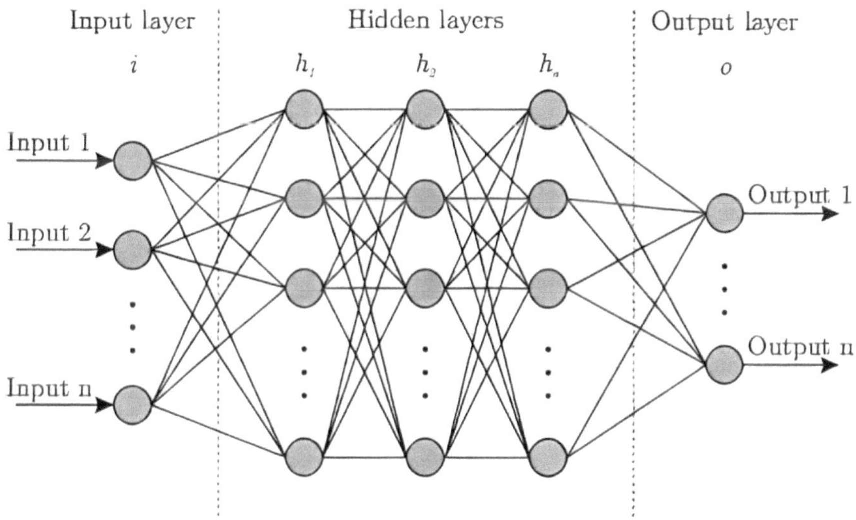

FIGURE 10.2

Architecture of a simple neural network. The connections shown are described in the text below.

Wikipedia commons; Facundo Bre.

determines if this value exceeds a threshold. If so, it is passed on to all the nodes in the next layer.

4. The model is synchronous in that all neurons in a given layer receive their inputs and calculate their activities at the same time. This means that all the calculations in the *N*th layer must be completed before the $(N + 1)$ layer can begin its calculations.

Because of this summing process, each layer receives increasingly complex inputs. At any given layer, the network creates an internal representation of the data that incorporates relationships among the feature vectors of each sample. Basically, a NN utilizes multiple levels of representation to model both simple and complex representations among data. They are powerful classifiers that can extract relationships among classes and make predictions and inferences from the data. Best of all, they can do this without the need to manually define the feature vectors for the set. When we were distinguishing between our dog images versus cat images, we no longer needed to input features such as eye-snout ratios, etc. We can simply present multiple known examples each to the program. For images, we would use a value of red, green, blue, and intensity for each pixel of the image without knowing explicitly what the image is describing. We start with a set of known examples; this is divided into a training set and a validation set.

Initially, the NN's weights are assigned randomly. When training begins, each output is compared to its (known) input. Unless we are extremely lucky, the output will not be correct. The trick is to adjust the weights until the model is accurate in its conclusions. During training, which involves a large number of samples, the output of the NN is compared to the known value of the input by means of a loss (or cost) function which is a quantitative measure of the difference. The weights are adjusted incrementally by minimizing the cost function using calculus, starting at the layer just under the output layer and working backwards toward the layer just after the input layer. This top–down process is called backpropagation, and the actual procedure is called gradient descent. Backpropagation assumes that tweaking all the weights independently actually leads to a reduction in overall error. This is not intuitive since changing one weight leads to a ripple effect throughout the rest of the network. However, if we make the changes small enough and do this repeatedly rather than in a single pass, backprop works reasonably well.

The NN is next presented with the validation step, which involves known samples that the network has not seen during training. The NN's performance on this validation set is essentially its final exam. There are a number of criteria for defining the reliability of the model, including overall accuracy, false positives, false negatives, and so on. Only when criteria for these have been met is the NN ready to process a sample for which we do not know the answer. The next step would be to present the same unknown to a group of human experts and compare their results to those of the machine. After this final exam, it is ready to graduate.

Adding more layers to the NN increases its power to create internal representations of increasing complexity. However, as we try to backpropagate, the gradients become flattened out as we get further and further away from the output layer. With too many hidden layers, this may actually cause the output to get worse. One way is to add "skip connections" that bypass one or more layers of the network. In this manner, each layer has to only fix the residual error remaining from another layer. This type of network is known as a ResNet.

Convolutional neural networks

The main categories of neural network applications are image analysis and natural language processing. For the moment, the discussion will focus on images (pun intended). Convolutional neural networks (CNNs) were designed to simplify and improve the classification ability of a NN when presented with huge feature sets and huge numbers of samples. An image composed of gigabytes of data is a good example of this. Again, inspiration came from the brain. Neurophysiologic studies of vision in the brain revealed that it does not process its view of the world all at once. It breaks the image into little chunks called visual fields, processes these, and then combines the separate representations of these into a visual whole. The major conceptual breakthrough that evolved from this observation is that the earliest stages of a NN need not be fully connected and can be divided into fully connected subunits, each of which is processed separately. Each such subset is subjected to a convolutional filtering step to extract local features. This is done by applying a small matrix to each field to extract important local features. These are then fed into pooling layers that down sample them by selecting the most important values from each. This is done repeatedly, and the final representations are passed to a fully connected part of the NN, which then integrates the compact set of representations. This is illustrated in Fig. 10.3.

The convolutional filters themselves are just matrices of numbers. By proper choice of values, they can detect specific features in the images, such as horizontal and vertical lines, edges, shapes, and so on. This also has the important feature of making the resulting representations more robust with respect to changes in the position of the feature in the image. Interestingly, we can start with random convolutional matrices and use backprop to optimize these just as we did for optimizing weights.

CNNs are typically used in applications such as image classification, object detection, facial recognition, natural language processing, and autonomous driving. One of the major advantages of CNNs is their ability to learn hierarchical representations of the input data. The early layers of the network learn to recognize simple features, such as edges and corners, while later layers learn to recognize more complex patterns that are combinations of these simple features. An important feature of a CNN is that since the early representations are relatively simple, they are similar for CNNs trained with similar data. This means that if we have an algorithm to distinguish benign from malignant breast lesions and want to create a new one to distinguish malignant from benign prostate lesions instead of starting from random weights, we can use the early stages of the neural network that has been trained on breast as the starting point for the new prostate NN, and let that output be the input of an untrained NN that will be trained with prostate images. In other words, we only start with random weights in the later layers

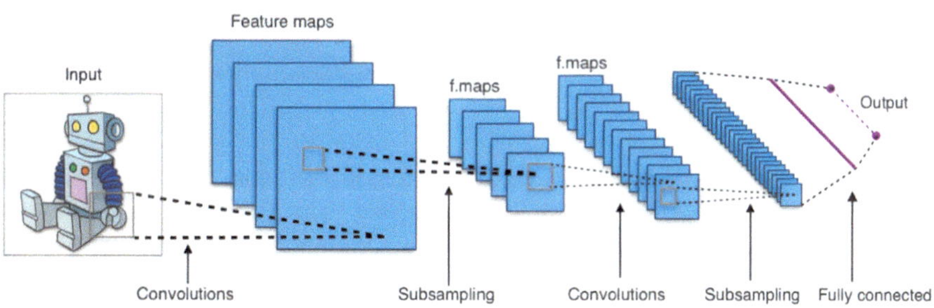

FIGURE 10.3

Schematic of a typical convolutional network illustrating convolutions and down sampling. The image representation of the final layer is passed to fully connected layers. And a classification prediction is made.

Creative commons attribution-share alike 4.0.

for the prostate network and begin with known weights taken from the breast model. This is known as transfer learning and allows for the use of a much smaller training set of samples than we would otherwise need.

To summarize thus far, the basics of deep learning include the architecture of a neural network, error correction via backpropagation, and the development of convolutional networks. The deep learning models described above are examples of supervised learning, since we have to train the system using many known examples. It is also possible to do unsupervised learning, where no information is given about the label of each sample. The program simply looks for patterns within this unlabeled data. For example, unsupervised learning can find clusters within the data. Thus, a small sampling from within each cluster will tell you the class of the objects in that cluster.

Subsequent advances discussed in the following sections include recurrent networks, generative models, and reinforcement learning algorithms, as well as novel implementations such as transformer attention models. The latest approaches include neuromorphic computing and quantum computing. These will be discussed in the following sections.

Recurrent neural networks

The deep learning strategies described above work well for stationary inputs (e.g., an image can be presented as input all at once), so they have great difficulty dealing with sequential or time series data. Natural language processing (NLP) is a good example of this type of problem. A given word may only have unambiguous meaning in the context of what has already gone before it in a sentence. There may also need to be a reference to information further back in the text sequence. Recurrent networks address this problem by giving the hidden nodes access to their own previous states. In other words, the value of a hidden node at time t is dependent both on its inputs at time t and its own state at time $(t-1)$. In some implementations, it also depends on the states of other hidden nodes at $(t-1)$. In the original recurrent networks, the previous states were treated just like any other input to a node. However, current networks make use of "gates" to make decisions about when to make use of the prior value of the node. There are also loops that determine how far back in the sequence one must go to maximize the capability of the network $(t-1)$, $(t-2)$, etc. Such gating decisions are the basis for algorithms known as long-short-term memory (LSTM).

IBM's Watson is an early example of the application of natural language processing. It incorporates NLP and data analytics to inform clinical decision-making. It is built on a modular architecture that includes multiple layers of algorithms, including software for semantic parsers and analyzers. The components work together to understand the meaning of text, identify relevant information, and generate responses. Its repertoire is limited to specific uses such as health care, finance, and customer service. Watson has been made obsolete by newer machine learning algorithms, some of which are described in the next paragraph.

The problem is that even in more modern LSTMs, the more recent events will have an outsize impact on the network. Attention mechanisms have therefore been introduced. In an LSTM, the entire sequence is shown to the network, and we rely on the network to retain all of the relevant information as it processes the inputs sequentially. A more efficient version known as a transformer network has led to major improvements. The key innovation of the transformer is the self-attention mechanism, which allows the network to selectively focus on different parts of the input sequence at each layer of the network.

Attention works by storing the state of the recurrent neural networks (RNN) after seeing each element of the input sequence. The weights are determined by the similarity between each pair of input elements. The sequential output is then used to train a separate network to assign "importance" weights to the elements of this output sequence. Training allows for optimization of both the representation and attention portions of the network.

As an example, this allows for language translation where the two languages may have different word ordering as well as different words to convey the same meaning. LTSM with attention models has become state-of-the-art for natural language processing (NLP).

An attention network is a more general concept that can be applied to a wide variety of neural architectures and not just transformers. In all such cases, the basic idea is to allow the network to selectively focus on different parts of the input sequence or different parts of the network itself during the computation. In fact, it has recently been discovered that the attention mechanism is so powerful that combining it with LSTM is not necessary and that transformer-attention models alone will suffice.

Generative models: autoencoders and generative adversarial networks

Generative models are ones that can generate novel representations, taking a set of examples and then generating new examples that are unique but are similar to the original examples. Deep Fakes are examples of the use of generative models.

Generative models can also be used to "process" examples to improve the efficiency of a NN. A typical example is an autoencoder. The first part of an autoencoder processes the sample just like a traditional NN to achieve a compact but complex representation. Contradistinction to a NN, it then reverses the process to reconstruct the original sample. When the reconstruction is perfect, we can use the representation rather than the original. In other words, the autoencoder gives us a lossless compression of the original data that can be used to train a conventional NN. Since we are comparing the output to the input, we need to provide no information as to what class the sample belongs to, and so the autoencoder is an example of unsupervised learning (Fig. 10.4).

A generative adversarial network (GAN) is a more sophisticated example of a generative model. GANs consist of two neural networks, a generator network and a discriminator network. The generator tries to generate new data samples, and the discriminator tries to distinguish between and the discriminator tries to distinguish between the generated samples and the original training data. The discriminator knows the original, and the generator is a forger trying to reproduce a perfect version of that original. The two networks are trained in a competitive game, where the generator tries to fool the discriminator that the samples are real, while the discriminator tries to correctly identify them. As in the case of the autoencoder, this gives rise to a compact internal representation of the original that can be used to train an NN. By suitable perturbations of the input, the GAN can be trained to produce similar but not identical representations of the original data and can generate realistic images, creating speech, and creating music. They have been used to create photorealistic images of celebrities, animals, and landscapes and have been a particular boon in lampooning politicians and pornography, in addition to more important real-life applications in the arts and sciences.

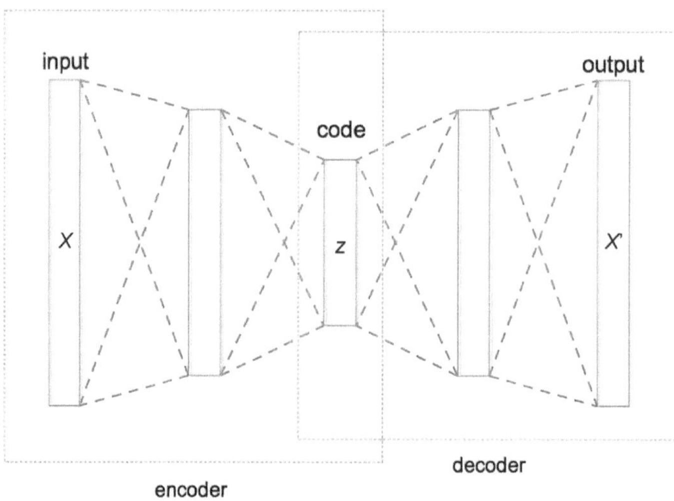

FIGURE 10.4

Autoencoder architecture. The first part (encoder) processes the sample like a traditional NN to achieve a compact representation of the original. Then the second part (decoder) reverses this process to reconstruct the original sample. When the original (*X*) and reconstructed sample (*X'*) are identical, we can use the compact representation (*Z*) rather than the original for subsequent AI applications.

Reinforcement learning

Most neural networks are trained in a supervised manner using a set of labeled input-output pairs. In some cases, unsupervised strategies are used as an initial step to increase both efficiency and accuracy. For certain kinds of tasks, this is not practical. For example, for certain sequential decision problems, we can only accurately assess the performance of the system at certain points. For example, in chess, we don't know whether the strategy being employed is a good one or not until the game is over. This means we can't train the network to decide what moves to make in the middle of a game using standard supervised learning. This is where **reinforcement learning** comes into play. Like many other AI approaches, reinforcement learning (RL) is based on biologic brain function and takes its inspiration from operant conditioning. RL works by providing some sort of "reward" signal that tells the machine when it is doing well. The system acts in such a manner as to maximize its reward through a process of trial and error. RL algorithms make use of deep neural networks, but there is now a reward function to be maximized rather than a loss function to be minimized.

In more detail, the RL learning mechanism consists of the following components:

- Agent: the decision maker that interacts with the environment.
- State: The representation of the environment that the agent observes at each time step
- Action: The agent makes a decision based on the current state of the environment.
- Reward: A scalar signal that the agent receives from the environment after taking an action.

The goal of the agent is to maximize the cumulative reward over time via trial and error. It starts by taking random actions, and as it interacts with the environment, it learns which actions lead to higher rewards. The agent updates its policy based on the rewards it receives, using backpropagation. The best-known examples are game playing and autonomous driving, although RL learning can be applied alone or in partnership with other deep learning strategies for a variety of other tasks, including those with applications to pathology. It can also be used for text summarization.

ChatGPT: everything, everywhere, all at once

As has been discussed above, LTSM and transformer models have been very competent at natural language processing. By adding a generative algorithm, modern LNP has taken a major lead forward. ChatGPT is one such generative pre-trained transformer model. It is designed to understand natural language and generate responses similar to those of a human being. It can be used for language translation, text summarization, and sophisticated question answering. If given prompts such as "define and describe GTP in less than 500 words," it can provide a discussion that is contextually relevant and explanatory. Moreover, ChatGTP can generate text that is similar in style and tone to that which it has been trained on. Basically, it captures relationships between words and their context in a sentence or sentence sequence based on probability distributions.

Since ChatGTP has been trained on a very large sampling of text data, including books, articles, and other sources of human knowledge, its dataset is massive and is continuing to grow as new iterations of ChatGPT appear. Additionally, the model can learn from new data and thus improve its performance over time. Since there can be biased and/or incorrect data in its training data, it may generate biased or inappropriate responses. The most recent version, ChatGTP-4, can use images as well as text as input and provide outputs based upon both these sources of information. Unfortunately, there is much misleading, incorrect, or biased data in the world literature, and care must be taken to see that errors based on this do not creep into the ChatGPT output.

Things to come

With the above overview of modern deep learning, we turn to two new developments in AI that have the potential to further accelerate the field, namely neuromorphic computing and quantum computing.

Neuromorphic computing (NC) is the easier of the two to understand. NC implements brain-like computation in hardware rather than software and is based on devices that can change properties due to input. In contrast to software implementation, NC is relatively energy efficient to operate. Unfortunately, the typical neural networks described in previous sections are not particularly good representations of an organic brain. In a NN, information between neurons is conveyed by changes in the amplitude of the connections between them that represent spikes of information traveling from synapse to nerve cell to axon. The brain uses both amplitude and frequency modulation, as well as relative arrival time between different sets of spikes, to convey interaction among neurons. Most neuromorphic computers implement spiking neural networks (SNN) that better represent what's going on in the brain. SNNs have been around for a long time but have not met with much enthusiasm within the AI crowd because they are not computationally efficient, and it is hard to define a continuous differentiable variable based upon spike arrival time or spike rate. However, by building neural networks in hardware, some of these difficulties can be overcome. In addition, a neuromorphic network can implement biologically inspired rules such as Hebbian learning, where the weight of a synapse is based on the degree of correlation with other connecting neurons (neurons that fire together wire together). This is local learning. In contrast, backpropagation is global in that all weights are adjusted together in a top–down manner. The use of neuromorphic computers represents an important direction for machine learning and AI to take.

Quantum computing (QC) is hard to understand intuitively. Most physicists accept that although the mathematics of quantum computing work beautifully, it is impossible, in contrast, to visualize or understand the underlying physical reality it represents. Fortunately, we can take advantage of some quantum properties for computation without this level of understanding. Recall that in classical computing, information is processed in binary form, using bits that can be either 0 or 1. In quantum computing, information is represented using quantum bits, or qubits, which can be in a superposition of all possible states between zero and one. These multiple super positional states need not be associated with equal probabilities. For example, a qubit can have a 25% probability of being zero and a 75% chance of being one when it is measured. The act of measurement causes the superpositions to collapse into a single defined value. The practical aspect of this is that a single qubit can encode much more information than a single bit. In terms of information content, 13 qubits could encode the equivalent of 8192 bits. From a processing time point of view, it would take a century to perform a 63-qubit operation in a classical computer. The problem is that qubits are very fragile and collapse not only when we want them to but also through interaction with their environment, which is why we don't see quantum effects in objects such as cats or pathologists. Since we want qubits to be stable enough so that they only take a value of 0 or 1, when we measure them, qubits must be generated and maintained at very low temperatures.

Qubit-based algorithms are powerful because of the ways in which multiple qubits can interact with each other when they are in their premeasured states. In fact, two or more qubits can be so tightly correlated (entangled) that we can't learn something about one without learning something about the others. Information is manipulated in a quantum computer by entangling and unentangling it. The specific steps by which these manipulations can lead to the construction of basic building blocks in quantum-based software are beyond the scope of this chapter. Because they can process large data sets in record time, a role for quantum computers in machine learning is obvious. Quantum machine learning has already been implemented in support vector machines and has begun to be explored for deep learning in the form of quantum neural nets.

Summary and overview

Artificial intelligence is based on machine learning. While much of current AI is based on advanced neural networks, there has also been much work using simpler geometric, probabilistic and stratification algorithms. While early work on AI in pathology has relied of image classification and analysis, advances in both natural language processing and multimodal deep learning algorithms allow for the incorporation of information derived from electronic health records including history, physical, and laboratory information to enable diagnosis, treatment, and prognosis. Another use would be to search for patterns across EHRs to help find previously unrecognized common features in diseases that might inform epidemiological and etiologic discovery. Unfortunately, because of the exponential growth in data acquisition and computation, both energy efficiency and processing become limiting. The fact that AI is becomingly increasingly important across all fields of human endeavor makes this even more of an issue. However, new forms of computation such as neuromorphic computing and quantum computing are coming to the rescue. In particular, quantum computing is likely to lead to a revolution in computation in general and AI specifically. This field is in its infancy but quantum versions of both SVM and NNs have already been described.

References for further reading

As this is a general introduction to the underpinnings of artificial intelligence, no specific references are presented for the architectures described.

Here are many books and reviews on this topic. The following are suggested, in order of complexity and detail:

1. Cohen, S. (ed.), (2021). *Artificial intelligence and deep learning in pathology*, Elsevier.
2. Glassner, A. (2021). *Deep learning: A visual approach*, Nostarch Press.
3. Russell, S. & Norvig, P. (2021). *Artificial intelligence: A modern approach*, (4th ed.), Pearson Press.
4. Johnston, E. R., Harrigan, N., & Gimeno-Segovia, M. (2019). *Programming quantum computers*, O'Reilly Press.

CHAPTER

11 Academic industry partnerships

Liron Pantanowitz

Department of Pathology, University of Pittsburgh Medical Center, Pittsburgh, PA, United States

Introduction

Practicing pathology informaticists typically work closely with industry since most technology designed for the laboratory is supplied by vendors. Similarly, academics working in the field of digital pathology are likely to partner with industry either as a consultant, serving on a medical advisory board, participating in sponsored research, or being involved in a clinical trial for regulatory approval. Furthermore, since traditional funding sources in academia are often lacking, industry-sponsored research has become attractive. Individuals in academia who develop novel technology may be involved in filing patents for their invention or spin-off a new company (Newco). With an academic industry partnership (AIP), there are often numerous agreements that need to be signed, challenging decisions to be made concerning intellectual property (IP), and potential conflicts of interest to be avoided. Therefore, for those involved in the field of digital pathology and artificial intelligence (AI), especially if there is an entrepreneurial incentive, it is imperative to understand what it takes to promote a successful AIP. Several AIPs have proven to be fruitful in other fields such as biotechnology and drug development (Kneller, 2001; Rose et al., 2015), including incubator and accelerator models (Srimathveeravalli et al., 2017). The Digital Pathology Association (DPA) published a practical white paper on this topic, which addresses many of the areas relevant to pathology and provides guidance on how to manage an AIP and mitigate potential risks (Pantanowitz et al., 2022).

Digital pathology clinical trials

Funded AIPs have been instrumental to facilitate clinical trials for manufacturers to get their digital pathology systems cleared by regulatory bodies such as the Food & Drug Administration (FDA) in the USA (Borowsky et al., 2020; Mukhopadhyay et al., 2018). These clinical trials solicited the participation of pathologists from several academic and private pathology practices. To date, at

Digital Pathology. DOI: https://doi.org/10.1016/B978-0-443-13809-6.00015-4

least three digital pathology companies have received FDA approval to market their digital pathology systems for primary diagnosis in surgical pathology; these include the Philips Intellisite Pathology Solution (PIPS) approved in April 2017, the Aperio AT2 DX from Leica approved in May 2019, and the Nanozoomer S360MD from Hamamatsu approved in October 2022. All parties involved in these trials were responsible for managing ethical (e.g., institutional review board approval) and regulatory issues (e.g., adhering to good clinical practice).

Academic industry partnership benefits

For academia, these partnerships help fund translational research and stimulate innovation (Gregory, 1997; Rao & Mulloth, 2017; Rybniceck & Konigsgruber, 2019; Van Norman & Eisenkot, 2017b). Industry collaboration is becoming a popular mechanism for additional revenue streams in academia. They also lower the barrier for academics to develop marketable technology they can commercialize. By working with industry, some institutions may gain early or low-cost access to newer, more sophisticated products. Additional benefits include fodder for research and publications, as well as the opportunity for faculty and trainees to keep ahead in their field and broaden their experience.

For industry, there are benefits to partnering with an academic medical center because they are an excellent source of novel technology and knowledge, provide access to scientific and medical experts, offer ethical access to clinical patients and data, fuel publications that help with marketing, and this collaboration presents an opportunity for companies to easily receive market feedback (Rasmussen et al., 2018; Schofield, 2013).

Risks associated with academic industry partnership

Unfortunately, there are risks associated with an AIP. These include conflict of interest (COI), possible litigation, and undermining of academic standards (e.g., publication bias) (Gregory, 1997; Hillerbrand & Werker, 2019; McDonnell et al., 2021; Meslin et al., 2015; Prigge, 2005; Singh & Singh, 2005). Bad press may erode public trust in research or cause others to shy away from similar AIPs. Today, with the need to access large patient datasets to develop AI algorithms, concern has emerged regarding ethics and data ownership (Birch et al., 2020).

Academia versus industry priorities

Academia and industry share traditional core missions such as research and education. However, both partners face different constraints to achieve these goals with divergent agendas and varying prioritization. For academic institutions, education and clinical care are of the highest institutional priority, followed by research and then innovation. For industry, to achieve a competitive advantage, there is a compelling need to prioritize innovation over research and education. Moreover, compared to corporate culture, which is more guarded about sharing propriety information because they want to monetize their innovations, in academia researchers are usually driven to share and publish their novel findings. It is important not to let these different agendas and competing priorities hinder a successful AIP.

Contracts

The **sponsored research agreement (SRA)** is the type of contract most executed with an AIP. This represents a formal contract between a university and the sponsor that intends to fund and conduct specific research during a defined timeframe at the university. Funds received from a company that do not demand receiving rights in return (e.g., licensed IP) are best considered a gift. SRAs can be supported by funding received from a nonprofit (e.g., federal government, foundation) or for-profit (e.g., private industry) sponsor. These agreements typically include the scope of work to be undertaken, deliverables (e.g., publication of results), budget to perform the research, payment obligations, options to license IP arising from this research, compliance requirements, and rights to terminate the project.

A **technology transfer agreement (TTA)** is used when an academic partner develops an innovation and registers it as IP and then seeks to transfer their IP to an industry partner. Licensing of IP to an industry partner (licensor) is best handled by an institution's technology transfer office (TTO). Academic inventors (licensees) in the USA who develop a start-up company and license their own IP from their university may seek Small Business Innovation Research (SBIR) funding from the federal government. Alternatively, a company may approach an academic partner via a Small Business Technology Transfer (STTR) agreement to develop technology in partnership. STTR grants are suitable for early-stage spin-off companies, whereas SBIR is best for self-sufficient companies.

An **industry consulting agreement (ICA)** is a personal contract established between an academic faculty member (i.e., consultant) and an industry client who wants to retain specific services from this consultant. This agreement should define the scope and financial terms of the consulting activity. In these situations, the consultant's university is typically not a party in these personal agreements. Of note, compensation is usually, but not always, paid directly to faculty. Albeit that rule may differ among institutions, in general faculty are permitted to spend a proportion (e.g., up to 20%) of their time consulting with industry.

Other agreements that may likely be needed to support an AIP include a **non-disclosure agreement (NDA)** (Sunderland, 2020) or **confidential disclosure**

agreement (CDA), and **material transfer agreement (MTA)** or **data use agreement (DUA)**. A clinical trial agreement (CTA) may combine elements of all these agreements. Signing an NDA or CDA is often necessary prior to any party sharing their proprietary information or other IP. At most universities, their TTO or sponsored program office (SPO) can provide a template to initiate NDA or CDA negotiations. Most universities will not permit investigators involved in an AIP to sign these agreements on behalf of the university. For industry-sponsored clinical trials, where an academic center only needs to receive confidential information relevant to execute a study, a one-way agreement will suffice. The MTA assists with the acquisition of various biological and research materials, including data. The purpose of a DUA is to govern the transfer (both outgoing and incoming) and use of nonpublic, restricted data (e.g., patient data with personalized health information) between parties. The DUA should provide details that define the subject matter of data, ownership, permitted use, limitations, rights (e.g., to publish and/or patent results), as well as information about data disposal and liability.

Innovation cycle

Innovation entails the invention of new products and/or services that may result in commercialization. By fueling the efficiency and effectiveness of innovation, a successful AIP can be an excellent mechanism to encourage entrepreneurship in a pathology department (Bruneel et al., 2010). For this reason, certain universities have created centers, institutes, or incubators to foster industry partnerships (e.g., MIT Media Lab). This mechanism allows faculty and students to connect with industry, as well as entrepreneurs. Of note, the digital pathology industry has numerous examples where AIPs have thrived, and companies were consequently spun out of academia (e.g., InterScope, Omnyx, Paige.AI, among others) (Pantanowitz et al., 2018).

Innovation begins with product discovery and ends with commercialization (Fig. 11.1). The innovation cycle can be further divided into four phases: discovery, project selection, product development, and commercialization (de Kluyver, 1977). With the discovery phase, a researcher or principal investigator (PI) typically invents a novel device, software, or AI algorithm. This is followed by the project selection phase, when a decision needs to be made whether to invest in commercializing this invention. Tactics such as advocacy and market analysis are often employed at this time to gage the likelihood that an invention will lead to commercial success. An AIP can be particularly useful at this stage because industry partners have better expertise and infrastructure for this process. Funding during this phase may involve seed/angel investment (Mas & Hsueh, 2017). The subsequent product development phase involves developing a prototype or pilot within the academic system, often with the help of local business developers to

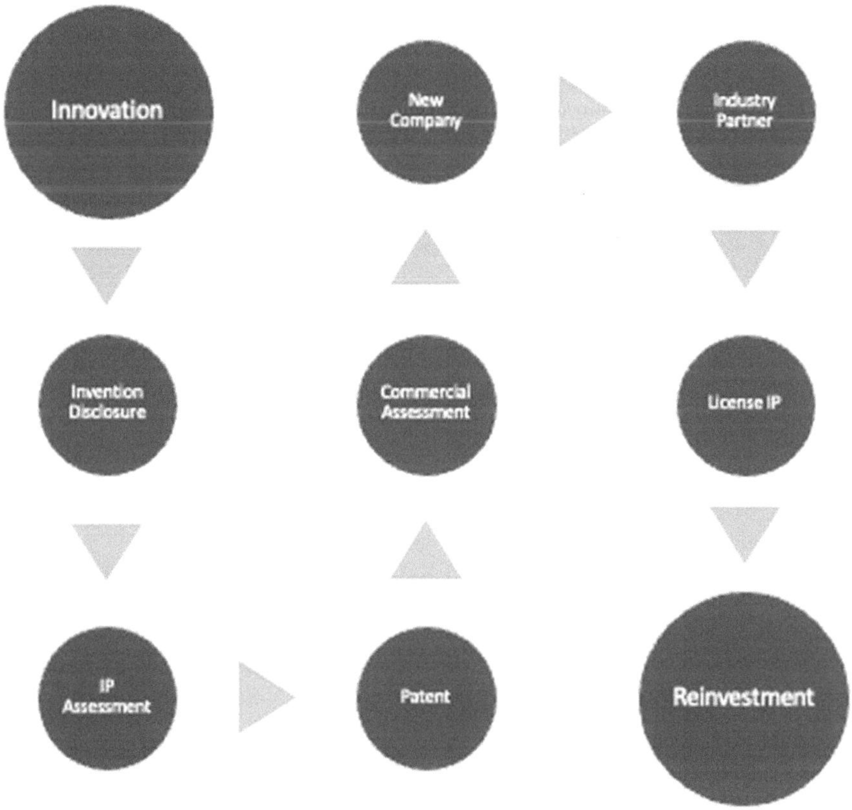

FIGURE 11.1

Innovation cycle demonstrating how an invention can go from the lab to market.

translate the derisked (validated) technology into an advanced product. Finally, there is a commercialization phase where the product is brought to market and adapted to meet customer needs. This may necessitate licensing the product to a company and/or formation of a NewCo (i.e., spin-off or start-up company). Again, an AIP can be helpful because success is dependent on a sales pipeline and building a customer base.

Once a final product is successfully launched, revenue gained can ideally be reinvested back into the academic medical center to develop related or new products. This, in turn, can support the development of further inventions. In addition, if the faculty involved remain engaged, they can help provide valuable feedback to improve the end product and offer assistance with regulatory approval. They may also decide to act as an individual consultant for this company or serve on their advisory board.

The pathologist's role

Pathologists can be involved at any stage of an AIP. With digital pathology, pathologists are often summoned by industry to serve as subject-matter experts (SME) to provide scientific guidance on product development regarding feasibility, design, workflow impact, and clinical utility. When working with industry, pathologists need to pay attention to ethical (e.g., IRB) and regulatory issues. It is advisable for academic pathologists to consult with their university TTO or technology licensing office (TLO), who will help protect the institution's IP. Faculty may be obligated to assign their inventions to their university.

Disclosure and conflict of interest

Conflict of interest (COI) is best avoided at every step of the innovation process. The safest policy is to disclose everything, every time, especially when there are significant financial conflicts involved (e.g., ownership of stock, licensing revenues, etc.). It is important for academic leaders to avoid drawing negative attention if they are offered equity, board seats, or receive other industry perks. For pathologists, it is imperative that they disclose their COI when conversing with key decision makers about making a financial recommendation related to a company partner or their industry. Disclosures often also apply to family members. If an academic pathologist owns equity in a company, they may be prohibited from taking on a high-ranking executive title (e.g., CEO). In the USA, the Physician Payments Sunshine Act requires manufacturers of medical products to publicly disclose any payments given to physicians or teaching hospitals (Richardson, 2014). This information is published annually in a publicly searchable database.

Intellectual property

IP refers to an invention to which one has legal rights and may accordingly apply for a patent, copyright, trademark, or trade secret protection (Trerise, 2016; Van Norman & Eisenkot, 2017a). In this way, inventors can earn recognition and receive financial benefits (e.g., monopoly rights) from their invention. It is important to note that IP rights at academic institutions usually do not fall automatically to the individual inventor but rather to their institution. Patents rely on mandatory public disclosure, whereas trade secrets rely on secrecy from the public. Patent protection typically lasts 20 years from filing, whereas trade secret protection can be perpetual. However, trade secret protection is lost if someone reverse engineers an invention, whereas a patented invention is protected against reverse engineering. Patents are also restricted to a specific country. Hence, filing a patent in the USA does not automatically confer protection for inventions internationally.

Since the advent of digital pathology, there has been a steady increase in related patents (Cucoranu et al., 2014). The most requested patents in digital pathology are associated with product utility and design. For digital pathology, most patents can be classified into the following technological categories: scanners, telepathology, pathology/diagnostic consultation networks, image analysis, and computer-aided diagnosis (CAD) tools.

Conclusion

With the introduction of digital pathology and more recently in the era of AI (Jain et al., 2014), there has been increasing interest for industry to partner with pathology laboratories, especially those working at academic institutions. The benefits of an AIP outweigh the risks (Ankrah & AL-Tabbaa, 2015; Cyert & Goodman, 1997). Establishing a successful and sustainable AIP requires stakeholders to understand all the aforementioned factors that may influence this partnership. An AIP may fail due to conflict among stakeholders, overpromising, improper budgeting, unclear or changing aims, poor communication, inadequate governance, failed deliverables, arguments over IP, and a damaging scandal (Breault & Knafl, 2020; Tierney et al., 2016). To be successful, there needs to be trust, integrity, as well as compatibility of values and organizational work cultures. Seeking legal counsel including a TTO early on is recommended, especially to help execute contracts, address data sharing and ownership, as well as navigate IP issues, licensing, and revenue opportunities upfront. A prosperous AIP can be fortified by establishing clear goals, a governance structure, a responsible budget, providing project management, carefully managing resources, regular communication, and having a champion on both sides to lead. It is equally important to manage disclosures and comply with all pertinent institutional policies and regulations. Finally, AIPs are to be encouraged because, by working together, academic institutions and industry can create new digital pathology solutions to solve problems and improve the field of pathology.

Disclosures

Liron Pantanowitz consults for Hamamatsu and NTP, serves on the medical advisory board for Ibex and AIxMed, and is a coowner of LeanAP and Placenta AI.

References

Ankrah, S., & AL-Tabbaa, O. (2015). Universities–industry collaboration: A systematic review. *Scandinavian Journal of Management*, *31*(3), 387–408. Available from https://doi.org/10.1016/j.scaman.2015.02.003.

Birch, K., Chiappetta, M., & Artyushina, A. (2020). The problem of innovation in technoscientific capitalism: Data rentiership and the policy implications of turning personal digital data into a private asset. *Policy Studies*, *41*(5), 468–487. Available from https://doi.org/10.1080/01442872.2020.1748264.

Borowsky, A. D., Glassy, E. F., Wallace, W. D., Kallichanda, N. S., Behling, C. A., Miller, D. V., Oswal, H. N., Feddersen, R. M., Bakhtar, O. R., Mendoza, A. E., Molden, D. P., Saffer, H. L., Wixom, C. R., Albro, J. E., Cessna, M. H., Hall, B. J., Lloyd, I. E., Bishop, J. W., Darrow, M. A., ... Bauer, T. W. (2020). Digital whole slide imaging compared with light microscopy for primary diagnosis in surgical pathology. *Archives of Pathology & Laboratory Medicine*, *144*(10), 1245–1253. Available from https://doi.org/10.5858/arpa.2019-0569-oa.

Breault, J. L., & Knafl, E. (2020). Pitfalls and safeguards in industry-funded research. *Ochsner Journal*, *20*(1), 104–110. Available from https://doi.org/10.31486/toj.19.0093, http://www.ochsnerjournal.org/content/ochjnl/20/1/104.full.pdf.

Bruneel, J., D'Este, P., & Salter, A. (2010). Investigating the factors that diminish the barriers to university–industry collaboration. *Research Policy*, *39*(7), 858–868. Available from https://doi.org/10.1016/j.respol.2010.03.006.

Cucoranu, I., Parwani, A. V., Vepa, S., Weinstein, R., & Pantanowitz, L. (2014). Digital pathology: A systemic evaluation of the patent landscape. *Journal of Pathology Informatics*, *5*(1), 2014.

Cyert, R. M., & Goodman, P. S. (1997). Creating effective university-industry alliances: An organizational learning perspective. *Organizational Dynamics*, *25*(4), 45–57. Available from https://doi.org/10.1016/s0090-2616(97)90036-x.

de Kluyver, C. A. (1977). Innovation and industrial product life cycles. *California Management Review*, *20*(1), 21–33. Available from https://doi.org/10.2307/41164735.

Gregory, E. H. (1997). University–industry strategic partnerships: Benefits and impediments. *Industry and Higher Education*, *11*(4), 253–254. Available from https://doi.org/10.1177/095042229701100413, https://uk.sagepub.com/en-gb/eur/industry-and-higher-education/journal202558.

Hillerbrand, R., & Werker, C. (2019). Values in university–industry collaborations: The case of academics working at universities of technology. *Science and Engineering Ethics*, *25*(6), 1633–1656. Available from https://doi.org/10.1007/s11948-019-00144-w.

Jain, S. H., Rosenblatt, M., & Duke, J. (2014). Is big data the new frontier for academic–industry collaboration? *JAMA: The Journal of the American Medical Association*, *311*(21), 2171–2172. Available from https://doi.org/10.1001/jama.2014.1845, http://jama.jamanetwork.com/data/Journals/JAMA/930297/jvp140032.pdf.

Kneller, R. (2001). Technology transfer: A review for biomedical researchers. *Clinical Cancer Research*, *7*(4), 761–774.

Mas, J. P., & Hsueh, B. (2017). An investor perspective on forming and funding your medical device start-up. *Techniques in Vascular and Interventional Radiology*, *20*(2), 101–108. Available from https://doi.org/10.1053/j.tvir.2017.04.003, http://www.elsevier.com/inca/publications/store/6/2/3/1/9/9/index.htt.

McDonnell, J. M., Dalton, D. M., Ahern, D. P., Welch-Phillips, A., & Butler, J. S. (2021). Methods to mitigate industry influence in industry sponsored research. *Clinical Spine Surgery*, *34*(4), 143–145. Available from https://doi.org/10.1097/BSD.0000000000001098, http://journals.lww.com/jspinaldisorders/pages/default.aspx.

Meslin, E. M., Rager, J. B., Schwartz, P. H., Quaid, K. A., Gaffney, M. M., Duke, J., & Tierney, W. M. (2015). Benchmarks for ethically credible partnerships between industry and academic health centers: Beyond disclosure of financial conflicts of interest. *Clinical and Translational Medicine*, *4*(1). Available from https://doi.org/10.1186/s40169-015-0077-y.

Mukhopadhyay, S., Feldman, M. D., Abels, E., Ashfaq, R., Beltaifa, S., Cacciabeve, N. G., Cathro, H. P., Cheng, L., Cooper, K., Dickey, G. E., Gill, R. M., Heaton, R. P., Kerstens, R., Lindberg, G. M., Malhotra, R. K., Mandell, J. W., Manlucu, E. D., Mills, A. M., Mills, S. E., … Taylor, C. R. (2018). Whole slide imaging versus microscopy for primary diagnosis in surgical pathology: A multicenter blinded randomized noninferiority study of 1992 cases (pivotal study). *American Journal of Surgical Pathology*, *42*(1), 39–52. Available from https://doi.org/10.1097/PAS.0000000000000948, http://journals.lww.com/ajsp/pages/default.aspx.

Pantanowitz, L., Bui, M. M., Chauhan, C., ElGabry, E., Hassell, L., Li, Z., Parwani, A. V., Salama, M. E., Sebastian, M. M., Tulman, D., Vepa, S., & Becich, M. J. (2022). Rules of engagement: Promoting academic-industry partnership in the era of digital pathology and artificial intelligence. *Academic Pathology*, *9*(1). Available from https://doi.org/10.1016/j.acpath.2022.100026, https://www.sciencedirect.com/journal/academic-pathology.

Pantanowitz, L., Sharma, A., Carter, A. B., Kurc, T., Sussman, A., & Saltz, J. (2018). Twenty years of digital pathology: An overview of the road travelled, what is on the horizon, and the emergence of vendor-neutral archives. *Journal of Pathology Informatics*, *9*(1)40. Available from https://doi.org/10.4103/jpi.jpi_69_18.

Prigge, G. W. (2005). University–industry partnerships: What do they mean to universities?: A review of the literature. *Industry and Higher Education*, *19*(3), 221–229. Available from https://doi.org/10.5367/0000000054300486, https://uk.sagepub.com/en-gb/eur/industry-and-higher-education/journal202558.

Rao, B., & Mulloth, B. (2017). The role of universities in encouraging growth of technology-based new ventures. *International Journal of Innovation and Technology Management*, *14*(4). Available from https://doi.org/10.1142/S0219877017500146, http://www.worldscinet.com/ijitm/ijitm.shtml.

Rasmussen, K., Bero, L., Redberg, R., Gøtzsche, P. C., & Lundh, A. (2018). Collaboration between academics and industry in clinical trials: Cross sectional study of publications and survey of lead academic authors. *BMJ*, *363*, k3654. Available from https://doi.org/10.1136/bmj.k3654.

Richardson, E. (2014). The physician payments sunshine act. *Health Affairs*. Available from https://doi.org/10.1377/hpb20141002.272302.

Rose, D. M., Marshall, R., & Surber, M. W. (2015). Pharmaceutical industry, academia and patient advocacy organizations: What is the recipe for synergic (win-win-win) collaborations? *Respirology*, *20*(2), 185–191. Available from https://doi.org/10.1111/resp.12458, http://www.blackwell-science.com/res.

Rybniceck, R., & Konigsgruber. (2019). What makes industry–university collaboration succeed? A systematic review of the literature. *Journal of Business Economics*, *89*, 221–250, 2019.

Schofield, T. (2013). Critical success factors for knowledge transfer collaborations between university and industry. *Journal of Research Administration*, *44*, 38–56, 2013.

Singh, A., & Singh, S. (2005). The connection between academia and industry. *Mens Sana Monographs*, *3*(1), 5. Available from https://doi.org/10.4103/0973-1229.27876.

Srimathveeravalli, G., Balesh, E., Cheng, C. P., & Chen, D. (2017). If you build it, they will come: How to establish an academic innovation enterprise. *Techniques in Vascular and Interventional Radiology*, *20*(2), 121–126. Available from https://doi.org/10.1053/j.tvir.2017.04.005, http://www.elsevier.com/inca/publications/store/6/2/3/1/9/9/index.htt.

Sunderland, J. J. (2020). The academic NDA: Justification, process, and lessons learned. *Journal of Nuclear Medicine*, *61*(4), 480–487. Available from https://doi.org/10.2967/JNUMED.119.238287, http://jnm.snmjournals.org/content/61/4/480.full.pdf + html.

Tierney, W. M., Meslin, E. M., & Kroenke, K. (2016). Industry support of medical research: Important opportunity or treacherous pitfall? *Journal of General Internal Medicine*, *31*(2), 228–233. Available from https://doi.org/10.1007/s11606-015-3495-z, http://www.springerlink.com/content/120414/.

Trerise, J. (2016). The influence of patents on science. *Politics, Philosophy & Economics*, *15*(4), 424–450. Available from https://doi.org/10.1177/1470594x16666022.

Van Norman, G. A., & Eisenkot, R. (2017a). Technology transfer: From the research bench to commercialization: Part 1: Intellectual property rights—Basics of patents and copyrights. *JACC: Basic to Translational Science*, *2*(1), 85–97. Available from https://doi.org/10.1016/j.jacbts.2017.01.003, http://www.journals.elsevier.com/jacc-basic-to-translational-science.

Van Norman, G. A., & Eisenkot, R. (2017b). Technology transfer: From the research bench to commercialization: Part 2: The commercialization process. *JACC: Basic to Translational Science*, *2*(2), 197–208. Available from https://doi.org/10.1016/j.jacbts.2017.03.004, http://www.journals.elsevier.com/jacc-basic-to-translational-science.

CHAPTER

Ethical aspects of using artificial intelligence in digital and computational pathology 12

Chhavi Chauhan
American Society for Investigative Pathology, Rockville, MD, United States

Introduction

Pathology is the backbone of medicine that guides clinical practice, patient outcomes, as well as overall patient wellbeing. It is also a very data-centric branch of medicine that has a huge potential to leverage artificial intelligence (AI)-, deep learning (DL)-, and machine learning (ML)-based tools to improve overall healthcare delivery. The responsible use of these tools in pathology can help improve clinical decision-making by (1) enhancing access to diverse annotated pathology datasets, (2) significantly decreasing the time to access and analyze clinical samples, (3) allowing faster second consultations, (4) enabling global collaboration on difficult cases, especially in resource-limited settings, (5) improving workflow efficiencies to decrease turnaround times, (6) improving training and pathology education as well as knowledge sharing, (7) improving accuracy of diagnoses by reducing human error, (8) supporting complex predictive analysis to enable precision medicine, and ultimately guiding overall faster and better healthcare delivery to patients for their betterment (Hamilton et al., 2012).

The branch of pathology that leverages information from digitized specimen slides (whole slide imaging (WSI)) is referred to as digital pathology (DP), and it heavily relies on data management for expedited and improved patient outcomes. DP has been around for several years, and WSI has paved the way for computational pathology. However, adoption of DP has moved at a glacial pace prior to the Coronavirus Disease 2019 (COVID-19) pandemic that finally accelerated the adoption of both teleconsultation and DP. Despite the recent shifts, we are still far from the anticipated mass adoption of DP, primarily due to the concerns surrounding the costs associated with the transition, the magnitude of needed new infrastructure and trained personnel, and quite remarkably, a full understanding of the needed AI- and ML-based tools that are transparent, trustworthy, interoperable, explainable, and adhere to the key ethical principles to not only support their development and deployment but also enable them to remain sustainable and

Digital Pathology. DOI: https://doi.org/10.1016/B978-0-443-13809-6.00016-6

scalable. This chapter will focus on a few of the ethical principles surrounding the use of digital pathology, especially bias and fairness, patient consent and patient data privacy and safety, and equity, and highlight some successful use cases of AI/ML in DP and computational pathology.

The shift from pathology to digital and computational pathology

The discipline of pathology allows diagnoses of diseases by analysis of tissue, cell, or body fluids. DP specifically refers to the digitization of glass slides with patient samples into whole slide images (WSI), replacing the need for microscopes with computer monitors. DP makes image analysis more convenient and ergonomic in ways while increasing the AI-enabled analysis capability. Computational pathology encompasses computationally analyzing an array of methods to analyze patient specimens to study diseases. Computational pathology focuses on the extraction of information from digitized pathology images with associated metadata, usually leveraging artificial intelligence (AI) methods such as image analysis, ML, and DL (Abels et al., 2019). Both DP and computational pathology have the potential to improve patient disease management by improving and expediting disease detection to enable better patient care for more positive outcomes.

DP requires a broad range of resources that range from infrastructure, specialized instruments and tools, trained personnel, as well as processes pertaining to digitization, whereas computational pathology relies heavily on the quality and accuracy of the training datasets and multistakeholder input for efficient model development. It is imperative that both DP and computational pathology and the associated processes conform to ethical guidelines and frameworks for desired positive outcomes. The future of DP and computational pathology will likely be augmented intelligence empowered by responsible and ethical use of AI (Colling et al., 2019).

Ethical concerns in leveraging artificial intelligence in digital pathology and computational pathology

The ethical concerns leveraging AI in DP and computational pathology largely reflect the concerns surrounding the use of AI in healthcare broadly. Some of the thoughts and questions that constantly surface include: What does ethical use of patient data look like, especially when trust, transparency, and explainable AI are not yet mainstream? Inequitable access to healthcare, health disparities, and socioeconomic disparities have finally come to the forefront in the light of the recent pandemic. How should we address technological disparity to enable access to

AI-empowered, equitable healthcare? What level and kind of vigilance should we envision, develop, and deploy when we are talking about patient data, especially in the light of ransomware and cyber-attacks? How do we establish trust in AI in DP and computational pathology?

There is a huge need for large sets of patient data with adequate representative populations to train AI/ML algorithms, which raises issues pertaining to patient consent, data privacy and security, as well as data de-identification. It is imperative that these models are ethically trained, tested, and deployed (and scaled) after a careful review by human experts to ensure that they are safe, reliable, trustworthy, explainable, and that they perform as intended and do not result in unintended consequences. Safeguards need to be built in to identify any unintended performances with detailed procedures in place to circumvent them as well as manage the outcomes with proper, responsible action taken to protect the interests of all stakeholders.

There are several ethical aspects, including but not limited to (1) algorithmic accountability, (2) quality and accuracy of data used for training AI datasets, (3) data security and privacy, (4) engineering of AI systems, (5) safety of AI systems, (6) explainability and interoperability, (7) trustworthiness, (8) transparency, (9) accessibility, (10), data safety, (11) governance, etc., that need to be adequately considered while developing and deploying AI-based tools and to ensure their scalability and sustainability. The next sections will focus on some of these fundamental ethical aspects needed for us to embrace digital pathology to make it mainstream. There are several ethical aspects that have been comprehensively discussed previously by myself (Chauhan & Gullapalli, 2021; Fig. 12.1) and others (Jackson et al., 2021).

Bias and fairness

From time immemorial, humans and their behaviors have been impacted by biases. In fact, the cognitive bias codex lists 180 different kinds of biases (Fig. 12.2) that can impact human decision-making.

Even small biases or personal identifiers persistent in the training datasets can be used by heuristic models to arrive at undesirable and unintended outcomes with deep irrevocable impacts on the health and safety of patients (Obermeyer et al., 2019).

Addressing bias and equity in healthcare

Though AI/ML has the potential to transform healthcare and enable equitable access and treatment, its selective use has the potential to exacerbate the existing healthcare inequities and potentially introduce new disparities if bias is not addressed properly.

Metrics have been proposed to analyze bias, and strategies have been proposed to mitigate bias to ensure health equity by following the total product lifecycle

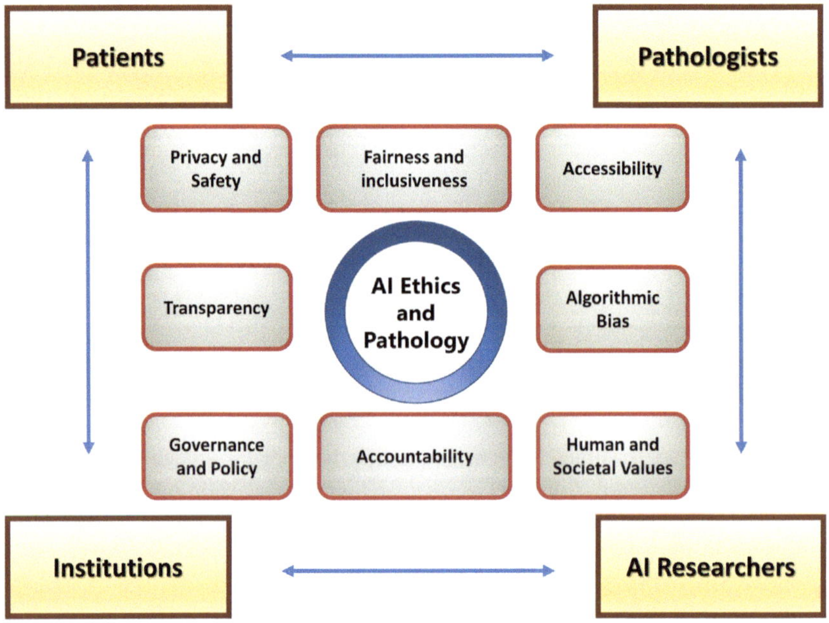

FIGURE 12.1

Key participants and issues of emerging importance in the study of artificial intelligence (AI) ethics in pathology. Understanding the ethical issues shown at the center of the image is critical to enabling a robust framework of AI implementation within pathology practice.

(TPLC) framework for healthcare AI/ML (Abràmoff et al., 2023) by including all pertinent stakeholders. A 5-year Canadian-Australian research program under the IMPACT (Innovative Models Promoting Access to Care Transformation) program may also help to improve access to public healthcare for vulnerable populations.

Some bias mitigation strategies include data preprocessing to clean data to reduce bias, model selection to prioritize fairness, postprocessing to adjust model outputs to account for fairness, enhancement of transparency and accountability in the AI tools used, and interdisciplinary multistakeholder collaboration among a diverse set of experts to address bias effectively (Ferrara, 2023).

Ethical concerns pertaining to patient privacy and choice

While deploying AI infrastructure in pathology, several patient-facing ethical concerns immediately come to mind. First and foremost, it is critical to ensure patient privacy when using AI to analyze pathology images. The rules are quite strict in the US; however, in some other countries there is no direct comparison for the

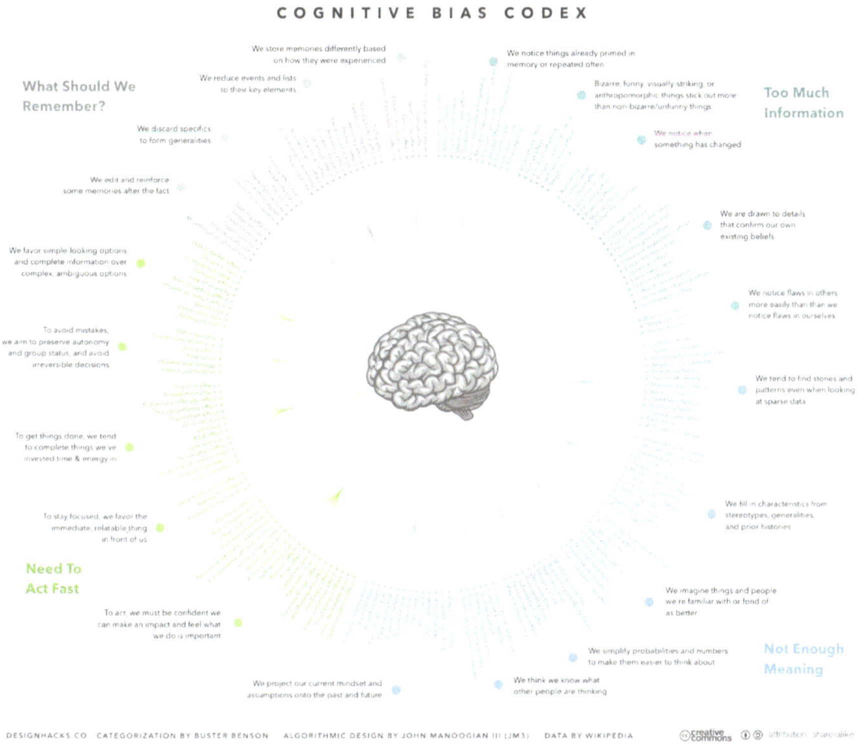

FIGURE 12.2

Bias codex created by John Manoogian III and Buster Benson. A visual representation of biases reported to date. The biases are represented in a circle and are divided into four quadrants, each of which depicts a specific group of cognitive biases.

Source*: Cognitive Bias Codex with Definitions, an Extension of the work of John Manoogian by Brian Morrissette.jpg.*

Health Insurance Portability and Accountability Act (HIPPA), which may make the patient data vulnerable to malicious use and the trained models susceptible to misperformance due to the presence of additional identifiers, which may lend the heuristic models to change the analysis course to arrive at the desired outcome, but with previously unintended consequences.

We are still far from providing a choice to the patients to seek their consent for using their data in AI research or to consent to a treatment regimen that was augmented by the use of AI. There still remain deep fears about undetected underlying bias in training datasets that may disproportionately impact specific patient populations.

We have a long way to go to educate patients to build trust and transparency surrounding the use of AI/ML in pathology as well as institute clear consent procedures both for data sharing and for data analysis. Shared decision-making involving both the patients and pathologists is the only trustworthy and transparent approach that will enable full-scale adoption of AI/ML in the realm of pathology. It is needless to say that we will have to rely on robust national and international governance mechanisms to completely transform this sector.

Ethical concerns pertaining to ownership and control of patient data

There seems to be a tendency for AI technologies to be owned and controlled by private entities, such as corporations, large clinics, or even public bodies, despite the original source of the training dataset and primary research being academic institutions or healthcare facilities. It is critical to implement systemic oversight to ensure data privacy safeguards and patient agency (Murdoch, 2021). There is no denying that AI/ML tools can easily compromise patient data privacy if the training datasets have hidden personal identifiers. There are tools available that can successfully reidentify deidentified or even anonymized data, which can compromise patient data privacy protection. Such risks may increase the vulnerability of patient data for malicious use. Therefore, privacy implications and data security should be addressed in tandem to allow safe technological advancement while safeguarding patient information.

Regulation of artificial intelligence and machine learning and oversight

Given the inherent black box nature of AI and its critical use in DP and computational pathology, it is crucial to institute standards and to regulate its use for continued intended performance with high accuracy. Any changes or even marginal shifts in the training datasets can force these models to deviate in unexplainable ways and compromise their performance and accuracy.

There is an urgent need for an equitable regulatory framework for ML to clearly define (1) requirements based on risk, (2) best practices for developing, training, deploying, and scaling AI tools with a focus on bias assessment and mitigation, (3) auditing procedures to constantly verify performance, (4) frameworks to identify and mitigate any unintended performance, and (5) key responsibility for issues as well as responsible individuals to monitor and mitigate these issues when they arise.

The United Kingdom was the first to launch the AI Act (AIA), which is a cross-sector outcome-based framework to regulate the use of AI. The AIA is based on the foundation of five core principles: safety, security, and robustness; transparency and explainability; fairness, accountability, and governance; and contestability and redress. Though there is still time for AIA to fully become effective, cross-sector efforts are already under way to comply with the listed frameworks for using AI in all domains in the United Kingdom (https://www.gov.uk/government/publications/ai-regulation-a-pro-innovation-approach/white-paper; last accessed 05/15/24; ISBN: 978-1-5286-4009-1).

The President of the United States of America issued the AI Bill of Rights, which is a set of principles designed to ensure transparent development of AI tools and to monitor inaccurate and biased data and other related concerns. The goal of the AI Bill of Rights is to protect privacy and civil rights for all citizens (https://www.whitehouse.gov/ostp/ai-bill-of-rights/; last accessed on 05/15/24). The FDA itself has launched guidance for the use of software as a medical device (SaMD), with specific focus on AI/ML-enabled medical devices (https://www.fda.gov/medical-devices/software-medical-device-samd/artificial-intelligence-and-machine-learning-aiml-enabled-medical-devices; last accessed on 05/15/24). However, there continues to be a gap in concrete guidance, approved methods, as well as best practices for individuals and organizations in deploying ML tools.

Successful artificial intelligence case studies in digital pathology and computational biology

Predicting patient outcomes with multimodal data integration

An ML model can now predict patient outcomes from WSI from histology tissue and genomic sequencing data from The Cancer Genome Atlas for many cancer types. This model also independently identified *IDH1* mutations as a prognostic factor in low-grade gliomas, suggesting the use of machine learning models for better prognosis for patients (Chen et al., 2022). The continued enrichment of the training dataset for this model will perhaps further improve its performance as well as utility.

Stratifying prostate cancer risk

A new light microscopy approach allows imaging prostate biopsies in 3D and leverages AI to accurately predict cancer recurrence as well as stratify high-risk versus low-risk patients (Xie et al., 2022). After validation in larger patient cohorts, this technology may augment pathologists' interpretation of traditional 2D sections.

Augmenting pathologists in prostate cancer detection

Paige Prostate, an algorithm aimed to assist pathologists in cancer detection in WSIs derived from prostate biopsies, was the first AI product in DP to receive clearance from the US Food and Drug Administration (FDA). Paige Prostate was reported to improve cancer detection on individual WSIs by around 7.3% as compared to pathologists' unassisted reads of WSIs of individual biopsies (FDA Authorizes Software that Can Help Identify Prostate Cancer | FDA; last accessed on 05/14/24). To date, more than 500 healthcare AI algorithms have received FDA clearance, with four so far for pathology (FDA has now cleared more than 500 healthcare AI algorithms, four of which are for pathology—Pathology news; last accessed on 15.05.24).

Achieving ethical use of artificial intelligence in digital pathology

The responsible and ethical use of AI in DP and computational pathology can be achieved by building trust for AI by leveraging key ethical frameworks (Chauhan & Gullapalli, 2021). Federated learning is a good approach to training models on-device or on locally available data within the safety network parameters. There is an urgent need to implement algorithmic assessment and auditing into the compliance process with periodic checks on the intended performance of AI and defined frameworks to impede any unintended outcomes. Once instituted, these auditing frameworks should be constantly updated and fine-tuned to comply with internal as well as external industry best practices.

Conclusions

AI continues to prove to be a powerful tool that can augment pathologists to improve disease diagnosis, augment a physician's ability to better serve the needs of a patient, help the physicians realize their passion for healing patients better, expedite patient care with more positive outcomes, and allow extension of high-quality diagnosis globally, even in resource-limited settings with a paucity of trained pathologists. However, it is critical for us to continue to focus on safe, ethical, and responsible use of AI in DP and computational pathology to realize its full potential. In the healthcare domain, there is no circumventing "humans in the loop," not only to ascertain accurate and timely disease diagnoses but also to continue to strive to improve patient care and wellbeing. The future of DP and computational pathology will likely be augmented intelligence empowered by responsible and ethical use of AI.

Disclosures

None declared.

Funding

None.

References

Abels, E., Pantanowitz, L., Aeffner, F., Zarella, M. D., van der Laak, J., Bui, M. M., Vemuri, V. N. P., Parwani, A. V., Gibbs, J., Agosto-Arroyo, E., Beck, A. H., & Kozlowski, C. (2019). Computational pathology definitions, best practices, and recommendations for regulatory guidance: A white paper from the Digital Pathology Association. *Journal of Pathology*, *249*, 286–294.

Abràmoff, M. D., Tarver, M. E., Loyo-Berrios, N., Trujillo, S., Char, D., Obermeyer, Z., Eydelman, M. B., Foundational Principles of Ophthalmic Imaging and Algorithmic Interpretation Working Group of the Collaborative Community for Ophthalmic Imaging Foundation, Washington, D.C., & Maisel, W. H. (2023). Considerations for addressing bias in artificial intelligence for health equity. *Digital Medicine*, *6*, 170. Available from https://doi.org/10.1038/s41746-023-00913-9.

Chauhan, C., & Gullapalli, R. R. (2021). Ethics of AI in pathology: Current paradigms and emerging issues. *The American Journal of Pathology*. Available from https://doi.org/10.1016/j.ajpath.2021.06.011.

Chen, R. J., Lu, M. Y., Williamson, D. F. K., Chen, T. Y., Lipkova, J., Noor, Z., Shaban, M., Shady, M., Williams, M., Joo, B., & Mahmood, F. (2022). Pan-cancer integrative histology-genomic analysis via multimodal deep learning. *Cancer Cell*, *40*(8), 865–878, e6. Available from https://www.cell.com/cancer-cell/fulltext/S1535-6108(22)00317-8, e6.

Colling, R., Pitman, H., Oien, K., Rajpoot, N., Macklin, P., Snead, D., Sackville, T., & Verrill, C. (2019). Artificial intelligence in digital pathology: A roadmap to routine use in clinical practice. *The Journal of Pathology*, *249*, 143–150.

Ferrara, E. (2023). *Fairness and bias in artificial intelligence: A brief survey of sources, impacts, and mitigation strategies*. https://arxiv.org/pdf/2304.07683

Hamilton, P. W., Wang, Y., & McCullough, S. J. (2012). Virtual microscopy and digital pathology in training and education. *Journal of Pathology, Microbiology and Immunology*, *120*(4), 305–315. Available from https://doi.org/10.1111/j.1600-0463.2011.02869.x.

Jackson, B. R., Ye, Y., Crawford, J. M., Becich, M. J., Roy, S., Botkin, J. R., de Baca, M. E., & Pantanowitz, L. (2021). The ethics of artificial intelligence in pathology and laboratory medicine: Principles and practice. *Academic Pathology*, *8*. Available from https://doi.org/10.1177/2374289521990784.

Murdoch, B. (2021). Privacy and artificial intelligence: Challenges for protecting health information in a new era. *BMC Medical Ethics*, *22*, Article number: 122.

Obermeyer, Z., Powers, B., Vogeli, C., & Mullainathan, S. (2019). Dissecting racial bias in an algorithm used to manage the health of populations. *Science*, *366*(6464), 447–453. Available from https://doi.org/10.1126/science.aax234.

Xie, W., Reder, N. P., Koyuncu, C., Leo, P., Hawley, S., Huang, H., Mao, C., Postupna, N., Kang, S., Serafin, R., Gao, G., Han, Q., Bishop, K. W., Barner, L. A., Fu, P., Wright, J. L., Keene, C. D., Vaughan, J. C., Janowczyk, A., … Liu, J. T. C. (2022). Prostate cancer risk stratification via nondestructive 3D pathology with deep learning-assisted gland analysis. *Cancer Research*, *82*(2), 334–345. Available from https://doi.org/10.1158/0008-5472.CAN-21-2843, 34853071, PMCID: PMC8803395.

CHAPTER

Next-generation imaging in pathology

13

Richard Torres[1,2]

[1]Applikate Technologies Inc., Fairfield, CT, United States
[2]Laboratory Medicine, Yale School of Medicine, New Haven, CT, United States

Why next-gen imaging matters

Imaging technologies have seen a renaissance of innovation over the past few decades, fueled by key discoveries in optics and supported by continuing rapid advancement in computational hardware and analysis techniques. Superficially, the appeal of many new imaging methods for pathology largely revolves around speed and ease, while some offer avenues for expanding the amount and type of data that can be extracted from a given specimen. An exciting perspective is that some emerging techniques could fully supplant preparation of physical slides and, in doing so, engender easier and faster practical implementation of digital pathology. This would increase access and use of powerful image analysis tools while creating opportunities for efficient workflows that yield improvements in diagnostic precision and, by consequence, quality of care. In this regard, next-generation imaging technologies represent a path for pathology to rapidly produce tangible benefits to patients through widespread digital adoption.

An entry point for evaluating many new imaging techniques has been intraoperative assessment (i.e., frozen section), which also encompasses the concept of rapid onsite evaluation (ROSE). A frozen section is a prominent pain point for pathologists due to the degradation of histology by artifacts of freezing (Desciak & Maloney, 2000; Peters, 2010), making critical decisions dependent on inferior morphologic information. Therefore, there is ample recognition of the need for better intraoperative assessment processes (Taxy, 2009). Also, there is a lower regulatory burden for using new imaging methods in the rapid onsite/intraoperative setting. Two reasons for this are: (1) the frozen tissue standard has known severe limitations, and (2) the final definitive diagnosis must always be rendered separately with standard paraffin embedded glass slides. These reduce the overall risk of incomplete or inadequate information produced by alternative imaging techniques. While this makes early implementation easier, the high variability of sample types submitted for intraoperative assessment introduces challenges for generalizability. It also only represents a minority of the total of pathology

Digital Pathology. DOI: https://doi.org/10.1016/B978-0-443-13809-6.00022-1

practice and therefore a small fraction of the areas of pathology in which new imaging modalities could have a profound impact.

The intraoperative use helped evolve the misnomer of new imaging techniques as "ex vivo" microscopy (EVM), which in fact narrowly refers to applications where freshly excised tissue is microscopically visualized prior to being chemically fixed (Wells et al., 2018). However, as discussed below, tissue fixation and clarification can greatly enhance image quality while still accelerating time to digital image review, so methods that are not technically "ex vivo" may be grouped under the EVM moniker (Krishnamurthy et al., 2019).

Regardless, the prospect of fully eliminating the need for preparation of physical slides, with all its associated labor, time, and technical challenges, is a driving motivation for developing new imaging modalities for pathology. The analogies are the transition from physical radiologic films to digital data, as well as from physical photographic prints to digital cameras. In this respect, use of digital whole slide imaging (WSI) is akin to processing physical x-rays or photograph negatives, then scanning the resulting film so that it can be stored more reliably, shared widely, and analyzed by image processing tools. Transition to pure digital workflow has the potential to have a similarly meaningful impact.

The desire to use direct digital imaging modalities is relatable for pathology, but compared to the transformation in radiology, histology poses additional challenges. Light transmission microscopy of thin physical wax-embedded tissue sections has a proven record of providing reliable information on the complex genetic, protein, and biochemical make-up that corresponds to specific disease states. It is a gold standard with a far higher amount of informational content than that used in radiology—for example, a comprehensive 3D MRI data set collected at moderately high resolution may be $512 \times 512 \times 256 \times 16$ bits ~ 1 billion bits (Dandu, 2008), whereas a standard 15×15 mm slide at $40\times$ equivalent resolution (250 nm pixels) translates into about $60{,}000 \times 60{,}000 \times 24$ bits ~ 86 billion - bits, more than 80 times a 3D MRI. Furthermore, capturing the full range of that information with direct imaging of whole tissue is not trivial. A key reason is the complexity surrounding the achievement of adequate contrast coupled with high microscopic resolution in varied biological specimens and doing so in a relevant time frame. The imaging modalities described in this chapter demonstrate how innovations have emerged that present a range of solutions to this challenge with the potential to fuel the full digital transformation of pathology.

Looking beyond whole slide imaging

At its core, digital pathology promises to have an enormous impact on clinical care in two key aspects: (1) reduction in interpretative errors and variability by facilitating routine sharing with expert subspecialists for consultation (Hanna et al., 2020; Zarella et al., 2018) and (2) enabling computerized image analysis tools that can theoretically further improve consistency, sensitivity, and accuracy of interpretation (Bera et al., 2019; Lee, 2023). More generally, it also has potential to improve review and analysis efficiency (Moscalu et al., 2023). However, WSI implementations have considerable cost and effort downsides, which have affected adoption (Kumar et al., 2020). Compared to standard physical slide review for histology, WSI scanners increase technical personnel needs and management effort, require additional expenditures, and can delay initial review. And although technical developments continue to improve color variability and focusing errors, these characteristics can degrade human and machine interpretation (Jain, 2024).

Another critical aspect is that WSI does not address problems that are related to standard physical slide preparation. Formalin fixation with paraffin embedding (FFPE) and hematoxylin and eosin staining (H&E) is a process that has remained largely unchanged for more than a century. The reasons for its endurance as a gold standard can be characterized as being:

- forgiving chemistry that does not depend on highly precise, timed incubations
- wide applicability, making it useful in the majority of clinical specimens
- use of inexpensive reagents
- visualization at submicron resolutions
- accumulated experience of more than 100 years

On the other hand, the process suffers from key limitations, including:

- consumption of tissue from small samples, which can negatively impact downstream testing (Thunnissen et al., 2012)
- limited provision of dimensional perspective
- highly manual and technically complex processes dependent on skilled staffing
- artifacts from physical tissue cutting and staining (McInnes, 2011; Taqi et al., 2018)
- variable yield of nucleic acid recovery (Turashvili et al., 2012)
- costly asset management with associated errors and high risk of samples being lost (Hanna et al., 2019; Heher et al., 2016; Morelli et al., 2013)
- reduction in efficiency of care through long process times

Beyond these aspects, physical slides present challenges for obtaining expert remote review and cannot be readily analyzed by computerized methods. WSI can address these latter limitations, which are key to improving the quality of diagnoses. But by adding cost, effort, time, and potential artifacts, WSI poses barriers to adoption. In short, *the need for prior preparation of physical slides caps the range of gains in efficiency and precision achievable by WSI.*

New imaging techniques offer opportunities to markedly improve workflow and quality of diagnoses. Furthermore, by avoiding the need for complex processing, they may lower the barriers to digital adoption, expanding access to digital benefits (including to remote and low resource settings without local pathology expertise), thereby multiplying the contribution to care that digital pathology can achieve.

What makes a new imaging tech relevant to pathology?

As a starting point, it is worth understanding the context of emerging digital pathology technologies. Notably, imaging advances have occurred in the field of “in vivo” microscopy, particularly for ocular, dermatologic, and gastrointestinal screening applications, many employing methodologies described in this chapter and detailed in other reviews (Wells et al., 2018). However, contrast and resolution are severely limited in vivo by physics constraints that are difficult to overcome, namely, poor intrinsic cellular contrast and high light scattering properties of tissue. The variety of techniques described below for improving contrast, clarity, and resolution in tissue imaging are not generally accessible for use in living subjects due to the risk of tissue damage and patient injury. As such, in vivo imaging cannot currently serve as an adequate substitute for excised tissue evaluation.

Additionally, not all new imaging developments in “ex vivo” microscopy are equally likely to contribute to pathology practice in the foreseeable future. Clinical arenas (both human and veterinary) have specific performance requirements that often differ from those in research. Aspects of distinct relevance to clinical work are high throughput, fast turnaround, low cost, and low effort, as well as compatibility with the standard of care. Imaging techniques that require dedicated personnel performing technically complex processes that take extensive time to complete are unlikely to find broad adoption in clinical laboratories. Workflow changes aside, any new method that (1) results in deviation from established interpretative approaches (how pathologists recognize disease patterns), (2) negatively affects existing protocols for diagnosis and management, or (3) limits common ancillary tests naturally faces a more difficult path in attaining wide usage.

In principle, tissue imaging without paraffin embedding or thin slicing could be achieved without sacrificing quality, creating opportunities for enormous gains in time efficiency, reduced cost, and improved diagnostic precision, while also enabling an expanding role of image analysis and engendering an energized new era of digital pathology. Below we describe emerging technologies being applied for microscopic examination of tissue. We focus specifically on those being developed as either a complement or potential substitute for human clinical, veterinary, and standard research histology.

It may be noted that, within research histology, distinctions can also be made between techniques for facilitating established routine histology (i.e., studies employing hematoxylin and eosin and other standard stains) and those expressly for augmenting existing methods of visualization of pathophysiologic processes to further our understanding of disease (e.g., characterizing the distribution of a specific epitope or genetic pattern). Investigative projects benefit immensely from imaging approaches that identify specific molecular details or elucidate spatial relationships that are not able to be visualized with standard microscopy slides. This new data can eventually serve as the basis for changes in clinical practice, but the risk to human health of misinterpreting unfamiliar data forms mandates that a considerable body of evidence be acquired before achieving broad acceptance. And, as already stated, clinical use necessitates minimal workflow disruptions for adoption. For its part, routine H&E histology in research (e.g., toxicology) may have similar needs to clinical practice as the goal is more likely to be that of matching existing standards while improving efficiency.

Finally, it is worth considering that the transformation of a new imaging modality into a commercial product is a rather significant endeavor. Hardware engineering development is a costly process replete with hurdles and mines. For starters, robotic systems are difficult to get working reliably. It can take years of refinement to automate a single task from an expert technical operator. The steps for physical slide preparation are highly technique dependent, making them supremely challenging for robotic systems, as evidenced by how rather modest and piecemeal the adoption of existing automation systems for standard histology has been. For the development to be executed correctly, engineering standards require detailed planning and constant quality monitoring aimed at ensuring robust performance but at the cost of added time and cost of development. Associated regulatory guardrails meant to ensure this process is done correctly mean added documentation and effort. And, critically, biological samples are highly varied, making them a particular challenge for automating systems. Given this reality, it is prudent to be wary of efficiency, quality, and cost claims of new techniques until they have been tested under representative scenarios.

As such, the most immediately relevant techniques for clinical practice are those either commercialized or far down a path to commercialization, particularly those with testing for routine histology in human and veterinary clinical settings, as well as for research endeavors that employ standard histology techniques. Together, these are expected to influence the largest components of histology work in the years to come.

What is the new tech and how does it work?

There is a constant and seemingly endless stream of new imaging modalities being generated throughout the world, and any description of new technology is fated to become rapidly outdated. The sheer volume of literature on the topic of optics and imaging for histology also makes it possible for interesting approaches to go unnoticed for years. Nonetheless, many new developments are variations on known themes. Plus, given the long developmental cycle of these hardware instruments and the already substantial literature on many alternative imaging methods, some salient trends can be discerned.

Underpinning all slide-free techniques for microscopic imaging of tissue in pathology are two key concepts: (1) the ability to produce thin optical sections drives image quality, and (2) there are trade-offs in terms of speed and image quality for every approach, although the combined theoretical capabilities are intrinsically higher for some methodologies over others, based on the corresponding physical principles.

Optical sectioning refers to the ability to isolate signals to a specific plane or surface without undue interference from other portions of a thick specimen. Not doing so would otherwise cloud images, rendering them uninterpretable, akin to looking at uncut tissue under transmitted light microscopy. It is the reason histology evolved to wax-embedding with thin physical sectioning: to limit visual analysis to a controlled portion of tissue without being clouded from the remainder of the sample. Whatever means are employed to derive image contrast in tissue, be it fluorescence, light reflectance, frequency absorption, etc., interpreting the composition and architecture at a cellular level requires the isolation of the signal to a plane that is at most a few micrometers thick.

The speed versus quality balance is a consideration for any imaging modality (and for tissue sectioning as well!). Generally, noise reduction and higher resolution can be achieved with longer preparation and/or slower imaging. However, there are intrinsic limits to speed and quality for each imaging modality, and understanding the goal for a given pathology use-case (e.g., intraoperative evaluation versus primary diagnosis) can help narrow compatible options.

From the perspective of intraoperative review, the image quality requirements are somewhat lower. As mentioned previously, this is because the comparative standard is of low quality: freezing damages tissue and creates artifacts that are difficult to tease out (Desciak & Maloney, 2000; Peters, 2010). Therefore, imaging approaches that favor speed (either preparation, imaging, or both) over image quality are attractive for intraoperative microscopic evaluation of tissue. The direct imaging approaches that have been most thoroughly explored in this arena include (1) stimulated Raman scattering (SRS) (Corden et al., 2023; Lu et al., 2016; Movahed-Ezazi et al., 2023; Orringer et al., 2017); (2) microscopy with ultraviolet surface excitation (MUSE) (Xie et al., 2019); (3) confocal microscopy (Au et al., 2023; Baas et al., 2023; Togawa et al., 2023; Villarreal et al., 2021); and (4) multiphoton microscopy (aka nonlinear microscopy) (Ching-Roa et al., 2022; Matsui et al., 2022; Sehm et al., 2020; Treacy et al., 2023; Villarreal, Pal, et al., 2023; Xi et al., 2023). Additionally, microcomputed CT (microCT) (Albers et al., 2018; Khimchenko et al., 2016; Senter-Zapata et al., 2016) and optical coherence microscopy (OCM) (van Manen et al., 2018; Wang et al., 2022) have been examined. These methods vary in the amount of tissue preparation, including contrast enhancement or staining. Some, such as multiphoton, can be performed with or without the use of extrinsic dyes. Additionally, the various techniques have different practical limits in speed of data acquisition, and their features are counterbalanced by the quality level of the images produced.

While the threshold for matching frozen section image quality is relatively low, competing with gold-standard light microscopy of formalin-fixed paraffin embedded (FFPE) tissue slides is a more challenging goal. Methods being actively evaluated for the concept of true "slide-free" pathology to replace FFPE are lightsheet/selective plane illumination microscopy (SPIM) (Glaser et al., 2019; Poola et al., 2019), confocal (Guida et al., 2021; Titze et al., 2021), and multiphoton microscopy (Olson et al., 2016; Torres et al., 2021). In principle, lightsheet has the highest collection speed potential, while multiphoton has the highest image quality potential in thick tissue, but the balance depends highly on the specific implementation. For example, the speed of lightsheet microscopy becomes most apparent when looking to collect finely spaced three-dimensional images, for which noncommercial experimental systems report data collection rates on the order of 100–140 megapixels (or megavoxels for 3D) per second (Prince et al., 2023; Yang et al., 2022). Even so, fast implementations of confocal and multiphoton can match or surpass these overall imaging speeds, with experimental reports showing multiphoton microscopy running at 50–200 megapixels per second (Ching-Roa et al., 2023; Olson et al., 2018). Importantly, effective imaging speed is affected by a multitude of factors, including how much blank space is imaged, image processing time (which can be considerable for large 3D datasets), stage movement, data upload, sample loading, etc. The tradeoff with speed and quality during collection is also a critical variable. In addition, complexity and cost affect the practicality of certain approaches, so all must be considered when comparing techniques for use in clinical or other high-volume environments.

Parenthetically, evaluation of published data of the imaging modalities described in this review should include consideration of the tested specimen source. Shorter processing times usually produce lower image quality, so comparing techniques should take note of that trade-off. And while there is considerable investigative value in analyzing samples that have been previously embedded, dewaxed, stained, and imaged in 3D, the image quality obtained from such specimens may not be representative of what can be expected from routine unembedded specimens. This is because sectioning provides a very flat surface, which improves the consistency of optical sections at depth, and the clarification is improved by the multistep process. In other aspects, consideration for imaging clinical specimens should include assessment of total image processing times. Prolonged computation could impact turnaround, throughput, and/or overall cost (if more computing resources are needed). In addition, there are distinctions within this group in terms of the expected capability to provide a definitive *primary* diagnosis based on ultimate image quality. Confocal and multiphoton have the highest intrinsic resolution capability.

Notably, the image quality of the different techniques is also affected to different degrees with increasing depth. Regardless of approach, these techniques typically require the use of optimized tissue clearing (normalizing light path through refractive index matching) to achieve the high threshold of quality needed to compete with physical slide microscopy. Clearing reduces scattering of light, enabling more sensitive detection of signals from a specific imaging plane. In terms of clearing methods, there are also a multitude of approaches (Richardson et al., 2021; Weiss et al., 2021). These are described in more detail below, but they all seek to counteract the opacity created by differences on how light passes through water versus membranes and proteins. When these optical properties are adequately tuned, it has the effect of rendering the tissue extremely transparent, considerably more than the typical "clearing" agent xylene does, thereby allowing incident laser light and the consequent emitted fluorescence to traverse unobstructed through samples.

Given this complex interplay of trade-offs, the comparative evaluation of slide-free methods has many parameters (Table 13.1). Each of the most relevant current imaging modalities is discussed in detail below. Given active trends in *research* environments, we touch upon several other relevant methodologies, including expansion microscopy, superresolution microscopy, and deep-learning-based image enhancement.

Stimulated Raman histology

Raman spectroscopy is a technique capable of evaluating the composition of biological specimens through the identification of types of chemical bonds. It is based on a physical mechanism of light scattering known as Raman scattering, which occurs when a photon interacts with a molecule and a portion of the light energy gets absorbed into vibration of molecular bonds, yielding light with lower

Table 13.1 Considerations in comparing slide-free methods.

Overall image quality (optical sectioning)
Pixel resolution
Sample types tested: fresh, fixed, FFPE dewaxed, and FFPE slide
Accessible imaging depth
Speed of acquisition
Time to digital image
Sample throughput
Remote or local viewing
Cost of sample preparation (e.g., reagents)
Cost of instrumentation
Complexity of sample setup
Sample type restrictions
Stage of commercial development

energy. The lower energy corresponds to shifts of the light to longer wavelengths, which can be quantitatively measured.

The standard Raman scattering effect produces a very weak signal, which means a long time is required to collect reliable data. Over the past few decades, advances in several technical areas have improved sensitivity as well as achievable resolution. As illustrated in Fig. 13.1, a typical modern stimulated Raman histology (SRH) setup includes two pulsed lasers (like those used in multiphoton microscopy described below), which are carefully aligned and coordinated in the timing of their pulses (Lee et al., 2021). Light is focused through a microscope objective to excite the sample in a spot. The lasers are coordinated in such a way that they promote and magnify the vibrational absorption of light energy by specific molecules. For example, the frequency difference can be tuned to quantify light energy absorption predominantly from amino acids, lipids, or pigments. When the method is applied across a specimen surface, a map showing the spatial localization of components can be created (Fig. 13.2).

Given that standard H&E morphology is largely a map of chemical composition, it is no surprise that SRH images are highly reminiscent of what is visible on H&E. The SRH signal can even be recolored to use similar hues as H&E (Fig. 13.2), making it look more recognizable by human pathologists. A key difference is that whereas H&E is a map of protein content and nucleic acid distribution (specifically chemical bases and acids), variability in compounds within

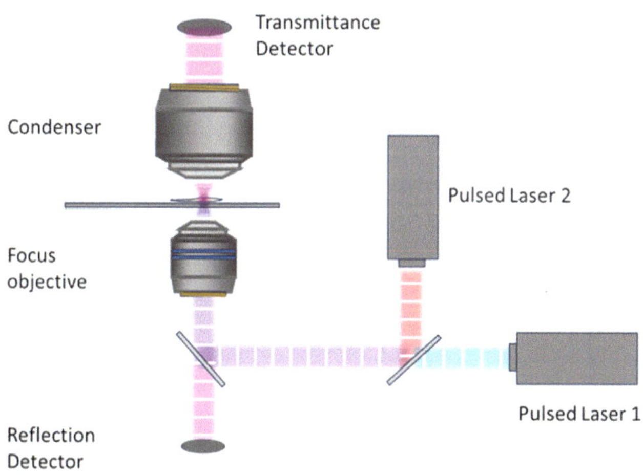

FIGURE 13.1 Typical stimulated Raman histology optical setup.

Two different pulsed lasers (like in multiphoton/two-photon) are aligned and coordinated to simultaneously focus on a specific location within tissue. The difference in frequencies is selected to match the vibrational frequency of specific bonds (e.g., CH_2 and CH_3 carbon-hydrogen bonds), so that the signal can be preferentially amplified. Detection can occur in reflectance or in transmittance.

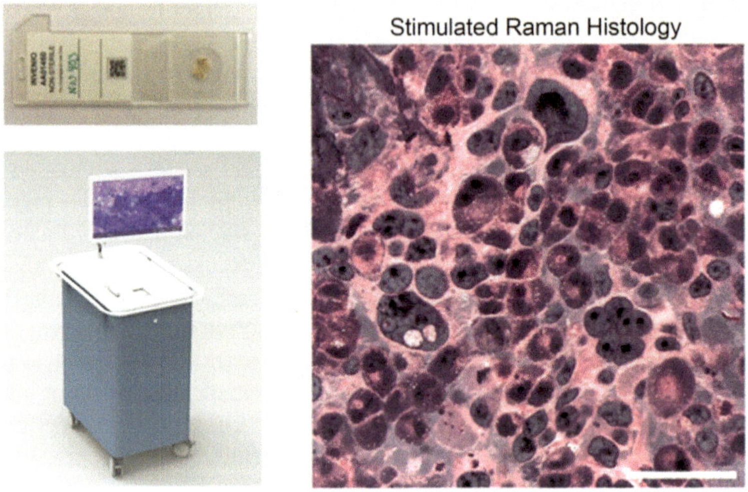

FIGURE 13.2 Stimulated Raman histology (SRH) example image and instrument.

Clockwise from top left: fresh sample of brain tissue placed in container for imaging, resulting pseudocolorized image, instrument used for SRH imaging.

Modified from Orringer, D.A., Pandian, B., Niknafs, Y.S., Hollon, T.C., Boyle, J., Lewis, S., Garrard, M., Hervey-Jumper, S.L., Garton, H.J.L., Maher, C.O., Heth, J.A., Sagher, O., Wilkinson, D.A., Snuderl, M., Venneti, S., Ramkissoon, S.H., McFadden, K.A., Fisher-Hubbard, A., Lieberman, A.P., ... Camelo-Piragua, S. (2017). Rapid intraoperative histology of unprocessed surgical specimens via fibre-laser-based stimulated Raman scattering microscopy. Nature Biomedical Engineering, *1(2). https://doi.org/10.1038/s41551-016-0027. Used under Creative Commons License.*

structures creates SRH signals that differ intrinsically from an H&E image, so they are not fully equivalent.

A prominent advantage of this technique is that virtually no tissue preparation is needed. Fresh specimens can be analyzed immediately without undergoing further processing. The method can be completely nondestructive, meaning that the same tissue can be subsequently processed in any other manner, ensuring gold standard analysis is accessible. It is also appealing that SRH can provide composition information. The latter aspect is an avenue for expanding analysis, theoretically providing information difficult to discern on H&E alone. For example, a spectroscopic pattern may be associated with oncoproteins such as alpha fetoprotein (AFP), providing potential avenues for enhancing the recognition of neoplastic tissue (Huang, 2023).

Several publications detail testing of the SRH technique in clinical specimens, including for margin assessment (Koya et al., 2020; Lizio et al., 2021), measuring cartilage degradation (Gaifulina et al., 2021), characterizing breast cancer (Li et al., 2021), skin lesions (Nguyen et al., 2021), gastrointestinal biopsies (Liu et al., 2022), bladder cancer (Osterberg et al., 2014), prostate (Mannas et al.,

2023), and brain cancer (Di et al., 2021; Riva et al., 2021). Some Raman analyses have focused on identifying discriminatory signatures with numerical outputs, while others have sought to create visual data for interpretation (the histology aspect of SRH), which may then be combined with machine learning for classification, addressing the need for pathologists to undergo retraining for interpretation. A recent study illustrates the potential for rapid tumor detection (at low resolution) in fresh breast cancer specimens (Pandey et al., 2023).

Numerous companies offer Raman spectroscopy instrumentation. A few companies currently advertise systems specifically designed for clinical tissue evaluation using SRS, including Invenio Imaging (Santa Clara, CA), Lightcore Technologies (Marseille, France), and CytoVeris (Farmington, CT).

With respect to limitations, a principal challenge for SRH is imaging speed. When compared to scanning fluorescence microscopy (e.g., confocal and multiphoton), Raman imaging is at least an order of magnitude slower for equivalent levels of resolution, even with experimental fast systems (Wang et al., 2024). Reasons include limits related to the physics of the various phenomena—vibrational processes take considerably longer than fluorescence excitation and emission. Additionally, measurement typically requires combinations of light pulses, and importantly, there is a tradeoff in terms of the amount of compositional information that can be obtained (based on the number of frequency bands detected) and speed. Standard resolution analysis of tissue is typically measured in minutes per square millimeter. As such, the pace of typical clinical workflows currently limits detailed microscopic evaluation of specimens by Raman to only a few millimeter. On the other hand, as described above and illustrated in Fig. 13.3, low-resolution approaches meant to roughly map larger specimens are a tactic that could produce data that is complementary to detailed microscopic examination.

Because Raman is not exactly duplicative of H&E, it is generally coupled with an image analysis algorithm. Dependence on the performance of "black-box" machine learning systems for which there is limited experience, both in terms of accumulated use data and of covering the full extent of biological variability, introduces an element of diagnostic uncertainty in any given case. Time and experience are expected to reduce this barrier, although how much is unknown.

Another aspect of SRM is that it is most effectively implemented in specimens thin enough to allow detection of signals through a sample, such as those mounted between a slide and coverslip. This reflects the benefit of having separate paths for excitation and detection and that the effect is more efficiently detected in transmission. However, surface imaging systems are more practical in high-throughput environments.

Efforts continue at the research level to speed collection and visualize more depth. The current stage of development makes the technique most amenable to evaluation of small and thin specimens or surfaces, with more complicated use in routine clinical samples, particularly for thorough or detailed morphologic evaluation. Nonetheless, the relative ease in preparation for imaging and ability to detect

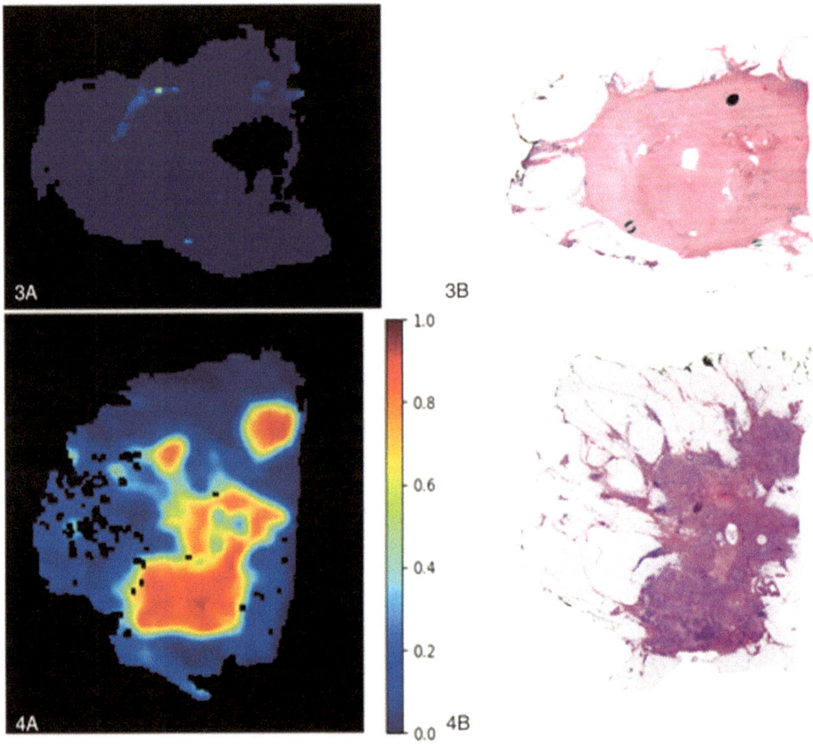

FIGURE 13.3 Low-resolution mapping of breast cancer using SRH.

Example cancer probability maps (left panels) generated using Raman spectroscopy on fresh breast tissue at resection with corresponding physical slide H&E (right panels). The top sample shows no/low cancer probability. Bottom sample shows high cancer probability in areas corresponding to cancer on H&E. *SRH*, Stimulated Raman histology.

Adapted from Pandey, R., Fournier, D., Root, G., Riccio, M., Shirvalkar, A., Zamora, G., Daigneault, N., Sapack, M., Zhong, M. & Harigopal, M. (2023). AI-powered biomolecular-specific and label-free multispectral imaging rapidly detects malignant neoplasm in surgically excised breast tissue specimens. Archives of Pathology & Laboratory Medicine, 147*(11), 1298–1306. https://doi.org/10.5858/arpa.2022-0228-oa.*

compositional aspects highlight a potential complementary role for SRH in pathology, especially in focused intraoperative or rapid-evaluation settings.

Photoacoustic microscopy

In photoacoustic microscopy, light-absorbing molecules in tissue are stimulated in such a manner that it elicits momentary thermal expansion at a molecular scale. The vibration results in the transfer of energy as sound waves that are externally detectable using an acoustic sensor—in essence, it is akin to light-excited ultrasound (Fig. 13.4).

The light excitation may be nanosecond pulses of laser light, similar to that used in multiphotons. The advantage of this approach over fluorescent methods is that acoustic waves travel more easily through tissue than emitted light, so signal detection can be efficient. On the other hand, signal strength and consistency with depth are counterchallenges, especially in tissues without a high quantity of intrinsically absorbing molecules. Relative to standard ultrasound, enhanced precision is achieved via excitation with focused pulsed laser light, improving contrast and resolution. The light absorption in unstained specimens comes primarily from natural chromophores, with hemoglobin and its breakdown products being principal contributors; melanin, lipid, and water can also be absorbers. Logically, vasculature, red blood cells, and liver have been usual targets. Exogenous contrast agents have also been explored.

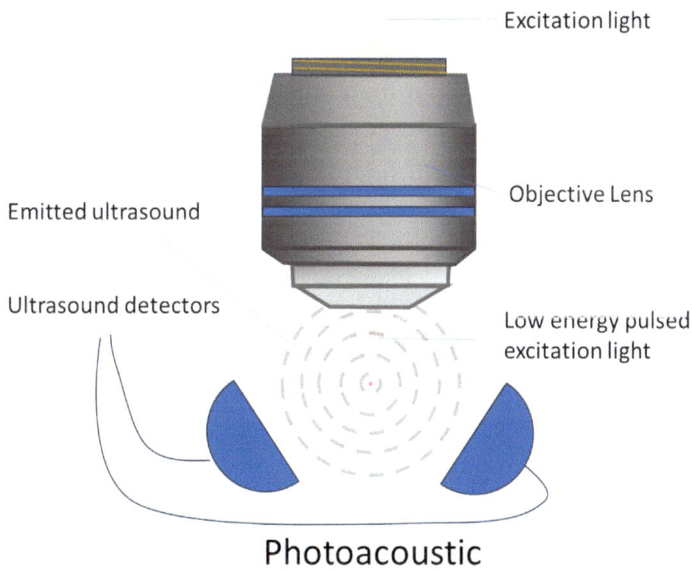

FIGURE 13.4 Schematic of photoacoustic microscopy technique.

This hybrid light/sound method uses pulsed laser light such as that used in multiphoton/two-photon microscopy to excite molecules within a sample. The excitation light is tuned to produce expansion/vibration in chromophores within tissue (e.g., hemoglobin). The energy is converted into sound waves, which are detected using ultrasound acoustic detectors.

Apart from it being exploitable as a finer resolution variant of ultrasound, this hybrid of light and sound can be used to recreate histology. When using focused excitation with high numerical aperture objectives, the technique can be called optical resolution photoacoustic microscopy (OR-PAM). A variant that uses ultraviolet excitation has been referred to as UV-PAM (Cheng et al., 2022). In a particular variant combining pulsed short-wavelength and infrared lasers, resulting in UV photoacoustic effect and reflective microscopy, it is called UV-PARS (Restall et al., 2021) (being commercialized by illumiSonics, Waterloo, Ontario, CA). By employing the use of ultraviolet range excitation, these latter methods excite nucleic acids, rendering nuclear contrast, but are then also largely restricted to surface imaging. The combined reflectance modality aids in the recognition of protein, improving the mimicking of standard H&E stains, at the expense of system complexity and limited depth.

Because the signal sources in PAM, UV-PAM, and UV-PARS are not fully analogous to standard H&E (which is based on differential staining of acidic and basic molecules), these techniques have been paired with pseudocolor staining based on neural network/deep learning algorithms. When applied to unstained FFPE slides, the UV photoacoustic methods have demonstrated close prediction of the resulting H&E stain (Martell et al., 2023). As discussed below, nonlinear transformations using machine learning to mask differences in signal sources pose risks of false representation, which will need to be addressed with detailed testing in fully representative diagnostic settings. And as with other techniques at this stage of development, evolutionary steps to look out for are further demonstration in unsectioned and unembedded specimens and testing of robustness to clinical workflows and environments.

Microscopy with ultraviolet surface excitation and fluorescence imitating brightfield illumination

With a descriptive name and a catchy acronym, MUSE (being developed by MUSE microscopy, Dana Point, CA) refers to an imaging technique devised for rapid visualization of unembedded tissue (Fereidouni et al., 2017). In its basic implementation (Fig. 13.5), MUSE produces a fluorescent topographic image of tissue by taking advantage of the limited depth of penetration of short-wavelength ultraviolet light. Whereas infrared (low energy) light can traverse on average a couple hundred microns into most tissues, short UV can excite fluorophores only within the first 20 μm or so (Yoshitake et al., 2018). This produces optical sectioning of the exposed surface layer.

Key features of the MUSE approach are potentially low cost and minimal tissue preparation. The method typically employs the use of an ultraviolet LED (between approximately 240 and 290 nm) to illuminate the tissue at an angle from the viewing objective. A multitude of fluorescent dyes are excited by this short wavelength. Fresh or fixed tissue can be dipped for seconds or minutes in

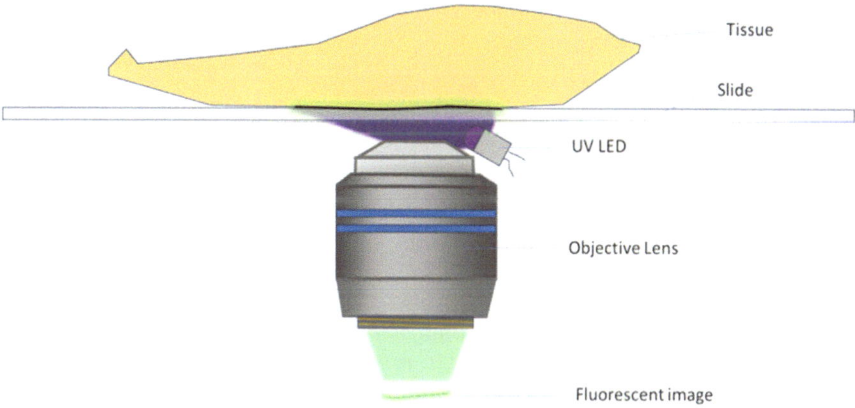

FIGURE 13.5 Diagrammatic representation of standard MUSE setup.

Fresh tissue is stained briefly with fluorescent dyes and is illuminated by an ultraviolet LED. Emitted surface fluorescence is collected via a microscope objective, and a fluorescent image is generated at a detector camera. *MUSE*, Microscopy with ultraviolet surface excitation.

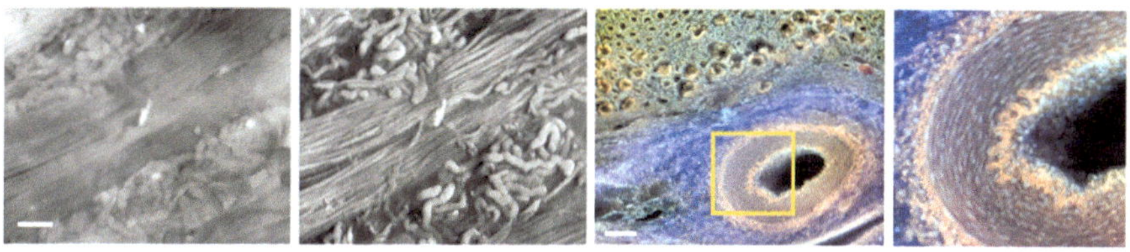

FIGURE 13.6 Representative MUSE images.

Native view of several tissues showing topological images of fluorescently stained samples. *MUSE*, Microscopy with ultraviolet surface excitation.

Adapted from Fereidouni, F., Harmany, Z. T., Tian, M., Todd, A., Kintner, J. A., McPherson, J. D., Borowsky, A. D., Bishop, J., Lechpammer, M., Demos, S. G., & Levenson, R. (2017). Microscopy with ultraviolet surface excitation for rapid slide-free histology. Nature Biomedical Engineering, *1(12), 957–966. https://doi.org/10.1038/s41551-017-0165-y under Creative Commons License.*

nuclear and protein dyes and then directly viewed under the fluorescent microscope.

Without image processing, the images adopt the look of something like surface terrain (see Fig. 13.6), which is different enough from standard slides that pathologic distinctions require experience. The degree to which the images can be used to render accurate and precise diagnoses in specific clinical instances has not yet been widely reported. What has been shown is that with image processing, the images can also be made to look like H&E slides, making recognition more immediate and expected to require less retraining of pathologists.

There are specific technical challenges that have been described in the implementation of MUSE, such as variable illumination across specimens and difficulties with the geometrical constraints of high numerical aperture objectives (Ching-Roa et al., 2021). Because high resolution usually means a short working distance of the objective, little room is available to evenly illuminate tissue. And if the tissue is irregular, then angled illumination can result in shadows.

There are also theoretical limits imposed by the physics of the basic MUSE technique in terms of image clarity and definition. The fact that the effective optical tissue "section" is governed by the depth of light penetration results in the appearance of slices that are considerably thicker and more cellular than that of a typical physical slide. Compared to the typical 5 μm thick physical slide section, the 20 μm imaging depth reduces clarity and definition by allowing out-of-plane cellular components to cloud the view. Investigative projects continue to develop strategies to ameliorate the illumination variability and optical sectioning challenges of MUSE by employing advanced imaging strategies. Examples include deconvolution (Xie et al., 2019) (a computational sharpening based on obtaining images at various levels, which slows imaging but can make results clearer) and speckle illumination (Wong et al., 2022) (an optical technique that produces light excitation patterns from which fluorophore location can be more precisely determined, adding complexity and time but improving overall quality). Both approaches are generally applicable to other fluorescent imaging modalities like confocal and multiphoton. One study evaluated the benefit of using immersion fluid for image improvement in MUSE (Yoshitake et al., 2018), comparing resulting image quality with multiphoton microscopy (nonlinear microscopy, NLM) and physical slide histology, providing a basis for evaluation of image quality differences between approaches (Fig. 13.7).

A related technique is called Fluorescence Imitating Brightfield Imaging (FIBI) (Au Yeung et al., 2023; Borowsky et al., 2023; Engel et al., 2022). In this variant (being developed by Histolix, Sacramento, CA), the excitation wavelength is slightly lower energy/longer wavelength, about 405 nm. As reported and illustrated in Fig. 13.8, surface staining is performed by brief sequential exposure to hematoxylin and eosin. The stains need not be fluorescent, as the penetration depth of the blue excitation light is meant to generate tissue autofluorescence below the surface of the specimen, creating a type of back-illumination. This autofluorescence emitted by tissue just below the surface is variably absorbed by the dyes on the surface in a manner analogous to a physical slide transillumination. The optical sectioning is created by the limited depth of dye penetration coupled by the limited depth of deep violet light excitation for generation of the backlight. Surface tissue irregularities that would preclude obtaining complete tissue cross sections can be counteracted by collecting data at different focus levels, at the obvious expense of imaging time. Pseudocoloring (converting non-H&E data to an H&E color profile) produced via machine learning can closely mimic H&E images. As more data on this

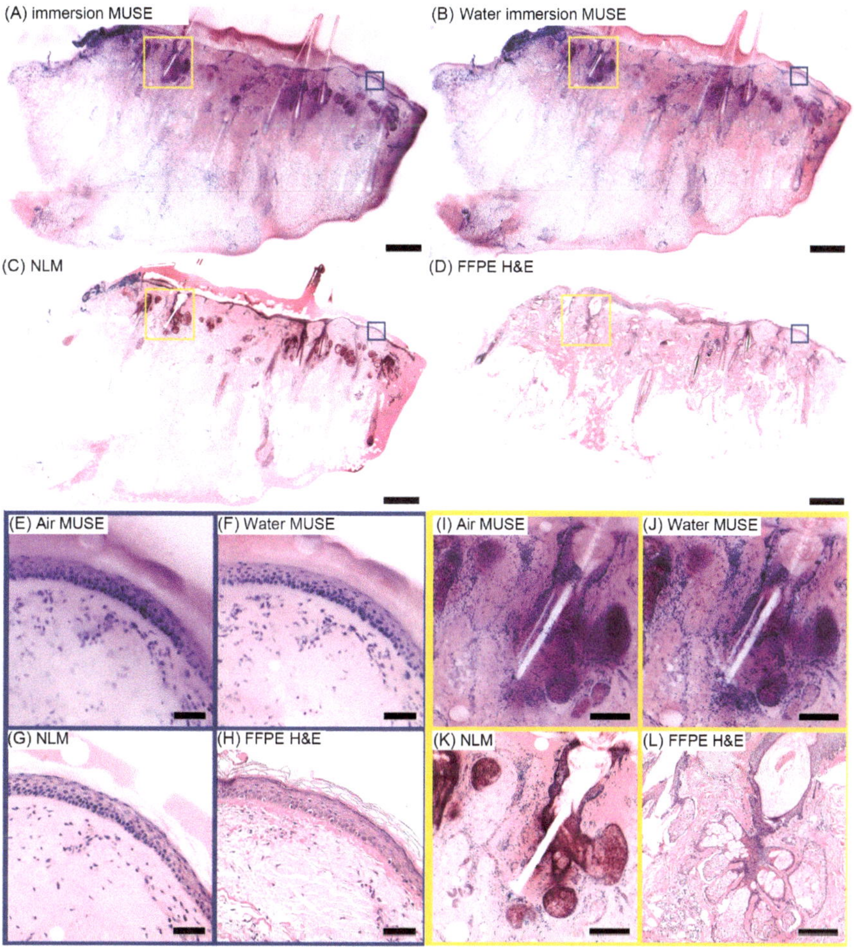

FIGURE 13.7 Paired comparison of fresh tissue imaging.

Images of fresh skin samples imaged by MUSE (A, E, and I), MUSE with water immersion (B, J, and L), multiphoton/nonlinear microscopy (C, G, and K), and standard physical slide preparation with H&E staining (D, H, and L). *MUSE*, Microscopy with ultraviolet surface excitation.

From Yoshitake, T., Giacomelli, M. G., Cahill, L. C., Schmolze, D. B., Vardeh, H., Faulkner-Jones, B. E., Connolly, J. L., & Fujimoto, J. G. (2016). Direct comparison between confocal and multiphoton microscopy for rapid histopathological evaluation of unfixed human breast tissue. Journal of Biomedical Optics, 21*(12). https://doi.org/10.1117/1.jbo.21.12.126021. Used under Creative Commons License.*

approach emerges, key aspects to look out for are robustness to routine use in a clinical environment and the level of overall imaging efficiency and resolution that can be achieved.

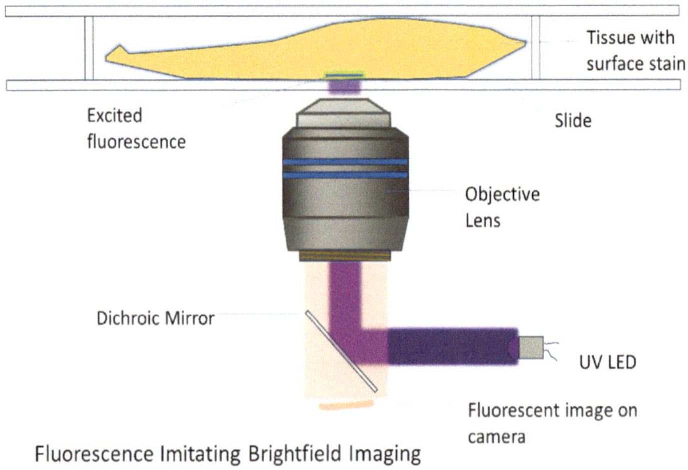

FIGURE 13.8 Diagrammatic representation of the FIBI approach.

Tissue is surface stained sequentially with hematoxylin and eosin (or other dyes). Ultraviolet excitation of tissue generates fluorescence from intrinsic tissue fluorophores up to a limited depth in a sample. The fluorescence emitted is absorbed variably by the surface dyes, imitating transmitted light imaging. The light absorption pattern is collected by a detector camera. A computer-generated image of a machine. Description automatically generated. *FIBI*, Fluorescence imitating brightfield illumination.

If the quality improvements of MUSE or the FIBI approach can be economically, efficiently, and reliably implemented, the speed and simplicity of the approaches make them attractive options for rapid tissue evaluation. Comparative clinical studies on direct comparison with physical slides will be critical steps toward the determination of the true potential.

Optical coherence tomography/optical coherence microscopy

Optical coherence tomography is by no means new and primarily an in vivo technique but is included here owing to the capability to produce slide-free microscopic images, which may be reviewed by pathologists. Optical coherence tomography (OCT) is likely familiar to readers as an imaging modality based on detecting light reflectance from tissue structures, extracted from interference patterns that are generated. It is akin to sonography, where the energy source is ultrasound rather than light. This light reflection has a limited correlation to microscopic structures seen in a standard stained slide and is a rather weak signal that, under normal circumstances, represents a complex mixture of structures at different depths. In optical coherence microscopy, the detection of light reflectance patterns incorporates the use of a pinhole to reduce the contribution of signal from outside a specific focus plane, enabling some "sectioning" to be visualized. In this regard, optical coherence microscopy (OCM) is a combination of OCT and confocal microscopy, although standard confocal microscopy relies on fluorescence whereas OCM is based on light reflectance, usually without staining.

Like SRH, a natural advantage of OCM is that it does not require any tissue preparation. Fresh tissue can be visualized, making it applicable to direct visualization of tissue in vivo, for example, skin and gastrointestinal tract, as well as retinal evaluation. In certain OCM implementations, the data is isotropic, meaning that the lateral (x–y) and axial (z) resolutions are equal, producing cubes of data that can make 3D visualizations look more natural, unlike the majority of other optical sectioning techniques, which produce data elongated in the z-direction.

For low-power applications, the technique can produce highly informative data with little preparation (Fig. 13.9). With slower imaging, image quality can improve. Techniques have evolved to enhance contrast in the ex vivo setting;

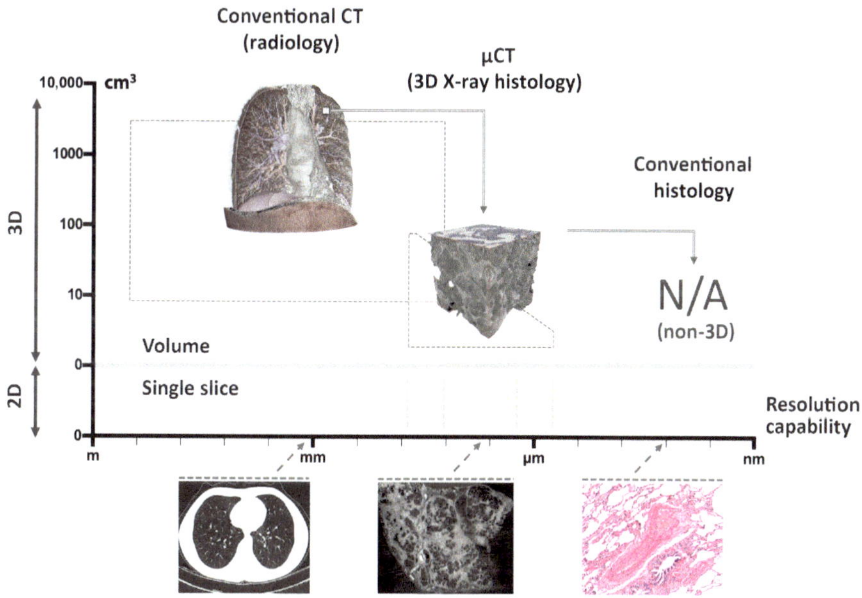

FIGURE 13.9 Resolution and 3D role of microCT.

The microCT approach provides three-dimensional characterization of specimens and tissue blocks at a resolution level that is lower than slide histology but higher than traditional magnetic resonance imaging and computed tomography.

From Katsamenis, O. L., Olding, M., Warner, J. A., Chatelet, D. S., Jones, M. G., Sgalla, G., Smit, B., Larkin, O. J., Haig, I., Richeldi, L., Sinclair, I., Lackie, P. M., & Schneider, P. (2019). The American Journal of Pathology, 189*(8), 1608–1620. https://doi.org/10.1016/j.ajpath.2019.05.004 under Creative Commons License.*

however, they come at the cost of reduced speed and increased preparation effort. And quality downsides remain in that cellular contrast is limited and practical subcellular resolution is still unattainable. It is possible that OCM methods evolve to provide supplemental views in the intraoperative setting, although the physical principle that underlies OCM makes it more amenable to rapid, complementary evaluation of larger specimens with less resolution and contrast than physical slides.

Confocal microscopy

Confocal microscopy is a well-known fluorescence imaging technique touted for its ability to generate high-resolution microscopic images with fine sectioning in the axial (z) direction, thereby capable of producing 3D reconstructions. Although there are multipoint imaging variants designed to increase speed, these also generally degrade image quality, so confocal for histology can be considered primarily a scanning-type of microscopy where each point in space is independently measured; there is usually no camera for imaging, just a detector collecting emitted light for a given spot.

Focusing a laser beam into a biological sample through a microscope objective lens creates a biconical ray of light (Fig. 13.10). As seen in the figure, the cone excites many fluorophores above and below the focus point of interest, each emitting light in all directions, so images reconstructed from scanning through the sample with a laser would look very fuzzy and bright. The confocal part refers to a physical pinhole being placed between the objective lens and a detector to block light that is coming from above and below the plane of interest. This works because *most* of the spherical emission of light from a fluorescent molecule above or below the focus point takes a different path toward a detector than *most* of the light from molecules at the focus point, so much of the extraneous signal can be blocked.

Numerous confocal microscopes are commercially available, and they have been extensively evaluated for use in pathology settings. There are a multitude of practical reasons why research confocal microscopes have not become part of the standard pathology analysis repertoire. Generally speaking, research applications are diverse, and imaging systems benefit from having high flexibility, which translates into accommodating a choice of objective lenses, fluorescent filters, specimen configurations, embedded image analysis controls, etc. This conflicts with the need in clinical settings for simple user interfaces, high speed at high image quality, and simplicity and ease of sample mounting and tracking.

In response to this, a few companies have produced confocal systems with designs and interfaces directed to histology, including clinical applications. Two examples are the VivaScope (caliber imaging and diagnostics) and the Histolog Scanner (SamanTree Medical, Lausanne, Switzerland). Such systems are designed to be more straightforward to use than a pure research microscope and offer basic supplies for preparing excised tissue for imaging. Rapid staining with fluorescent

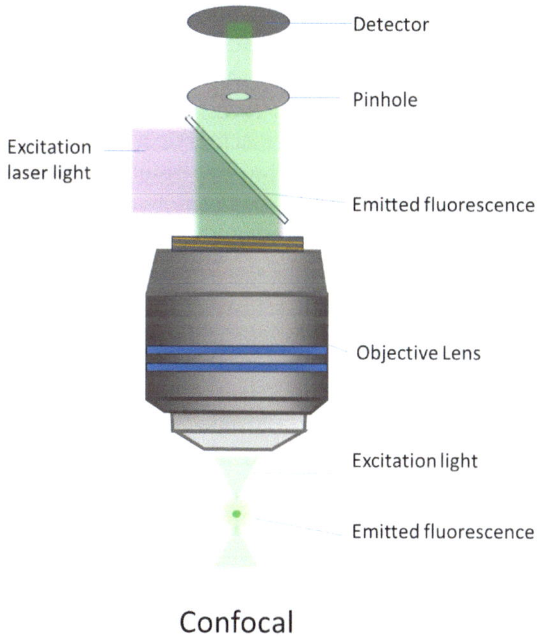

FIGURE 13.10 Schematic of confocal microscopy approach.

Laser light is directed onto a specimen, which is focused by a microscope objective. Biconical excitation light excites fluorophores throughout the specimen. The most intense fluorescent emission comes from the focal spot. The emitted light waves are collected through the same objective lens. As the light rays that pass through take different paths, much of the light coming from areas outside the focus point is blocked, whereas the light from the focus point passes through to a detector. Moving the point of focus throughout a specimen allows reconstruction of the fluorescent pattern.

dyes to mimic H&E can be applied in minutes, and fast imaging confocal can produce recognizable images in a few minutes more. Consequently, there has been interest in employing this technique for rapid review.

Confocal systems that image fluorescence can also integrate the use of a related technique termed reflectance confocal microscopy (RCM). Reflection occurs at points where there are sharp changes in the optical properties of tissue components (i.e., refractive index), such as interfaces between cells and extracellular fluid. The principle of reflectance with the use of a confocal pinhole is like that of optical coherence microscopy, with the difference here being that the detection is purely reflected light detection rather than light interference patterns. A practical difference is that OCM can image deeper than RCM; when used in vivo for dermatologic applications, the reviewable depth in RCM is typically only a few tens of micrometers. Apart from the rapid drop in intensity with depth, RCM produces patterns that are considerably different from H&E. But with experience and for specific scenarios, it provides information reported to be complementary to fluorescence staining of standard confocal.

At present, these confocal systems are being evaluated for use in a variety of clinical scenarios. Most experience exists in ex vivo assessment of skin, particularly as part of the Mohs' surgical approach (Forchhammer et al., 2023; Gareau et al., 2008; Longo et al., 2019), with imaging results exemplified in Fig. 13.11. Other organ systems that have been evaluated by confocal include the breast (Conversano et al., 2023; Nackenhorst et al., 2022), gastrointestinal tract (Guerrero et al., 2021), liver (Titze et al., 2022), kidney (Villarreal, Pérez-Anker, et al., 2023), prostate (Bertoni et al., 2020; Titze et al., 2021), and brain.

A fundamental challenge for confocal systems as it applies to clinical histology is that confocal rejection of out-of-plane light is not 100% effective. As a result, fluorescent confocal systems also lose clarity quickly with depth. Formalin fixation magnifies this effect. And even firm samples cut with clean, high-quality blades have surface irregularities in a scale that makes it difficult to obtain clear, full sections of tissue samples using confocal imaging. Compression of specimens can help alleviate this limitation. A study on fresh breast resection specimens showed that training of pathologists (and surgeons) can improve accuracy of interpretation (Conversano et al., 2023). Performance in fresh skin specimens has been variable, particularly sensitivity for detection of disease, but high specificity in assessment of basal cell carcinoma margins has been reported (Grizzetti & Kuonen, 2022; Longo et al., 2019).

As in SRH, OCM, and MUSE, there are computational and imaging strategies that can be applied to further improve image quality in confocal microscopy, but (also as in the other modalities), they present trade-offs in terms of time and complexity. Of note, the use of clearing techniques described below can improve deep tissue detection of excited fluorophores, albeit to a limited degree. The signal from tissue outside the plane of focus remains high with clearing in confocal, blurring the images obtained beyond the surface. In practical terms, this means that for the time being confocal is an option for review of surface histology in the rapid-evaluation setting, and which becomes considerably slower when trying to approach the image quality of physical slides.

Selective plane illumination microscopy (lightsheet)

The selective plane illumination microscopy (SPIM) technique, also known evocatively as lightsheet microscopy (LSM), is a variant of laser optical sectioning microscopy. It evolved as a means to speed up point-scanning for fluorescent three-dimensional imaging of tissues. The method involves reshaping the double cone created by focused laser beams (as used in confocal and standard fluorescent microscopy) into a thin sheet. Specially crafted lenses widen and flatten the double cone into an approximation of a plane of excitatory light. Specimens are fluorescently stained and made transparent by one of several clearing techniques. A camera is placed behind lenses at 90 degrees to the sheet of light. When a

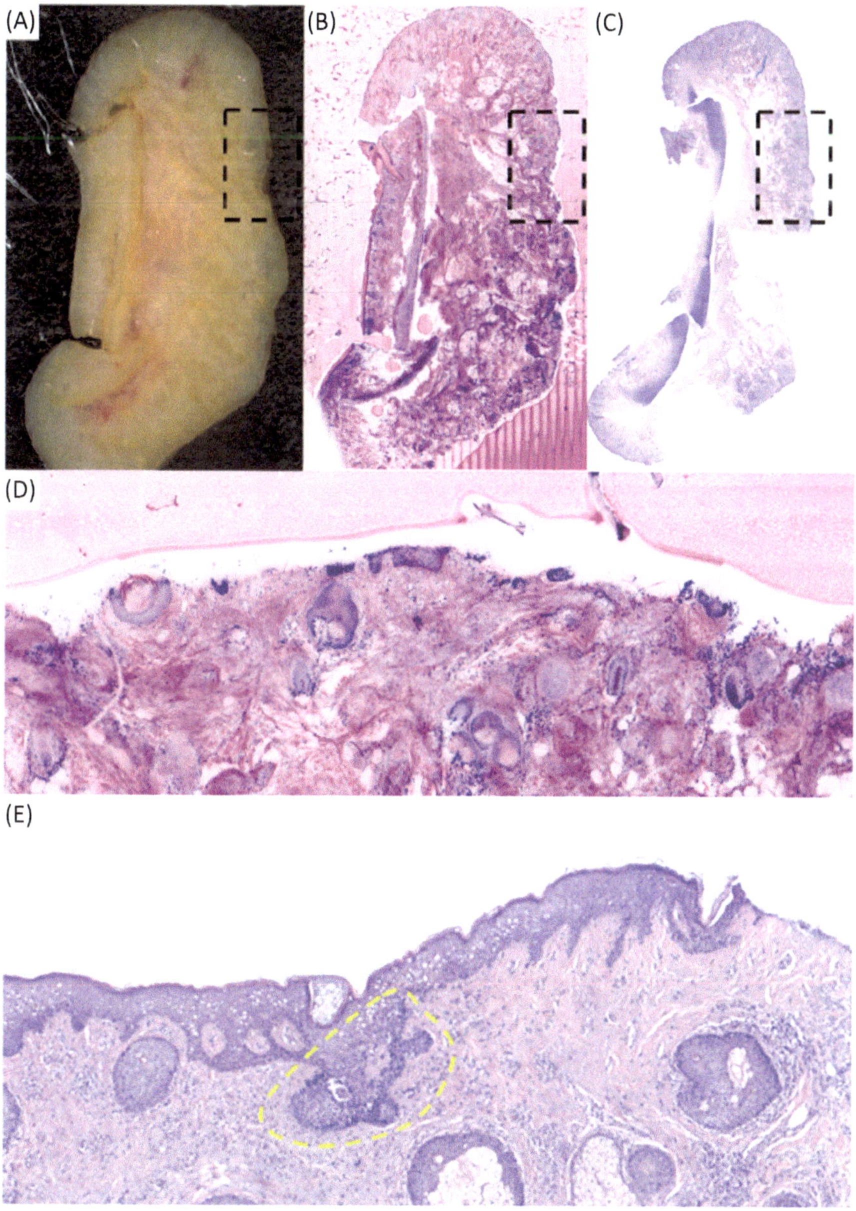

FIGURE 13.11 Confocal microscopy on fresh, stained skin, compared to physical slide.

Sample of basal cell carcinoma imaged in an intraoperative setting using confocal microscopy. (A) macro photo of specimen; (B) pseudocolored, ex vivo confocal image; (C) physical slide; (D) magnified confocal image; and (E) magnified physical H&E slide.

From Forchhammer, S., Grunewald, S., Möhrle, M., Metzler, G., Eigentler, T., Münch, A. -K., & Ogrzewalla, H. (2023). Diagnosis of basal cell carcinoma with ex-vivo confocal laser scanning microscopy in a real-life setting. Acta Dermato-Venereologica, *103. https://doi.org/10.2340/actadv.v103.4859, under Creative Commons lincense.*

specimen is transparent enough to allow the lightsheet to go deeply, the entire plane can be imaged at once. Sweeping the sheet of light through a sample allows imaging at video camera rate speeds. Notably, this can produce massive amounts of three-dimensional data quickly.

As illustrated in Fig. 13.12, multiple variants of the optical configuration for lightsheet microscopy exist (Liu et al., 2023). Common research implementations of lightsheet microscopy with commercial systems depend on complex sample mounting arrangements in between dual microscope objectives perpendicular to each other to allow simultaneous laser sheet excitation and plane detection, making them less applicable to high-throughput settings. An approach known as open-top lightsheet (OTLS) microscopy (Glaser et al., 2019) (being commercially developed by Alpenglow, Seattle, WA) uses an angled geometry that facilitates flat sample arrangement with simple orientation, more amenable to efficiently placing samples. A range of applications to histology using this system have been described, including prostate (Reder et al., 2019), esophagus (Reddi et al., 2023), and sentinel lymph nodes (Barner et al., 2022). Beyond simplifying the sample setup, progress has been achieved in addressing challenges of image processing speed and computational burden (Gao et al., 2023) as well as trade-offs in terms

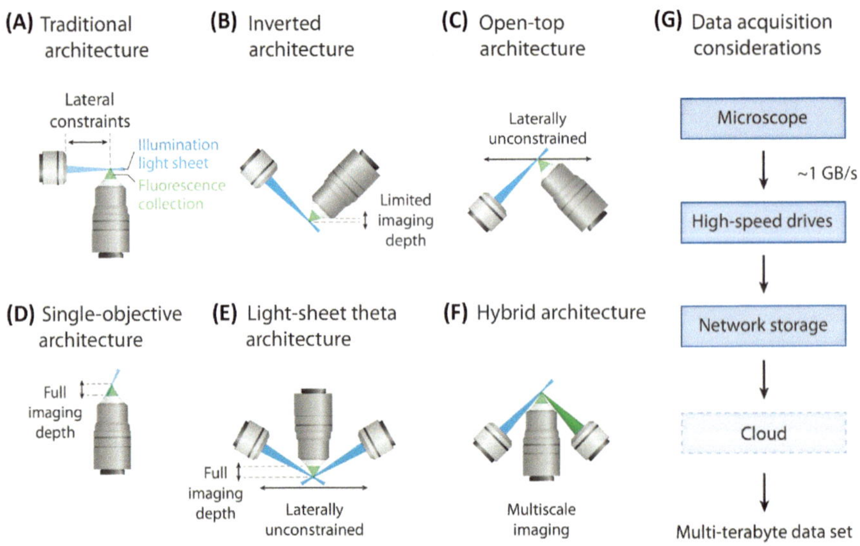

FIGURE 13.12 Variant geometries for selective plane illumination microscopy (SPIM)/ lightsheet microscopy (LSM).

A diagram of a structure is automatically generated with medium confidence.

From Liu, J. T. C., Glaser, A. K., Poudel, C., & Vaughan, J. C. (2023). Nondestructive 3D pathology with light-sheet fluorescence microscopy for translational research and clinical assays. Annual Review of Analytical Chemistry, 16, *231–252. https://doi.org/10.1146/annurev-anchem-091222–092734, used under Creative Commons license.*

of resolution and acquisition speed. Notably, several published works have explored the potential clinical value of three-dimensional analysis, particularly for prostate grading (Koyuncu et al., 2023) and for assessment of esophageal dysplasia (Barner, 2023; Reddi et al., 2023).

There are several salient characteristics of lightsheet microscopy relevant to the evaluation of tissue specimens in comparison to other ex vivo imaging techniques. While the preparation process of tissue is more extensive than for some other new imaging technologies such as Raman and Optical Coherence Microscopy, lightsheet offers capabilities for higher speed and deeper imaging. Lightsheet microscopy presents a different trade-off profile for resolution, depth, and speed for consideration relative to those modalities.

While research imaging continues to exploit lightsheet's volumetric imaging capabilities, there are some key aspects that relate to the use of lightsheet microscopy in clinical scenarios. A principal one is that the clarity and quality of images are dependent on the consistent thinness of the sheet of light. Inconsistent intensity and variable thinness of the lightsheets throughout a specimen result in distortions of imaging. And even highly transparent specimens have optical inhomogeneities derived from biological variability, which degrade image quality with depth. These affect all optical sectioning modalities, but the mechanism for optical sectioning in lightsheet results in it being impacted by this variability more so than other methods. As in other techniques, strategies employing tools like multiphoton excitation, computational adjustments, restricted fields-of-view, and patterned illumination (Chen et al., 2022; Daetwyler & Fiolka, 2023) have been experimentally pursued, rebalancing speed, complexity, and image quality trade-offs with lightsheet. Regardless, maintaining thin and consistent illumination throughout biologically diverse specimens is an important goal for maximizing resolution at depth with these methods and a key parameter for determining applicability for a specific application.

Finally, it is worth noting that clinical users need to weigh their appetite for additional data for review. High speed three-dimensional imaging is a strong suit for lightsheet microscopy, but the expectation of having pathologists review hundreds of levels for a given biopsy could impose an excessive burden. As discussed in the note on 3D imaging below, if clinical need for this type of data becomes established, a high number of levels is an aspect that would benefit greatly from automated computerized screening, provided that too becomes clinically acceptable.

The applicability of the technique to routine clinical work continues to be assessed. Imaging of deparaffinized, stained, and subsequently cleared samples in updated systems illustrates capacity for cellular resolution and clarity in these specimens (Fig. 13.13) (Liu et al., 2024). Developments to date highlight a desire to see what level of image quality consistency can be produced in less processed specimens. In addition, as in other modalities, practical implementation in high-throughput workflows and the effect of biological variability and inhomogeneities

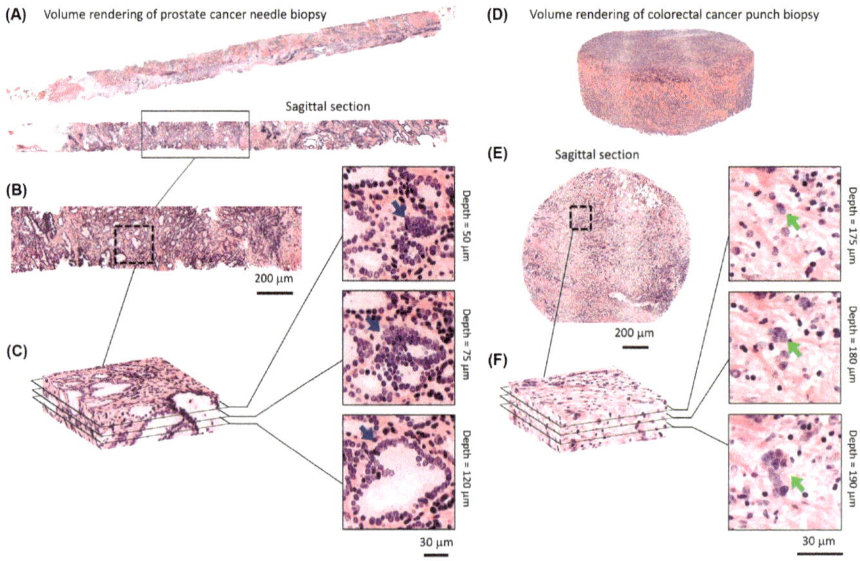

FIGURE 13.13 Open-top lightsheet microscopy on extracted FFPE tissue blocks.

Examples of 3D histology derived from using SPIM/LSM on prostate tissue, which has been deparaffinized from wax-embedded tissue blocks, stained, and cleared. Details are in the caption following the image. *FFPE*, Formalin fixation with paraffin embedding; *LSM*, lightsheet microscopy; *SPIM*, selective plane illumination microscopy.

From Liu, J. T. C., Chow, S. S. L., Colling, R., Downes, M. R, Farré, X., Humphrey, P., Janowczyk, A., Mirtti, T., Verrill, C., Zlobec, I., & True, L. D. (2024). Engineering the future of 3D pathology. The Journal of Pathology: Clinical Research, 10*(1). https://doi.org/10.1002/cjp2.347. Used under Creative Commons license.*

will be critical considerations. Nonetheless, progressive improvement in speed and quality supports a potential role in future histology.

Multiphoton microscopy

Multiphoton microscopy refers to a methodology that uses the property of simultaneous absorption of multiple pulses of light energy (photons) by fluorescent molecules. When a rapidly pulsed laser is focused by a standard microscope objective, the meaningful probability that multiple pulses of light reach a fluorophore at approximately the same time is restricted to a very small point in space (Fig. 13.14). Since the pulses of light combine to excite a fluorophore, lower-energy light (lower frequency/longer wavelength) is needed to achieve this fluorophore excitation than with more standard (continuous-wave) lasers. This presents an advantage for microscopy in that longer wavelengths of light can penetrate deeper into tissue without being scattered and weakened. And the fact that only a

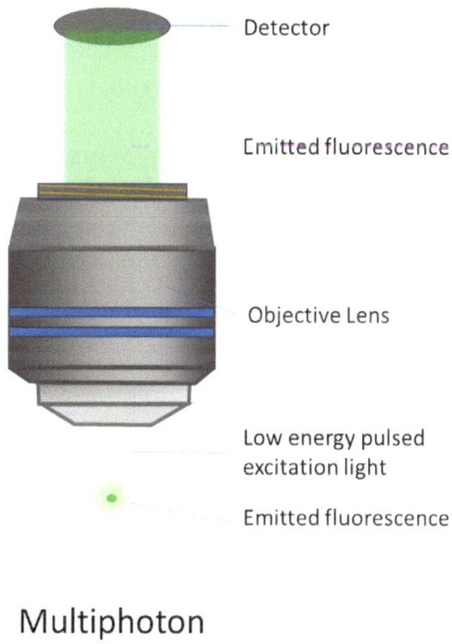

FIGURE 13.14 Schematic of multiphoton/two-photon microscopy setup.

Pulsed laser light of low power/long wavelength is used to excite fluorophores within tissue. Emitted fluorescence is detected through the same objective lens. Scanning the beam through a sample allows reconstruction of the desired sections. Tissue samples may be fresh unstained, fresh stained, and fixed/stained/cleared samples, each producing a different trade-off in terms of image quality and time to result. A close-up of a machine. Description automatically generated.

point in space is excited at a time simplifies fluorescence collection and guarantees optical sectioning. There are two-photon as well as three-photon variants of the technique referring to when two photons versus three (or more) combine to excite a fluorophore. In practice, multiphoton and two-photon are terms often used interchangeably.

The distinguishing feature of multiphoton microscopy for pathology is the capability for high-quality imaging at depth (Figure). Of the methodologies described to this point, confocal and multiphoton microscopy are the optical sectioning modalities that produce the highest resolution images. Superresolution techniques (described below) are capable of even higher resolution but achieve that through higher technical complexity and/or employing methods that require considerably more time for data collection, making them impractical for high-throughput histology work, among other challenges. Compared to confocal microscopy and high-energy systems, multiphoton has the advantage of being able to image clearer at depth and reducing the risk of damage to tissue from the incident laser light due to lower overall light energy exposure.

Several research-focused commercial multiphoton microscopes are available from major microscope manufacturers. These specific implementations continue to produce valuable pathophysiologic insights in investigative settings, but, like confocal systems, they are not designed to be readily implemented in clinical or other high-throughput environments. Research multiphoton systems are associated with one or more of: (1) time-consuming specimen preparation; (2) slow imaging; (3) high expense of equipment; (4) complex nonautomated operation; and (5) the need for pathologist retraining of raw fluorescent images.

Advances in the various aspects of imaging hardware, specimen preparation, and image processing have resulted in new systems with capabilities better suited for histology. For example, investigative rapid assessment approaches using multiphotons have been described that employ surface staining of fresh or fixed tissue and rapid imaging systems for intraoperative/intraprocedure evaluation (Ching-Roa et al., 2022; Yoshitake et al., 2016).

In contrast to Raman and OCM, and like confocal and lightsheet microscopy, the multiphoton-based method involves fluorescent staining and benefits from optical clearing of tissue. And, also like confocal and lightsheet, deep imaging is affected by biological inhomogeneities such as those produced by crystals, inorganic calcium, and carbon particles in tissue. The use of longer wavelengths for excitation and the intrinsic optical sectioning determined by the physics of the multiphoton technique give it an advantage in clarity at depth over other approaches that are clearing-based.

The method known as Clearing Histology with MultiPhoton Microscopy (CHiMP) (being developed by Applikate Technologies, Fairfield, CT) involves abbreviated tissue processing that is a variant of standard approaches including formalin fixation, dehydration, and clearing, but without the need for the most time-consuming and technically challenging steps of embedding in paraffin or physical sectioning. The processing can be implemented in some standard tissue processors. Imaging is then performed in an automated multiphoton scanner, akin to a whole slide imager. Subsequently, the tissue can be stored in clearing fluid or processed further for paraffin embedding and downstream analysis.

This implementation seeks a particular balance of preparation speed and effort with image quality. Unlike the rapid intraoperative techniques described, this approach employs longer preparation times (on the order of hours) to achieve better contrast and resolution, recognizing that pathologist interpretation accuracy (and image analysis performance) benefit proportionally to image quality and that benefits of speed and ease of processing with digital image production can be maintained. Notably, more rapid preparation methods resulting in lower overall image quality but faster results (such as for intraoperative analysis) are also accessible with this approach.

Published studies describe its use in a variety of tissues, including prostate and renal biopsy evaluation (Olson et al., 2016; Torres et al., 2014, 2021). As with standard tissue processing, the small size of core needle biopsies permits rapid

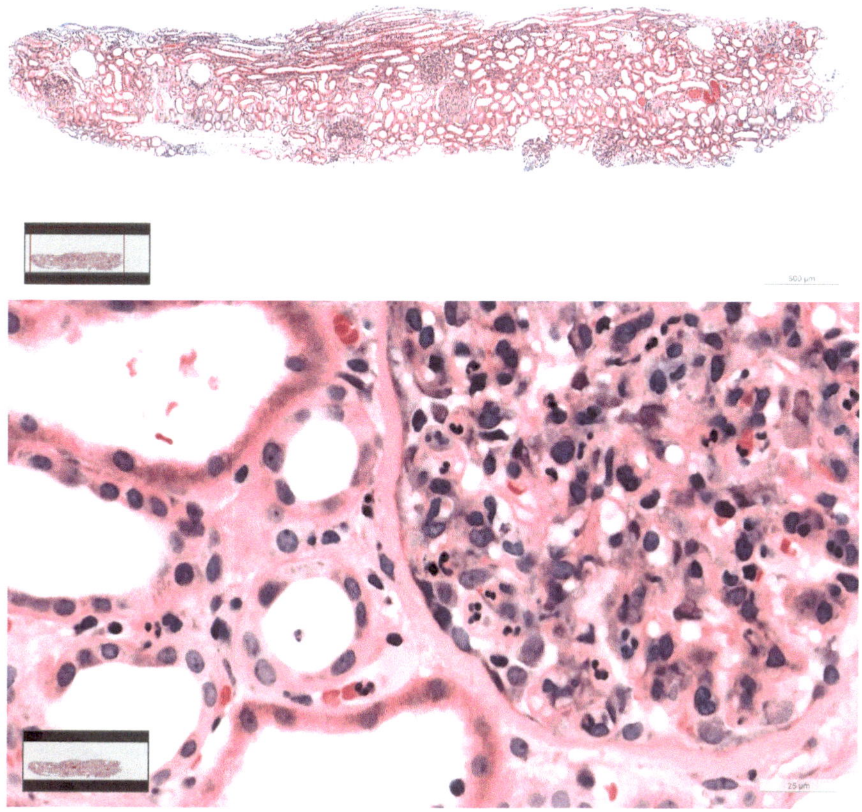

FIGURE 13.15 Clearing histology with MultiPhoton (CHiMP) microscopy example images.

A pseudocolored image of an unembedded clinical renal biopsy sample collected using an automated multiphoton microscope system approximately 3 h after tissue arrival to the laboratory. The sample was subsequently embedded in paraffin, cut into physical slides, and processed with standard stains for clinical diagnosis.

Courtesy: Sudhir, P., MD, PhD (Yale School of Medicine) and Applikate Technologies, Inc. (Fairfield, CT).

tissue processing to occur on the scale of a couple of hours, yielding access to images throughout the tissue samples with fine detail (Fig. 13.15).

A note on 3D imaging

Beyond the utility of visualizing 3D patterns of immunostaining in research specimens, publications have proposed increased diagnostic precision derived from analyzing 3D growth patterns using general protein and nucleic acid stains and relative single physical slides. A basis for this expectation comes from prior studies analyzing comprehensive sectioning through blocks. For example, incomplete sectioning of prostate biopsies has been shown to reduce sensitivity for detection of cancer foci and underestimate cancer grade relative to taking additional sections until a block is exhausted, with consequential therapeutic implications (Reyes & Humphrey, 1998).

The principal feature of lightsheet microscopy is the ability to produce finely spaced, detailed three-dimensional image maps of specimens with rapid accumulation of data. Both confocal and multiphoton microscopy can also provide three-dimensional imaging of tissue.

There are two key aspects that bear weighing in considering 3D imaging for clinical use. One is that establishing 3D imaging as superior to physical slides for diagnostic accuracy and, furthermore, providing improved prognostic accuracy, will require comprehensive analysis to confirm and quantify the benefit. It is reasonable to expect that it will become more possible to obtain that data once there is an established modality being used routinely. The second is that it is not yet clear what three-dimensional spacing is required for a given precision benefit. In other words, is histologic sampling every 1 μm any better than every 10, 20, or 50 μm? Based on the current use of architectural patterns in characterizing pathophysiologic processes, the benefit of three dimensionality is likely to be tissue-type-specific and dependent on the specific diagnostic scenario. Nonetheless, pathologist experience alone suggests there will be diagnostic benefits derived from accessing increased informational content from samples using deep imaging (Koyuncu et al., 2023).

Clearing techniques are also next-gen imaging tech

Techniques for making tissue transparent are as old as histology itself. Light scattering in tissue is due to differences in the interaction of light with components of membranes, proteins, and water, as a result of which light is redirected, reducing the ability to pinpoint the source. The refractive index quantifies how light interacts with a given material (inversely proportional to the speed of light in the medium). Normalizing the refractive index of different tissue constituents results in tissue being made transparent. In histology, xylene and its analogs produce this effect, as the refractive index of xylene is closer to that of proteins and membranes than that of water.

Over the past few decades, it has become evident that the use of clearing greatly enhances the capability of several optical sectioning microscopy methods to achieve high-quality imaging at depth (Dodt et al., 2007). Given the power of the technique, there now exists a dizzying array of clearing approaches, as characterized in detailed reviews (Richardson et al., 2021; Silvestri et al., 2016) (one listing more than 60 published protocols!) (Gómez-Gaviro et al., 2020), and many with clever acronyms such as Sca*l*e, FRUIT, ClearT, CUBIC, CLARITY, Cristal, SeeDB, DiSCO, etc. At a global level, they can be simply categorized as belonging to normalization of the refractive index with or without active lipid removal.

Lipid removal aids in clearing. Standard tissue processing results in a moderate amount of lipid extraction; dehydration post formalin fixation results in delipidation in the order of 50%. "Solvent-assisted clearing" refers to more comprehensive lipid removal using any of a variety of chemicals, such as detergents, to dissolve and remove lipids (and some proteins) from the tissue, thereby improving normalization of tissue optical properties.

With regards to normalizing the refractive index via fluid replacement, there are a multitude of published recipes. Some involve the use of compounds to directly increase the refractive index of water (e.g., high fructose concentrations—see DB); others reduce the refractive index of proteins (e.g., denaturation with urea); and others replace water with high refractive index nonaqueous fluids (e.g., organic solvents analogous to xylene). Several are combinations of these approaches.

The proliferation of clearing techniques has been rooted in the desire to achieve specific investigative goals such as preservation of fluorescence in genetically encoded fluorescent proteins (e.g., GFP) or to achieve efficient deep antibody staining for detailed compositional characterization. From a pathologist perspective, key considerations are more specifically the quality of imaging that they enable and the effects on established diagnostic pathways such as DNA/RNA and IHC testing, as well as cost, effort, and speed, the latter three relating to scalability for high throughput.

In this context, membrane changes from extensive lipid removal are likely problematic. Despite lipid extraction being part of standard tissue processing, compositional and morphologic alteration from extensive lipid extraction that is outside the range familiar to histology would adversely affect incorporation into clinical practice. Similarly, protein denaturation for refractive index reduction is bound to affect immunohistochemical studies, making it incompatible with current clinical workflows. In other aspects, derivative techniques such as that known as CLARITY involve gel stabilization with protein cross-linking followed by electrophoresis and/or solvent-based lipid removal, technically involved and time-consuming aspects that do not lend themselves well to high-throughput processing.

The compounds that are more immediately compatible with clinical workflows are those that serve as more direct xylene substitutes and follow analogous standard processing steps that preserve composition in an expected manner while still resulting in improved transparency. They also should be low toxicity, low cost, and stable. Examples are ethyl cinnamate, dibenzyl ether, and the combination of benzyl alcohol and benzyl benzoate (BABB). The latter is a modification of a clearing solution using methyl salicylate, originally developed more than a century ago (Spalteholz, 1914), and there is ample experience in its use for microscopic tissue preparation. BABB and its variants are considered inert with respect to dehydrated tissue specimens and have been used as storage media (Jacobson & Chapler, 1967; Orsini, 1962). Ethyl cinnamate and dibenzyl ether have been more recently recognized and achieve comparable clearing. The evolution of these protocols and their coupling with new optical sectioning modalities is an enabler for next-gen imaging in pathology.

There are other nonslide pathology techniques of interest

Beyond the next generation of digital methods actively being evaluated for pathology, there are several investigative techniques that, because of their current prominence and/or capabilities, are worth discussing in the context of clinical use.

Superresolution

One such next-gen imaging modality is superresolution imaging, which refers to what are now many related approaches that enable surpassing the theoretical resolution limits of standard microscopy. Without aiming to be comprehensive, methods in superresolution originate from two approaches: (1) using a laser to excite a fluorophore and using a second laser to suppress the excitation in a ring that encircles the excited region, effectively reducing the volume of fluorescence excitation, known as stimulated emission and depletion (STED) (Vicidomini et al., 2018), or (2) using fluorescent molecules that either blink naturally or are induced to blink, so that when imaged multiple times they produce data that can be analyzed statistically to improve the precision of localization, encompassing techniques known by the acronyms PALM (Betzig et al., 2006) (photoactivated localization microscopy) and STORM (Huang et al., 2008; Rust et al., 2006) (stochastic optical reconstruction microscopy). It is worth noting that some of the techniques already described to improve the resolution of other imaging modalities, such as deconvolution and structured illumination, can also be characterized as "superresolution" in that they computationally improve upon existing optical resolution limits. The STED and PALM/STORM methods and their numerous derivatives typically produce data at a resolution scale that falls somewhere between electron microscopy (a single nanometer) and confocal/multiphoton (a few hundred nanometers). The higher resolution is appealing from a clinical standpoint in as much as it could facilitate recognition of specific pathophysiologic processes at a cellular level. Historically, electron microscopy was once touted as potentially replacing much of standard histology, and yet it can be surmised that its use has been limited by the associated cost and complexity relative to its capability to add value. Predictably, compared to the other next-gen fluorescent imaging techniques, the resolution improvements in superresolution also come at the expense of one or more of: higher complexity of operation, higher equipment and reagent costs, longer sample preparation, and considerably slower imaging. Given that, commercial research microscopes that employ these and several related techniques are available, but clinical/high-throughput use would necessitate further advances in all these aspects.

A somewhat related and more recent technique that takes a different approach is expansion microscopy. As described in recent comprehensive reviews (Wen et al., 2023; Zhuang & Shi, 2023) the concept is to enable an effectively higher level of resolution by physically enlarging a specimen, taking advantage of the natural tendency of hydrogels to expand with water. In its original implementation, the simplified process involves several steps: (1) A specimen is exposed to fluorescent labels such as tagged antibodies, which are chemically activated for cross-linking; (2) the sample is permeated with (usually acrylamide) hydrogel components and allowed to polymerize and cross link; and (3) the gel is expanded with water. This results in physical separation of labeled components while preserving overall dimensional relationships, allowing nanometer reconstruction (on the order of 5–10 nm) while using more standard high-resolution imaging such as lightsheet microscopy or multiphoton microscopy. There are obvious analogies with hydrogel-based clearing techniques, and some nonhydrogel clearing techniques produce swelling that produces this effect, albeit to a much more limited degree. Expansion microscopy is devised to enable orders of magnitude expansion while preserving equidistance in all dimensions (isotropism).

In essence, expansion microscopy also broadens electron-microscopy-level capabilities with reduced needs for costly instrumentation and the added possibility of specificity in characterization. As with the more complex clearing protocols, the multistep process in current incarnations negatively impacts the generalizability. And the compositional changes induced also make it incompatible with existing ancillary test protocols. As such, these methods remain out of reach of routine analysis for clinical settings, but they present a foundation for future development with broader applicability and potential for further improvements in the quantity and quality of diagnostic information extractable from specimens.

Two other technical developments related to new imaging methods that have been explored for clinical histology include applications of machine learning (ML) for pseudocoloration of digital images and for prediction of immunohistochemical stains (IHC).

Pseudocoloration for H&E reproduction has been applied to a variety of imaging modalities, including unstained physical slides (Martell et al., 2023) and unembedded fresh or fixed tissue (Pillar et al., 2024). From a practical point of view, this reduces the burden on pathologists for retraining or simplifies or eliminates staining. A critical aspect for clinical use relates to the tendency of these approaches to "hallucinate," that is, create unpredictable transformations that deviate from the intended goal. In these instances, unknown patterns can potentially trick observers into seeing something different than what is present. In that regard, there remains a gap between apparently faithful recreations of H&E based on machine learning and the capability of pathologists to rely on it for rendering diagnoses.

The application of ML to slides has also been extended to predicting immunohistochemical staining from unstained or H&E-stained slides (Akbarnejad et al., 2023; Bai et al., 2022; Jackson et al., 2020). The importance lies in the opportunity to reduce the need to perform further processing in clinical cases that otherwise would need to undergo standard paraffin-embedding, slicing, and costly immunohistochemical staining. While similar arguments about "hallucination" could be brought forth, the difference lies in that, at least theoretically, these approaches could lead to the identification of recognizable features and patterns in the new digital modalities that correspond to the protein expression seen with IHC and, even more generally, with the classification of a diagnostic entity for which the IHC is being used. An analogy could be drawn between this development and the classification of tumor subtypes based on computerized clustering of RNA expression patterns, which can then be correlated to a few IHC stains. Complex analysis could result in simplified diagnostic algorithms. Although IHC use represents a minority of overall tissue processing volume in clinical environments, further reductions in labor and cost associated with physical slide preparation and expensive reagent use can only augment the value of pure digital histology.

A framework for validation is evolving

There has not been broad regulatory clearance for marketing of next-generation imaging systems as ex vivo diagnostic modalities in the clinical arena. Beyond satisfying regulatory requirements, histotechnologists, laboratory managers, and pathologists are likely to focus first on assuring themselves that any newly implemented technology will, at a minimum, not degrade quality of care.

Specific guidelines for WSI validation (Evans et al., 2022) produced by a collaborative group from leading professional organizations (Association for Pathology Informatics (API), College of American Pathologists (CAP), and American Society for Clinical Pathology (ASCP)) are based on quantitative analyses for how that goal can be achieved. The recommendations provided by such guidelines are applicable to images derived from alternative imaging modalities. Additionally, the CAP has been proactive in producing checklist items that refer specifically to implementing ex vivo microscopy (no date). These guidelines are valuable frameworks for the broad implementation of next-generation imaging approaches. Laboratory-derived procedures for new test development are also bound to be good starting points for any adoption of new imaging modalities, even for implementations as adjuncts to current practice.

In the case of alternative imaging modalities looking to supplant transmitted microscopy of physical slides, for an adequate risk/benefit profile, the overall clinical impact must be weighed as a whole. Doing so requires considering all the limitations of current practice and analyzing the specific benefits afforded by a particular imaging method.

Making sense of it all

While radiology and commercial cameras transitioned to full digital workflows years ago, pathology is undergoing its digital revolution moment. Whole slide imaging of standard physical slide-based histology specimens remains a valuable interim approach to improved care. However, its limitations highlight the need for complementary or alternative techniques to maximize information and improve diagnostic accuracy.

Several next-generation imaging modalities are nearing a stage of development where they can elevate the practice of pathology to a true digital enterprise. There are recurrent themes in terms of trade-offs between ease of preparation, speed of imaging, and the resolution and quality of resulting images. When evaluating a new pathology imaging system, the question will often be: Are both quality and speed satisfactory for a given pathologist task? Direct comparison of methods is challenging, and assessment should include overall throughput (including sample prep, loading, labeling, unloading, image acquisition, and image processing), labor, per sample cost, image quality, and robustness of the technique to expected biological variability. Imaging technologies have emerged that can enhance increasing aspects of pathology practice.

Overall, the new modalities offer considerable capabilities to complement existing workflows. Excitingly, specific implementations could provide marked benefits over standard physical slide preparation and WSI. Opportunities exist for extracting more and better data from tissue specimens to inform clinical decisions for improved patient outcomes, producing new standards for pathology, all while reducing turnaround times and with less work.

References

Akbarnejad, A., Ray, N., Barnes, P. J., & Bigras, G. (2023). Predicting Ki67, ER, PR, and HER2 statuses from H&E-stained breast cancer images. *arXiv*. Available from https://doi.org/10.48550/arXiv.2308.01982, https://arxiv.org.

Albers, J., Pacilé, S., Markus, M. A., Wiart, M., Vande Velde, G., Tromba, G., & Dullin, C. (2018). X-ray-based 3D virtual histology—Adding the next dimension to histological analysis. *Molecular Imaging and Biology*, *20*(5), 732–741. Available from https://doi.org/10.1007/s11307-018-1246-3.

Au, M., Almeida-Magana, R., Al-Hammouri, T., Haider, A., & Shaw, G. (2023). Accuracy of ex-vivo fluorescence confocal microscopy in margin assessment of solid tumors: A systematic review. *Journal of Histochemistry & Cytochemistry*, *71*(12), 661–674. Available from https://doi.org/10.1369/00221554231212948.

Au Yeung, S., Giaretta, P., Morningstar, T., Masuda, E., Questa, M., Fereidouni, F., Levenson, R. M., & Marsilio, S. (2023). Utility of fluorescence imitating brightfield imaging microscopy for the diagnosis of feline chronic enteropathy. *Veterinary Pathology*, *60*(1), 52–59. Available from https://doi.org/10.1177/03009858221131363.

Baas, D. J. H., Vreuls, W., Sedelaar, J. P. M., Vrijhof, H. J. E. J., Hoekstra, R. J., Zomer, S. F., van Leenders, G. J. L. H., van Basten, J. P. A., & Somford, D. M. (2023). Confocal laser microscopy for assessment of surgical margins during radical prostatectomy. *BJU International*, *132*(1), 40–46. Available from https://doi.org/10.1111/bju.15938, http://onlinelibrary.wiley.com/journal/10.1111/(ISSN)1464-410X.

Bai, B., Wang, H., Li, Y., de Haan, K., Colonnese, F., Wan, Y., Zuo, J., Doan, N. B., Zhang, X., Zhang, Y., Li, J., Yang, X., Dong, W., Darrow, M. A., Kamangar, E., Lee, H. S., Rivenson, Y., & Ozcan, A. (2022). Label-free virtual HER2 immunohistochemical staining of breast tissue using deep learning. *BME Frontiers, 2022*. Available from https://doi.org/10.34133/2022/9786242.

Barner, L. A. E. (2023). Artificial intelligence–Triaged 3-dimensional pathology to improve detection of esophageal neoplasia while reducing pathologist workloads. *Modern Pathology*, *36*, 2023.

Barner, L. A. E., Glaser, A. K., Mao, C., Susaki, E. A., Vaughan, J. C., Dintzis, S. M., & Liu, J. T. C. (2022). Multiresolution nondestructive 3D pathology of whole lymph nodes for breast cancer staging. *Journal of Biomedical Optics*, *27*(03). Available from https://doi.org/10.1117/1.jbo.27.3.036501.

Bera, K., Schalper, K. A., Rimm, D. L., Velcheti, V., & Madabhushi, A. (2019). Artificial intelligence in digital pathology—New tools for diagnosis and precision oncology. *Nature Reviews Clinical Oncology*, *16*(11), 703–715. Available from https://doi.org/10.1038/s41571-019-0252-y, http://www.nature.com/nrclinonc/archive/index.html.

Bertoni, L., Puliatti, S., Bonetti, L. R., Maiorana, A., Eissa, A., Azzoni, P., Bevilacqua, L., Spandri, V., Kaleci, S., Zoeir, A., Sighinolfi, M. C., Micali, S., Bianchi, G., Pellacani, G., Rocco, B., & Montironi, R. (2020). Ex vivo fluorescence confocal microscopy: Prostatic and periprostatic tissues atlas and evaluation of the learning curve. *Virchows Archiv*, *476*(4), 511–520. Available from https://doi.org/10.1007/s00428-019-02738-y.

Betzig, E., Patterson, G. H., Sougrat, R., Lindwasser, O. W., Olenych, S., Bonifacino, J. S., Davidson, M. W., Lippincott-Schwartz, J., & Hess, H. F. (2006). Imaging intracellular fluorescent proteins at nanometer resolution. *Science (New York, N.Y.)*, *313*(5793), 1642–1645. Available from https://doi.org/10.1126/science.1127344.

Borowsky, A. D., Levenson, R. M., Gown, A. M., Morningstar, T., Fleury, T. A., Henderson, G., Schaberg, K., Sybenga, A. B., Glassy, E. F., Taylor, S. L., & Fereidouni, F. (2023). A pilot validation study comparing fluorescence-imitating bright-field imaging, a slide-free imaging method, with standard formalin-fixed, paraffin-embedded hematoxylin-eosin–stained tissue section histology for primary surgical pathology diagnosis. *Archives of Pathology & Laboratory Medicine*, *148*(3), 345–352. Available from https://doi.org/10.5858/arpa.2022-0432-oa.

Chen, B., Chang, B.-J., Roudot, P., Zhou, F., Sapoznik, E., Marlar-Pavey, M., Hayes, J. B., Brown, P. T., Zeng, C.-W., Lambert, T., Friedman, J. R., Zhang, C.-L., Burnette, D. T., Shepherd, D. P., Dean, K. M., & Fiolka, R. P. (2022). Resolution doubling in light-sheet microscopy via oblique plane structured illumination. *Nature Methods*, *19*(11), 1419–1426. Available from https://doi.org/10.1038/s41592-022-01635-8.

Cheng, S., Zhou, Y., Chen, J., Li, H., Wang, L., & Lai, P. (2022). High-resolution photoacoustic microscopy with deep penetration through learning. *Photoacoustics*, *25*100314. Available from https://doi.org/10.1016/j.pacs.2021.100314.

Ching-Roa, V., Huang, C., & Giacomelli, M., (2023). Rapid large-volume tissue imaging with multiplane parallel strip-scanning two-photon microscopy, *Biophotonics Congress: Optics in the Life Sciences 2023 (OMA, NTM, BODA, OMP, BRAIN)*, Technical Digest Series (Optica Publishing Group, 2023), paper NW4C.2.

Ching-Roa, V. D., Huang, C. Z., & Giacomelli, M. G. (2021). Improved microscopy with ultraviolet surface excitation (MUSE) using high-index immersion illumination. *Biomedical Optics Express*, *12*(10), 6461–6473. Available from https://doi.org/

10.1364/BOE.435520, https://www.osapublishing.org/abstract.cfm?uri = boe-12-10-6461.

Ching-Roa, V. D., Huang, C. Z., Ibrahim, S. F., Smoller, B. R., & Giacomelli, M. G. (2022). Real-time analysis of skin biopsy specimens with 2-photon fluorescence microscopy. *JAMA Dermatology*, *158*(10), 1175–1182. Available from https://doi.org/10.1001/jamadermatol.2022.3628, http://archderm.jamanetwork.com/issues.aspx.

Conversano, A., Abbaci, M., van Diest, P., Roulot, A., Falco, G., Ferchiou, M., Coiro, S., Richir, M., Genolet, P.-M., Clement, C., Casiraghi, O., Lahkdar, A. B., Labaied, N., Ragazzi, M., & Mathieu, M.-C. (2023). Breast carcinoma detection in ex vivo fresh human breast surgical specimens using a fast slide-free confocal microscopy scanner: HIBISCUSS project. *BJS Open*, *7*(3). Available from https://doi.org/10.1093/bjsopen/zrad046.

Corden, C., Boitor, R., Dusanjh, P. K., Harwood, A., Mukherjee, A., Gomez, D., & Notingher, I. (2023). Autofluorescence-Raman spectroscopy for ex vivo mapping colorectal liver metastases and liver tissue. *Journal of Surgical Research*, *288*, 10–20. Available from https://doi.org/10.1016/j.jss.2023.02.014.

Daetwyler, S., & Fiolka, R. P. (2023). Light-sheets and smart microscopy, an exciting future is dawning. *Communications Biology*, *6*(1). Available from https://doi.org/10.1038/s42003-023-04857-4, https://www.springer.com/journal/42003.

Dandu, R. (2008). Storage media for computers in radiology. *Indian Journal of Radiology and Imaging*, *18*(4), 287–289. Available from https://doi.org/10.4103/0971-3026.43838, https://www.thieme-connect.com/products/ejournals/journal/10.1055/s-00050590.

Desciak, E. B., & Maloney, M. E. (2000). Artifacts in frozen section preparation. *Dermatologic Surgery*, *26*(5), 500–504. Available from https://doi.org/10.1046/j.1524-4725.2000.99246.x.

Di, L., Eichberg, D. G., Huang, K., Shah, A. H., Jamshidi, A. M., Luther, E. M., Lu, V. M., Komotar, R. J., Ivan, M. E., & Gultekin, S. H. (2021). Stimulated Raman histology for rapid intraoperative diagnosis of gliomas. *World Neurosurgery*, *150*, e135. Available from https://doi.org/10.1016/j.wneu.2021.02.122.

Dodt, H. U., Leischner, U., Schierloh, A., Jährling, N., Mauch, C. P., Deininger, K., Deussing, J. M., Eder, M., Zieglgänsberger, W., & Becker, K. (2007). Ultramicroscopy: Three-dimensional visualization of neuronal networks in the whole mouse brain. *Nature Methods*, *4*(4), 331–336. Available from https://doi.org/10.1038/nmeth1036.

Engel, T. N., Abraham, T. M., Morningstar, T., Fung, M. A., Rangchi, A., Kiuru, M., Fereidouni, F., & Levenson, R. (2022). Pilot study of fluorescence imitating brightfield imaging for rapid, slide-free dermatopathology. *Journal of Cutaneous Pathology*, *49*(12), 1060–1066. Available from https://doi.org/10.1111/cup.14308, http://onlinelibrary.wiley.com/journal/10.1111/(ISSN)1600-0560.

Evans, A. J., Brown, R. W., Bui, M. M., Chlipala, E. A., Lacchetti, C., Milner, D. A., Pantanowitz, L., Parwani, A. V., Reid, K., Riben, M. W., Reuter, V. E., Stephens, L., Stewart, R. L., & Thomas, N. E. (2022). Validating whole slide imaging systems for diagnostic purposes in pathology guideline update from the College of American Pathologists in collaboration with the American Society for Clinical Pathology and the Association for Pathology Informatics. *Archives of Pathology and Laboratory Medicine*, *146*(4), 440–450. Available from https://doi.org/10.5858/arpa.2020-0723-

CP, https://meridian.allenpress.com/aplm/article/146/4/440/464968/Validating-Whole-Slide-Imaging-Systems-for.

Fereidouni, F., Harmany, Z. T., Tian, M., Todd, A., Kintner, J. A., McPherson, J. D., Borowsky, A. D., Bishop, J., Lechpammer, M., Demos, S. G., & Levenson, R. (2017). Microscopy with ultraviolet surface excitation for rapid slide-free histology. *Nature Biomedical Engineering*, *1*(12), 957–966. Available from https://doi.org/10.1038/s41551-017-0165-y.

Forchhammer, S., Grunewald, S., Möhrle, M., Metzler, G., Eigentler, T., Münch, A.-K., & Ogrzewalla, H. (2023). Diagnosis of basal cell carcinoma with ex-vivo confocal laser scanning microscopy in a real-life setting. *Acta Dermato-Venereologica*, *103*, adv4859. Available from https://doi.org/10.2340/actadv.v103.4859.

Gaifulina, R., Nunn, A. D. G., Draper, E. R. C., Strachan, R. K., Blake, N., Firth, S., Thomas, G. M. H., McMillan, P. F., & Dudhia, J. (2021). Intra-operative Raman spectroscopy and ex vivo Raman mapping for assessment of cartilage degradation. *Clinical Spectroscopy*, *3*100012. Available from https://doi.org/10.1016/j.clispe.2021.100012.

Gao, G., Miyasato, D., Barner, L. A., Serafin, R., Bishop, K. W., Xie, W., Glaser, A. K., Rosenthal, E. L., True, L. D., & Liu, J. T. C. (2023). Comprehensive surface histology of fresh resection margins with rapid open-top light-sheet (OTLS) microscopy. *IEEE Transactions on Biomedical Engineering*, *70*(7), 2160–2171. Available from https://doi.org/10.1109/tbme.2023.3237267.

Gareau, D. S., Li, Y., Huang, B., Eastman, Z., Nehal, K. S., & Rajadhyaksha, M. (2008). Confocal mosaicing microscopy in Mohs skin excisions: Feasibility of rapid surgical pathology. *Journal of Biomedical Optics*, *13*(5), 054001. Available from https://doi.org/10.1117/1.2981828.

Glaser, A. K., Reder, N. P., Chen, Y., Yin, C., Wei, L., Kang, S., Barner, L. A., Xie, W., McCarty, E. F., Mao, C., Halpern, A. R., Stoltzfus, C. R., Daniels, J. S., Gerner, M. Y., Nicovich, P. R., Vaughan, J. C., True, L. D., & Liu, J. T. C. (2019). Multi-immersion open-top light-sheet microscope for high-throughput imaging of cleared tissues. *Nature Communications*, *10*(1). Available from https://doi.org/10.1038/s41467-019-10534-0, http://www.nature.com/ncomms/index.html.

Gómez-Gaviro, M. V., Sanderson, D., Ripoll, J., & Desco, M. (2020). Biomedical applications of tissue clearing and three-dimensional imaging in health and disease. *iScience*, *23*(8)101432. Available from https://doi.org/10.1016/j.isci.2020.101432.

Grizzetti, L., & Kuonen, F. (2022). Ex vivo confocal microscopy for surgical margin assessment: A histology-compared study on 109 specimens. *Skin Health and Disease*, *2* (2). Available from https://doi.org/10.1002/ski2.91.

Guerrero, J. A., Pérez-Anker, J., Fernández-Esparrach, G., Archilla, I., Diaz, A., Lopez-Prades, S., Rodrigo-Calvo, M., Lahoz, S., Camps, J., Puig, S., Malvehy, J., & Cuatrecasas, M. (2021). Ex vivo fusion confocal microscopy of colorectal polyps: A fast turnaround time of pathological diagnosis. *Pathobiology: Journal of Immunopathology, Molecular and Cellular Biology*, *88*(6), 392–399. Available from https://doi.org/10.1159/000517190, http://www.karger.com/journals/pat/pat_jh.htm.

Guida, S., Arginelli, F., Farnetani, F., Ciardo, S., Bertoni, L., Manfredini, M., Zerbinati, N., Longo, C., & Pellacani, G. (2021). Clinical applications of in vivo and ex vivo confocal microscopy. *Applied Sciences*, *11*(5), 1979. Available from https://doi.org/10.3390/app11051979.

Hanna, M. G., Parwani, A., & Sirintrapun, S. J. (2020). Whole slide imaging: Technology and applications. *Advances in Anatomic Pathology*, *27*(4), 251–259. Available from https://doi.org/10.1097/PAP.0000000000000273, http://www.anatomicpathology.com.

Hanna, M. G., Reuter, V. E., Samboy, J., England, C., Corsale, L., Fine, S. W., Agaram, N. P., Stamelos, E., Yagi, Y., Hameed, M., Klimstra, D. S., & Joseph Sirintrapun, S. (2019). Implementation of digital pathology offers clinical and operational increase in efficiency and cost savings. *Archives of Pathology and Laboratory Medicine*, *143*(12), 1545–1555. Available from https://doi.org/10.5858/arpa.2018-0514-OA, https://www.archivesofpathology.org/doi/pdf/10.5858/arpa.2018-0514-OA.

Heher, Y. K., Chen, Y., Pyatibrat, S., Yoon, E., Goldsmith, J. D., & Sands, K. E. (2016). Achieving high reliability in histology: An improvement series to reduce errors. *American Journal of Clinical Pathology*, *146*(5), 554–560. Available from https://doi.org/10.1093/AJCP/AQW148, http://ajcp.oxfordjournals.org/content/by/year.

Huang, B., Wang, W., Bates, M., & Zhuang, X. (2008). Three-dimensional super-resolution imaging by stochastic optical reconstruction microscopy. *Science (New York, N.Y.)*, *319* (5864), 810–813. Available from https://doi.org/10.1126/science.1153529.

Huang, L., Sun, H., Sun, L., Shi, K., Chen, Y., Ren, X., Ge, Y., Jiang, D., Liu, X., Knoll, W., Zhang, Q., & Wang, Y. (2023). label-free histopathological diagnosis of liver cancer based on Raman spectroscopy and deep learning. *Nature Communications*, *14*, 48.

Jackson, C. R., Sriharan, A., & Vaickus, L. J. (2020). A machine learning algorithm for simulating immunohistochemistry: Development of SOX10 virtual IHC and evaluation on primarily melanocytic neoplasms. *Modern Pathology*, *33*(9), 1638–1648. Available from https://doi.org/10.1038/s41379-020-0526-z, http://www.nature.com/modpathol/index.html.

Jacobson, H. N., & Chapler, F. K. (1967). Intrinsic innervation of the human placenta. *Nature*, *214*(5083), 103–104. Available from https://doi.org/10.1038/214103a0.

Jain, E. (2024). Whole slide imaging technology and its applications: Current and emerging perspectives. *International Journal of Surgical Pathology*, *32*(3), 433–448.

Khimchenko, A., Deyhle, H., Schulz, G., Schweighauser, G., Hench, J., Chicherova, N., Bikis, C., Hieber, S. E., & Müller, B. (2016). Extending two-dimensional histology into the third dimension through conventional micro computed tomography. *Neuroimage*, *139*, 26–36. Available from https://doi.org/10.1016/j.neuroimage.2016.06.005.

Koya, S. K., Brusatori, M., Yurgelevic, S., Huang, C., Werner, C. W., Kast, R. E., Shanley, J., Sherman, M., Honn, K. V., Maddipati, K. R., & Auner, G. W. (2020). Accurate identification of breast cancer margins in microenvironments of ex-vivo basal and luminal breast cancer tissues using Raman spectroscopy. *Prostaglandins & Other Lipid Mediators*, *151*106475. Available from https://doi.org/10.1016/j.prostaglandins.2020.106475.

Koyuncu, C., Janowczyk, A., Farre, X., Pathak, T., Mirtti, T., Fernandez, P. L., Pons, L., Reder, N. P., Serafin, R., Chow, S. S. L., Viswanathan, V. S., Glaser, A. K., True, L. D., Liu, J. T. C., & Madabhushi, A. (2023). Visual assessment of 2-dimensional levels within 3-dimensional pathology data sets of prostate needle biopsies reveals substantial spatial heterogeneity. *Laboratory Investigation*, *103*(12)100265. Available from https://doi.org/10.1016/j.labinv.2023.100265.

Krishnamurthy, S., Brown, J. Q., Iftimia, N., Levenson, R. M., & Rajadhyaksha, M. (2019). Ex vivo microscopy a promising next-generation digital microscopy tool for surgical pathology practice. *Archives of Pathology and Laboratory Medicine*, *143*(9),

1058–1068. Available from https://doi.org/10.5858/arpa.2019-0058-RA, https://www.archivesofpathology.org/doi/pdf/10.5858/arpa.2019-0058-RA.

Kumar, N., Gupta, R., & Gupta, S. (2020). Whole slide imaging (WSI) in pathology: Current perspectives and future directions. *Journal of Digital Imaging*, *33*(4), 1034–1040. Available from https://doi.org/10.1007/s10278-020-00351-z.

Lee, M. (2023). Recent advancements in deep learning using whole slide imaging for cancer prognosis. *Bioengineering*, *10*(8), 897. Available from https://doi.org/10.3390/bioengineering10080897.

Lee, M., Herrington, C. S., Ravindra, M., Sepp, K., Davies, A., Hulme, A. N., & Brunton, V. G. (2021). Recent advances in the use of stimulated Raman scattering in histopathology. *The Analyst*, *146*(3), 789–802. Available from https://doi.org/10.1039/d0an01972k.

Li, H., Ning, T., Yu, F., Chen, Y., Zhang, B., & Wang, S. (2021). Raman microspectroscopic investigation and classification of breast cancer pathological characteristics. *Molecules (Basel, Switzerland)*, *26*(4), 921. Available from https://doi.org/10.3390/molecules26040921.

Liu, J. T. C., Chow, S. S. L., Colling, R., Downes, M. R., Farré, X., Humphrey, P., Janowczyk, A., Mirtti, T., Verrill, C., Zlobec, I., & True, L. D. (2024). Engineering the future of 3D pathology. *The Journal of Pathology: Clinical Research*, *10*(1). Available from https://doi.org/10.1002/cjp2.347.

Liu, J. T. C., Glaser, A. K., Poudel, C., & Vaughan, J. C. (2023). Nondestructive 3D pathology with light-sheet fluorescence microscopy for translational research and clinical assays. *Annual Review of Analytical Chemistry*, *16*, 231–252. Available from https://doi.org/10.1146/annurev-anchem-091222-092734, http://www.annualreviews.org/journal/anchem.

Liu, Z., Su, W., Ao, J., Wang, M., Jiang, Q., He, J., Gao, H., Lei, S., Nie, J., Yan, X., Guo, X., Zhou, P., Hu, H., & Ji, M. (2022). Instant diagnosis of gastroscopic biopsy via deep-learned single-shot femtosecond stimulated Raman histology. *Nature Communications*, *13*(1). Available from https://doi.org/10.1038/s41467-022-31339-8.

Lizio, M. G., Boitor, R., & Notingher, I. (2021). Selective-sampling Raman imaging techniques for: Ex vivo assessment of surgical margins in cancer surgery. *Analyst*, *146*(12), 3799–3809. Available from https://doi.org/10.1039/d1an00296a, http://pubs.rsc.org/en/journals/journal/an.

Longo, C., Pampena, R., Bombonato, C., Gardini, S., Piana, S., Mirra, M., Raucci, M., Kyrgidis, A., Pellacani, G., & Ragazzi, M. (2019). Diagnostic accuracy of ex vivo fluorescence confocal microscopy in Mohs surgery of basal cell carcinomas: A prospective study on 753 margins. *British Journal of Dermatology*, *180*(6), 1473–1480. Available from https://doi.org/10.1111/bjd.17507.

Lu, F. K., Calligaris, D., Olubiyi, O. I., Norton, I., Yang, W., Santagata, S., Xie, X. S., Golby, A. J., & Agar, N. Y. R. (2016). Label-free neurosurgical pathology with stimulated Raman imaging. *Cancer Research*, *76*(12), 3451–3462. Available from https://doi.org/10.1158/0008-5472.CAN-16-0270, http://cancerres.aacrjournals.org/content/canres/76/12/3451.full.pdf.

Mannas, M. P., Jones, D., Deng, F. -M., Hoskoppal, D., Melamed, J., Orringer, D. A., & Taneja, S. S. (2023). Stimulated Raman histology, a novel method to allow for rapid pathologic examination of unprocessed, fresh prostate biopsies. *The Prostate*, *83*(11), 1060–1067. Available from https://doi.org/10.1002/pros.24547.

Martell, M. T., Haven, N. J. M., Cikaluk, B. D., Restall, B. S., McAlister, E. A., Mittal, R., Adam, B. A., Giannakopoulos, N., Peiris, L., Silverman, S., Deschenes, J., Li, X., & Zemp, R. J. (2023). Deep learning-enabled realistic virtual histology with ultraviolet photoacoustic remote sensing microscopy. *Nature Communications*, *14*(1). Available from https://doi.org/10.1038/s41467-023-41574-2, https://www.nature.com/ncomms/.

Matsui, T., Iwasa, A., Mimura, M., Taniguchi, S., Sudo, T., Uchida, Y., Kikuta, J., Morizono, H., Horii, R., Motoyama, Y., Morii, E., Ohno, S., Kiyota, Y., & Ishii, M. (2022). Label-free multiphoton excitation imaging as a promising diagnostic tool for breast cancer. *Cancer Science*, *113*(8), 2916–2925. Available from https://doi.org/10.1111/cas.15428.

McInnes, E. F. (2011). *Artifacts in histopathology. Background Lesions in Laboratory Animals* (pp. 93–99). Elsevier Inc.. Available from http://www.sciencedirect.com/science/book/9780702035197, http://doi.org/10.1016/B978-0-7020-3519-7.00008-5.

Morelli, P., Porazzi, E., Ruspini, M., Restelli, U., & Banfi, G. (2013). Analysis of errors in histology by root cause analysis: A pilot study. *Journal of Preventive Medicine and Hygiene*, *54*(2), 90–96.

Moscalu, M., Moscalu, R., Dascălu, C. G., Ţarcă, V., Cojocaru, E., Costin, I. M., Ţarcă, E., & Şerban, I. L. (2023). Histopathological images analysis and predictive modeling implemented in digital pathology—Current affairs and perspectives. *Diagnostics*, *13*(14), 2379. Available from https://doi.org/10.3390/diagnostics13142379.

Movahed-Ezazi, M., Nasir-Moin, M., Fang, C., Pizzillo, I., Galbraith, K., Drexler, S., Krasnozhen-Ratush, O. A., Shroff, S., Zagzag, D., William, C., Orringer, D., & Snuderl, M. (2023). Clinical validation of stimulated raman histology for rapid intraoperative diagnosis of central nervous system tumors. *Modern Pathology*, *36*(9)100219. Available from https://doi.org/10.1016/j.modpat.2023.100219.

Nackenhorst, M. C., Kasiri, M., Gollackner, B., & Regele, H. (2022). Ex vivo fluorescence confocal microscopy: Chances and changes in the analysis of breast tissue. *Diagnostic Pathology*, *17*(1). Available from https://doi.org/10.1186/s13000-022-01240-5, http://www.diagnosticpathology.org/.

Nguyen, H. T. M., Zhang, Y., Moy, A. J., Feng, X., Sebastian, K. R., Reichenberg, J. S., Fox, M. C., Markey, M. K., & Tunnell, J. W. (2021). Characterization of ex vivo nonmelanoma skin tissue using raman spectroscopy. *Photonics*, *8*(7), 282. Available from https://doi.org/10.3390/photonics8070282.

Olson, E., Levene, M. J., & Torres, R. (2016). Multiphoton microscopy with clearing for three dimensional histology of kidney biopsies. *Biomedical Optics Express*, *7*(8), 3089–3096. Available from https://doi.org/10.1364/BOE.7.003089, https://www.osapublishing.org/boe/viewmedia.cfm?uri = boe-7-8-3089&seq = 0.

Olson, E., Levene, M.J., & Torres, R. (2018). Fast, high-quality histology of cleared tissue samples by polygon-based multiphoton microscopy. *Optics InfoBase Conference Papers 2018* 10.1364/TRANSLATIONAL.2018.JW3A.18. https://www.osapublishing.org/beta/allconferences.cfm.

Orringer, D. A., Pandian, B., Niknafs, Y. S., Hollon, T. C., Boyle, J., Lewis, S., Garrard, M., Hervey-Jumper, S. L., Garton, H. J. L., Maher, C. O., Heth, J. A., Sagher, O., Wilkinson, D. A., Snuderl, M., Venneti, S., Ramkissoon, S. H., McFadden, K. A., Fisher-Hubbard, A., Lieberman, A. P., . . . Camelo-Piragua, S. (2017). Rapid intraoperative histology of unprocessed surgical specimens via fibre-laser-based stimulated

Raman scattering microscopy. *Nature Biomedical Engineering*, *1*(2). Available from https://doi.org/10.1038/s41551-016-0027, http://www.nature.com/natbiomedeng/.

Orsini, M. W. (1962). Technique of preparation, study and photography of benzyl-benzoate cleared material for embryological studies. *Reproduction (Cambridge, England)*, *3*(2), 283–287. Available from https://doi.org/10.1530/jrf.0.0030283.

Osterberg, E. C., Laudano, M. A., & Li, P. S. (2014). Clinical and investigative applications of Raman spectroscopy in urology and andrology. *Translational Andrology and Urology*, *3*(1), 84–88. Available from https://doi.org/10.3978/j.issn.2223-4683.2014.01.02, http://www.amepc.org/tau/article/download/3519/4439.

Pandey, R., Fournier, D., Root, G., Riccio, M., Shirvalkar, A., Zamora, G., Daigneault, N., Sapack, M., Zhong, M., & Harigopal, M. (2023). AI-powered biomolecular-specific and label-free multispectral imaging rapidly detects malignant neoplasm in surgically excised breast tissue specimens. *Archives of Pathology & Laboratory Medicine*, *147* (11), 1298–1306. Available from https://doi.org/10.5858/arpa.2022-0228-oa.

Peters, S. R. (2010). *Variables affecting the cutting properties of tissues and the resulting artifacts. A practical guide to frozen section technique* (pp. 97–115). Springer New York. Available from http://www.springerlink.com/openurl.asp?genre = book&isbn = 978-1-4419-1233-6, http://doi.org/10.1007/978-1-4419-1234-3_5.

Pillar, N., Li, Y., Zhang, Y., & Ozcan, A. (2024). Virtual staining of nonfixed tissue histology. *Modern Pathology*, *37*(5)100444. Available from https://doi.org/10.1016/j.modpat.2024.100444.

Poola, P. K., Afzal, M. I., Yoo, Y., Kim, K. H., & Chung, E. (2019). Light sheet microscopy for histopathology applications. *Biomedical Engineering Letters*, *9*(3), 279–291. Available from https://doi.org/10.1007/s13534-019-00122-y, http://www.springer.com/engineering/biomedical + eng/journal/13534.

Prince, M. N. H., Garcia, B., Henn, C., Yi, Y., Susaki, E. A., Watakabe, Y., Nemoto, T., Lidke, K. A., Zhao, H., Remiro, I. S., Liu, S., & Chakraborty, T. (2023). Signal improved ultra-fast light-sheet microscope (SIFT) for large tissue imaging. *bioRxiv*26928205. Available from https://doi.org/10.1101/2023.05.31.543002, https://www.biorxiv.org.

Reddi, D. M., Barner, L. A., Burke, W., Gao, G., Grady, W. M., & Liu, J. T. C. (2023). Nondestructive 3D pathology image atlas of barrett esophagus with open-top light-sheet microscopy. *Archives of Pathology and Laboratory Medicine*, *147*(10), 1164–1171. Available from https://doi.org/10.5858/ARPA.2022-0133-OA, https://allen.silverchair-cdn.com/allen/content_public/journal/aplm/147/10/10.5858_arpa.2022-0133-oa/5/i1543-2165-147-10-1164.pdf?.

Reder, N. P., Glaser, A. K., McCarty, E. F., Chen, Y., True, L. D., & Liu, J. T. C. (2019). Open-top light-sheet microscopy image atlas of prostate core needle biopsies. *Archives of Pathology and Laboratory Medicine*, *143*(9), 1069–1075. Available from https://doi.org/10.5858/arpa.2018-0466-OA, https://www.archivesofpathology.org/doi/pdf/10.5858/arpa.2018-0466-OA.

Restall, B. S., Haven, N. J. M., Kedarisetti, P., Martell, M. T., Cikaluk, B. D., Silverman, S., Peiris, L., Deschenes, J., & Zemp, R. J. (2021). Virtual hematoxylin and eosin histopathology using simultaneous photoacoustic remote sensing and scattering microscopy. *Optics Express*, *29*(9), 13864–13875. Available from https://doi.org/10.1364/OE.423740, https://www.osapublishing.org/oe/abstract.cfm?uri = oe-29-9-13864.

Reyes, A. O., & Humphrey, P. A. (1998). Diagnostic effect of complete histologic sampling of prostate needle biopsy specimens. *American Journal of Clinical Pathology, 109*(4), 416–422. Available from https://doi.org/10.1093/ajcp/109.4.416, http://ajcp.oxfordjournals.org/content/by/year.

Richardson, D. S., Guan, W., Matsumoto, K., Pan, C., Chung, K., Ertürk, A., Ueda, H. R., & Lichtman, J. W. (2021). Tissue clearing. *Nature Reviews Methods Primers, 1*(1). Available from https://doi.org/10.1038/s43586-021-00080-9, https://www.nature.com/nrmp/journal-information.

Riva, M., Sciortino, T., Secoli, R., D'Amico, E., Moccia, S., Fernandes, B., Nibali, M. C., Gay, L., Rossi, M., De Momi, E., & Bello, L. (2021). Glioma biopsies classification using Raman spectroscopy and machine learning models on fresh tissue samples. *Cancers, 13*(5), 1073. Available from https://doi.org/10.3390/cancers13051073.

Rust, M. J., Bates, M., & Zhuang, X. (2006). Sub-diffraction-limit imaging by stochastic optical reconstruction microscopy (STORM). *Nature Methods, 3*(10), 793–796. Available from https://doi.org/10.1038/nmeth929.

Sehm, T., Uckermann, O., Galli, R., Meinhardt, M., Rickelt, E., Krex, D., Schackert, G., & Kirsch, M. (2020). Label-free multiphoton microscopy as a tool to investigate alterations of cerebral aneurysms. *Scientific Reports, 10*(1). Available from https://doi.org/10.1038/s41598-020-69222-5.

Senter-Zapata, M., Patel, K., Bautista, P. A., Griffin, M., Michaelson, J., & Yagi, Y. (2016). The role of micro-CT in 3D histology imaging. *Pathobiology: Journal of Immunopathology, Molecular and Cellular Biology, 83*(2-3), 140–147. Available from https://doi.org/10.1159/000442387, http://www.karger.com/journals/pat/pat_jh.htm.

Silvestri, L., Costantini, I., Sacconi, L., & Pavone, F. S. (2016). Clearing of fixed tissue: A review from a microscopist's perspective. *Journal of Biomedical Optics, 21*(8), 081205. Available from https://doi.org/10.1117/1.jbo.21.8.081205.

Spalteholz, W. (1914). *Über das Durchsichtigmachen von menschlichen und tierischen Präparaten und seine theoretischen Bedingungen, nebst Anhang* (1914). Über Knochenfärbung. S. Hirzel.

Taqi, S. A., Sami, S. A., Sami, L. B., & Zaki, S. A. (2018). A review of artifacts in histopathology. *Journal of Oral and Maxillofacial Pathology, 22*(2), 279. Available from https://doi.org/10.4103/jomfp.JOMFP_125_15, http://www.jomfp.in/.

Taxy, J. B. (2009). Frozen section and the surgical pathologist a point of view. *Archives of Pathology and Laboratory Medicine, 133*(7), 1135–1138. Available from http://arpa.allenpress.com/pdfserv/10.1043%2F1543-2165-133.7.1135.

Thunnissen, E., Kerr, K. M., Herth, F. J. F., Lantuejoul, S., Papotti, M., Rintoul, R. C., Rossi, G., Skov, B. G., Weynand, B., Bubendorf, L., Katrien, G., Johansson, L., López-Ríos, F., Ninane, V., Olszewski, W., Popper, H., Jaume, S., Schnabel, P., Thiberville, L., & Laenger, F. (2012). The challenge of NSCLC diagnosis and predictive analysis on small samples. Practical approach of a working group. *Lung Cancer (Amsterdam, Netherlands), 76*(1), 1–18. Available from https://doi.org/10.1016/j.lungcan.2011.10.017.

Titze, U., Hansen, T., Brochhausen, C., Titze, B., Schulz, B., Gunnemann, A., Rocco, B., & Sievert, K.-D. (2021). Diagnostic performance of ex vivo fluorescence confocal microscopy in the assessment of diagnostic biopsies of the prostate. *Cancers, 13*(22), 5685. Available from https://doi.org/10.3390/cancers13225685.

Titze, U., Sievert, K.-D., Titze, B., Schulz, B., Schlieker, H., Madarasz, Z., Weise, C., & Hansen, T. (2022). Ex vivo fluorescence confocal microscopy in specimens of the liver: A proof-of-concept study. *Cancers*, *14*(3), 590. Available from https://doi.org/10.3390/cancers14030590.

Togawa, R., Hederer, J., Ragazzi, M., Bruckner, T., Fastner, S., Gomez, C., Hennigs, A., Nees, J., Pfob, A., Riedel, F., Schäfgen, B., Stieber, A., Lux, M. P., Heil, J., & Golatta, M. (2023). Imaging of lumpectomy surface with large field-of-view confocal laser scanning microscopy 'Histolog® scanner' for breast margin assessment in comparison with conventional specimen radiography. *The Breast*, *68*, 194–200. Available from https://doi.org/10.1016/j.breast.2023.02.010.

Torres, R., Olson, E., Homer, R., Martin, D. T., Levene, M. J., Perincheri, S., Sprenkle, P. C., & Humphrey, P. A. (2021). Initial evaluation of rapid, direct-to-digital prostate biopsy pathology. *Archives of Pathology & Laboratory Medicine*, *145*(5), 583–591. Available from https://doi.org/10.5858/arpa.2020-0037-oa.

Torres, R., Vesuna, S., & Levene, M. J. (2014). High-resolution, 2- and 3-dimensional imaging of uncut, unembedded tissue biopsy samples. *Archives of Pathology and Laboratory Medicine*, *138*(3), 395–402. Available from https://doi.org/10.5858/arpa.2013-0094-OA, http://www.archivesofpathology.org/doi/pdf/10.5858/arpa.2013-0094-OA.

Treacy, P. J., Khosla, A., Kyprianou, N., Falagario, U. G., Tsavaras, N., Wiklund, P., Tewari, A. K., & Durand, M. (2023). Value of multiphoton microscopy in uro-oncology: A narrative review. *Translational Andrology and Urology*, *12*(3), 508–518. Available from https://doi.org/10.21037/tau-21-973, https://tau.amegroups.com/article/view/110061/html.

Turashvili, G., Yang, W., McKinney, S., Kalloger, S., Gale, N., Ng, Y., Chow, K., Bell, L., Lorette, J., Carrier, M., Luk, M., Aparicio, S., Huntsman, D., & Yip, S. (2012). Nucleic acid quantity and quality from paraffin blocks: Defining optimal fixation, processing and DNA/RNA extraction techniques. *Experimental and Molecular Pathology*, *92*(1), 33–43. Available from https://doi.org/10.1016/j.yexmp.2011.09.013.

van Manen, L., Dijkstra, J., Boccara, C., Benoit, E., Vahrmeijer, A. L., Gora, M. J., & Mieog, J. S. D. (2018). The clinical usefulness of optical coherence tomography during cancer interventions. *Journal of Cancer Research and Clinical Oncology*, *144*(10), 1967–1990. Available from https://doi.org/10.1007/s00432-018-2690-9.

Vicidomini, G., Bianchini, P., & Diaspro, A. (2018). STED super-resolved microscopy. *Nature Methods*, *15*(3), 173–182. Available from https://doi.org/10.1038/nmeth.4593.

Villarreal, J. Z., Pérez-Anker, J., Puig, S., Xipell, M., Espinosa, G., Barnadas, E., Larque, A. B., Malvehy, J., Cervera, R., Pereira, A., Martinez-Pozo, A., Quintana, L. F., & García-Herrera, A. (2023). Ex vivo confocal microscopy detects basic patterns of acute and chronic lesions using fresh kidney samples. *Clinical Kidney Journal*, *16*(6), 1005–1013. Available from https://doi.org/10.1093/ckj/sfad019.

Villarreal, J. Z., Pérez-Anker, J., Puig, S., Pellacani, G., Solé, M., Malvehy, J., Quintana, L. F., & García-Herrera, A. (2021). Ex vivo confocal microscopy performs real-time assessment of renal biopsy in non-neoplastic diseases. *Journal of Nephrology*, *34*(3), 689–697. Available from https://doi.org/10.1007/s40620-020-00844-8, http://www.springer.com/medicine/nephrology/journal/40620.

Villarreal, P. P., Pal, R., Qiu, S., Coblens, O., Villasante-Tezanos, A., Resto, V., McCammon, S., & Vargas, G. (2023). Label-free imaging and histo-optical evaluation of head and neck cancers with multiphoton autofluorescence microscopy. *Cancers, 15* (4), 1302. Available from https://doi.org/10.3390/cancers15041302.

Wang, L., Fu, R., Xu, C., & Xu, M. (2022). Methods and applications of full-field optical coherence tomography: A review. *Journal of Biomedical Optics, 27*(5). Available from https://doi.org/10.1117/1.JBO.27.5.050901, http://www.spie.org/x866.xml.

Wang, Z., Han, K., Liu, W., Wang, Z., Shi, C., Liu, X., Huang, M., Sun, G., Liu, S., & Guo, Q. (2024). Fast real-time brain tumor detection based on stimulated raman histology and self-supervised deep learning model. *Journal of Imaging Informatics in Medicine*. Available from https://doi.org/10.1007/s10278-024-01001-4.

Weiss, K. R., Voigt, F. F., Shepherd, D. P., & Huisken, J. (2021). Tutorial: Practical considerations for tissue clearing and imaging. *Nature Protocols, 16*(6), 2732–2748. Available from https://doi.org/10.1038/s41596-021-00502-8, http://www.natureprotocols.com/.

Wells, W. A., Thrall, M., Sorokina, A., Fine, J., Krishnamurthy, S., Haroon, A., Rao, B., Shevchuk, M. M., Wolfsen, H. C., Tearney, G. J., & Hariri, L. P. (2018). In vivo and ex vivo microscopy: Moving toward the integration of optical imaging technologies into pathology practice. *Archives of Pathology & Laboratory Medicine, 143*(3), 288–298. Available from https://doi.org/10.5858/arpa.2018-0298-ra.

Wen, G., Leen, V., Rohand, T., Sauer, M., & Hofkens, J. (2023). Current progress in expansion microscopy: Chemical strategies and applications. *Chemical Reviews, 123* (6), 3299–3323. Available from https://doi.org/10.1021/acs.chemrev.2c00711.

Wong, I. H. M., Zhang, Y., Chen, Z., Kang, L., & Wong, T. T. W. (2022). Slide-free histological imaging by microscopy with ultraviolet surface excitation using speckle illumination. *Photonics Research, 10*(1), 120–125. Available from https://doi.org/10.1364/PRJ.440935, https://www.osapublishing.org/prj/abstract.cfm?uri = prj-10-1-120.

Xi, G., Huang, C., Lin, J., Luo, T., Kang, B., Xu, M., Xu, H., Li, X., Chen, J., Qiu, L., & Zhuo, S. (2023). Rapid label-free detection of early-stage lung adenocarcinoma and tumor boundary via multiphoton microscopy. *Journal of Biophotonics, 16*(11). Available from https://doi.org/10.1002/jbio.202300172, http://onlinelibrary.wiley.com/journal/10.1002/(ISSN)1864-0648.

Xie, W., Chen, Y., Wang, Y., Wei, L., Yin, C., Glaser, A. K., Fauver, M. E., Seibel, E. J., Dintzis, S. M., Vaughan, J. C., Reder, N. P., & Liu, J. T. C. (2019). Microscopy with ultraviolet surface excitation for wide-area pathology of breast surgical margins. *Journal of Biomedical Optics, 24*(02), 1. Available from https://doi.org/10.1117/1.jbo.24.2.026501.

Yang, B., Lang, M., Millett-Sikking, A., Zhao, X., Bragantini, J., VijayKumar, S., & Kamb, M. (2022). DaXi—high-resolution, large imaging volume and multi-view single-objective light-sheet microscopy. *Nature Methods, 19*, 461–469. Available from https://doi.org/10.1038/s41592-022-01417-2.

Yoshitake, T., Giacomelli, M. G., Cahill, L. C., Schmolze, D. B., Vardeh, H., Faulkner-Jones, B. E., Connolly, J. L., & Fujimoto, J. G. (2016). Direct comparison between confocal and multiphoton microscopy for rapid histopathological evaluation of unfixed human breast tissue. *Journal of Biomedical Optics, 21*(12), 126021. Available from https://doi.org/10.1117/1.jbo.21.12.126021.

Yoshitake, T., Giacomelli, M. G., Quintana, L. M., Vardeh, H., Cahill, L. C., Faulkner-Jones, B. E., Connolly, J. L., Do, D., & Fujimoto, J. G. (2018). Rapid histopathological imaging of skin and breast cancer surgical specimens using immersion microscopy with ultraviolet surface excitation. *Scientific Reports*, *8*(1). Available from https://doi.org/10.1038/s41598-018-22264-2.

Zarella, M. D., Bowman;, D., Aeffner, F., Farahani, N., Xthona, A., Absar, S. F., Parwani, A., Bui, M., & Hartman, D. J. (2018). A practical guide to whole slide imaging: A white paper from the digital pathology association. *Archives of Pathology & Laboratory Medicine*, *143*(2), 222–234. Available from https://doi.org/10.5858/arpa.2018-0343-ra.

Zhuang, Y., & Shi, X. (2023). Expansion microscopy: A chemical approach for super-resolution microscopy. *Current Opinion in Structural Biology*, *81*102614. Available from https://doi.org/10.1016/j.sbi.2023.102614.

Index

Note: Page numbers followed by "*f*" and "*t*" refer to figures and tables, respectively.

A

Academic industry partnership (AIP), 257
  academia *versus* industry priorities, 258–259
  benefits of, 258
  conflict of interest, 262
  contracts, 259–260
  digital pathology clinical trials, 257–258
  innovation cycle, 260–261, 261*f*
  intellectual property, 262–263
  risks, 258
  role of pathologist, 262
Academic inventors (licensees), 259
Accession, 80
Accuracy, 136
Acoustic sensor, 288–289
ADH. *See* Atypical duct hyperplasia (ADH)
AFP. *See* Alpha fetoprotein (AFP)
AIA. *See* Artificial Intelligence Act (AIA)
AIP. *See* Academic industry partnership (AIP)
AI. *See* Artificial intelligence (AI)
Alpha fetoprotein (AFP), 286
American National Standards Institute (ANSI), 220
Analytical Work Order Steps (AWOS), 237
Analytic step, 104, 106–112
  barcode quality considerations, 111–112
  scanner level quality checkpoint, 106–110, 108*f*
  scanner quality variability, 110–111
Anatomic pathology, 229*f*
ANN. *See* Artificial neural network (ANN)
ANSI. *See* American National Standards Institute (ANSI)
APIs. *See* Application programming interfaces (APIs)
Application programming interfaces (APIs), 222
Artifacts, 90–91
Artificial intelligence (AI), 257–258, 267–269, 270*f*, 272
  basics of, 243–244
  ChatGPT, 253–254
  in computational biology, 273–274
  in computational pathology, 268–272
  convolutional neural networks, 248–250
  deep learning, 243, 246–248, 249*f*
  in digital pathology, 274
  generative models, 251–252
    autoencoders, 251–252
    generative adversarial networks, 252
  machine learning, 272–273
  neuromorphic computing, 254–255
  for primary diagnosis of cytology, 174
  quantum computing, 255
  in rapid onsite evaluation (ROSE), 178–179
  recurrent neural networks, 250–251
  reinforcement learning, 253
  shallow learning, 244–246
Artificial Intelligence Act (AIA), 272–273
Artificial neural network (ANN), 246
Atypical duct hyperplasia (ADH), 140–141
Autoencoders, generative models, 251–252, 252*f*
Automated digital slide quality tools, 117–119
AWOS. *See* Analytical Work Order Steps (AWOS)

B

Backpropagation, 247–248
Barcodes, 111
  quality considerations, 111–112
Biopsy, 80–82
  protocol for, 84–85, 84*t*
Black-box, performance of, 287
Breast cancer, 140–141, 148–151
Business case, for digital pathology
  cost process
    acquisition and ongoing, 49–52
    associated with, 49–58
    avoidance and savings, 62–63
    calculators, 69
    capital acquisition, 50
    digital scanners, 52–53
    infrastructure, 58
    research and development, 53–54
    stand-alone slide acquisition, 48
  development of, 48–49
  expense of, 58
  financial benefits, 59–64
  financial indicators, 69–71
  healthcare systems, additional revenue to, 61–62
  hospital management, 72
  hybrid analog DP workflow, 63–64
  plans creation, 64–71
  required software solutions infrastructure, 57–58
  servers and data storage, 56–57
  space allocation, 50

Business case, for digital pathology (*Continued*)
specific software solutions infrastructure, 57–58
tangible and intangible benefits, 66*t*

C

CAD. *See* Computer-aided diagnosis (CAD)
Calculators, 69
CAMELYON challenge (CAncer MEtastases in LYmph nOdes challeNge), 143–144
Cancer Genome Atlas (TCGA), 119, 273
CAP Laboratory Accreditation Program (LAP), 208
CAP Laboratory General Checklist, 197–204
CCD. *See* Charge coupled device (CCD)
Cells, 3–4
Centers for Medicare and Medicaid Services (CMS), 137–139, 194
Charge coupled device (CCD), 3
ChatGPT, 253–254
Clearing Histology with MultiPhoton Microscopy (CHiMP), 304, 305*f*
Clearing techniques, next-gen, 306–307
Clinical Document Architecture (CDA), 222
Clinical Laboratory Improvement Amendments (CLIA), 137–139
CLIA'67, 191
CLIA'88, 191–192
history of, 191–192
Clinical Laboratory Standards Institute (CLSI), 121–122
Clinical tissue evaluation, 287
Clinical trial agreement (CTA), 259–260
Cloud-based digital storage, 36–37
CMOS. *See* Complementary metal-oxide semiconductor (CMOS)
CNNs. *See* Convolutional neural networks (CNNs)
COI. *See* Conflict of interest (COI)
College of American Pathologists (CAP), 80, 103, 197
Colorectal cancer, 148
Complementary metal-oxide semiconductor (CMOS), 3
Computer-aided diagnosis (CAD), 262–263
Confidential disclosure agreement (CDA), 259–260
Conflict of interest (COI), 258, 262
Confocal microscopy, 296–298, 297*f*, 299*f*
Conformité Européenne (European Conformity; CE marking), 188
Convolutional neural networks (CNNs), 118, 248–250, 249*f*
Coronavirus Disease 2019 (COVID-19), 267–268
pandemic, 187, 190, 193–194
Cost(s)
acquisition and ongoing, 49–52
associated with digital pathology, 49–58
avoidance and savings, 62–63
business case
calculators, 69
capital acquisition, 50
digital scanners, 52–53
infrastructure, 58
research and development, 53–54
stand-alone slide acquisition, 48
opportunity, 124
scanner annual maintenance, 53
scanner downtime, 53
Coverslipping method, 89
COVID-19. *See* Coronavirus Disease 2019 (COVID-19)
CPT. *See* Current procedural terminology (CPT)
CTA. *See* Clinical trial agreement (CTA)
Current procedural terminology (CPT), 221
Cytopathology, 171
whole slide imaging in, 171–176

D

Data use agreement (DUA), 259–260
DCIS (ductal carcinoma in situ), 140–141
Deep learning (DL) algorithm, 141–142, 149, 267–268
DeepLIIF, 145–146
Deep multimagnification network (DMMN), 142*f*
Dermatopathology, 141–142
DICOM. *See* Digital imaging and communication in medicine
DICOMweb, 230–231
Digital analysis algorithm, 235
Digital imaging, 229*f*, 230*f*
for education, 175–176
Digital imaging and communication in medicine (DICOM), 15, 41–42, 220, 222–223, 226*f*
annotation capabilities, 227
Integrating the Healthcare Enterprise, 231–239
Microscopy Bulk Simple Annotations, 228
network storage and retrieval, 230–231
in radiology imaging, 228–230
real-world mapping, 228
segmentation, 227
SNOMED CT, 231
standard microscope glass slide, 225
structured reporting, 227–228
supplement 122 to, 224–225
supplement 145 to, 225
supplement 222 to, 228
technical information, 223–224
very large image size and data requirements, 225–227

working group 13, 224
working group 26, 223
Digital pathology (DP), 103, 135, 171, 232
applications of, 1–2, 15–19, 16*t*
artificial intelligence in, 274
business plans creation, 64–71
challenges for, 19–21
clinical trials, 257–258
costs
acquisition and ongoing, 49–52
associated with, 49–58
avoidance and savings, 62–63
calculators, 69
capital acquisition, 50
digital scanners, 52–53
infrastructure, 58
research and development, 53–54
stand-alone slide acquisition, 48
development of, 48–49
digital imaging in, 2–3
downstream tools in, 6
ecosystem, 6, 20*f*
enhancing interoperability, 231–239
expense of, 58
financial benefits of, 59–64
financial indicators, 69–71
food and drug administration regulations in, 187–188
gross imaging, 7–8
healthcare systems, additional revenue to, 61–62
hospital management, 72
hybrid approach, 63–64
for immunohistochemistry positive control slides, 233
implementing, 19
importance of, 219–220
input devices for, 13
augmented reality and virtual reality, 14
mobile devices, microscope attachments for, 14
mouse and keyboard, 13
3D input devices, 13–14
touchscreen displays, 13
voice and gesture recognition, 14
interoperability, 219
interoperability and standards in, 14–15
key standards for, 41*t*
limitations of, 19*t*
milestones for, 4*t*
monitors/display devices, 11–12
noninferiority, 137–139
photomicroscopy, 6–7
primary diagnosis in, 192
QMS, 121–123, 123*f*
required software solutions infrastructure, 57–58
servers and data storage, 56–57
space allocation, 50
specific software solutions infrastructure, 57–58
stakeholders and roles in, 29*t*
standards development organization (SDO), 220
standards for, 40–43, 223–231
standards, types of, 220
tangible and intangible benefits, 66*t*
technology adoption, 23–25
timeline for, 3–5
U.S. Food and Drug Administration, 5
video streaming technology, 8–9
whole slide scanning, 9–10
Digital Pathology Association (DPA), 137–139, 257
Digital Pathology Evidence Creation (DPEC), 235, 237
Digital pathology image acquisition (DPIA), 235–236, 236*f*
Digital pathology ordering and workflow (DPOW), 235–237
Digital pathology systems (DPS), 1, 6, 257–258
Digital photomicroscopy, 6–7
Digital slide scanner, 52, 109
downtime cost, 53
preventative maintenance, 53
service and maintenance costs, 53
space allocation, 54
team staff, 54, 56
Digital workflow ergonomics, 28
Discriminator network, 252
Display devices, 11–12
Display monitors, 115–116
DL. *See* Deep learning (DL) algorithm
Down sampling, 225–227
DP. *See* Digital pathology (DP)
DPA. *See* Digital Pathology Association (DPA)
DPEC. *See* Digital Pathology Evidence Creation (DPEC)
DPIA. *See* Digital pathology image acquisition (DPIA)
DPOW. *See* Digital pathology ordering and workflow (DPOW)
DPS. *See* Digital pathology systems (DPS)
DUA. *See* Data use agreement (DUA)
Dynamic-robotic imaging devices, 10, 31–32

E

Education, digital imaging for, 175–176
Electronic health records (EHRs), 58, 221, 235
Eosin dye, 86–88

Ergonomics, digital workflow, 28
Estrogen receptor (ER), 144–146
Ethylenediaminetetraacetic acid (EDTA), 83
Ex vivo microscopy (EVM), 278, 280

F

Fast Healthcare Interoperability Resources (FHIR), 222
FDA-2020-D-1138, 191
Feasibility, 136
Federal Food, Drug, and Cosmetic (FD and C) Act, 187–188
FHIR. *See* Fast Healthcare Interoperability Resources (FHIR)
FIBI. *See* Fluorescence imitating brightfield illumination (FIBI)
Field of view (FOV), 7, 9
Financial benefits, 59–64
  Financial indicators, 69–71
FISH. *See* Fluorescence in situ hybridization (FISH)
Fixation, 80–83
Fluorescence imitating brightfield illumination (FIBI), 290–294, 294*f*
Fluorescence in situ hybridization (FISH), 18
Fluorescent confocal systems, 298
Fluorescent dyes, 290–291, 296–297
Fluorescent microscope, 290–291
Food and Drug Administration (FDA), 257–258, 273–274
  whole slide imaging, regulations, 187–188
    CAP Laboratory Accreditation Checklists, 197
    CAP Laboratory General Checklist, 197–204
    Class II, 189
    Class III, 188
    general guiding principles, 198–204
    good practice statements (GPSs), 198–202, 199*t*
    history of, 188–189
    legislation regulating LDTs, 214–216
    pixel pipeline, 189–190
    remote sign-out. *See* Remote whole slide imaging sign-out
    subsystems, 189
    three-tier medical device regulatory classification system, 188
    user training, 197
    validation/verification process, 198–204
Formalin-fixed paraffin embedded (FFPE) slides, 85, 279, 283
FOV. *See* Field of view (FOV)
Fresh tissue imaging, 293*f*

G

Gene expression, 149
General Data Protection Regulation (GDPR), 223
Generative adversarial networks (GAN), 118, 251–252
Generative models, AI and ML
  autoencoders, 251–252
  generative adversarial networks, 252
Generator network, 252
Good practice statements (GPSs), 198–202, 199*t*
Gradient descent, 247–248
Gross imaging, 7–8
Gross pathology, 7–8

H

Hallucination, 309–310
Hardware engineering development, 281
HDR displays. *See* High dynamic range (HDR) displays
Healthcare industry, standards in, 41
Healthcare Information and Management Systems Society (HIMSS), 220
Health Insurance Portability and Accountability Act (HIPPA), 222–223, 270–271
  and remote sign-out, 207
Health Level Seven (HL7), 42, 220
Hematein, 86–87, 188
Hematoxylin and eosin (H&E) stain, 81–82, 88*t*, 279, 285–286, 296–297
Hematoxylin dye, 86–87
High dynamic range (HDR) displays, 12
High power field (HPF), 113–114, 115*f*
HIMSS. *See* Healthcare Information and Management Systems Society (HIMSS)
HIPAA. *See* Health Insurance Portability and Accountability Act (HIPPA)
Histology, 79
  accession, 80
  digital pathology in laboratory, 88–92
  logistical challenges, 92–94
  microtomy, 85–86, 86*t*
  scanner, 296–297
  specimen grossing and fixation, 80–83
  staining, 86–88
  tissue embedding, 85
  tissue processing, 83–85, 84*t*
  workflow, 80–88
  WSI workflow, adoption, 94–95
Histology laboratory, 88–92
  preanalytic factors, in DP implementation, 79
    artifacts, 90–91
    coverslipping method, 89
    immunohistochemistry, 91–92

microscope slide labels, 90
special stains, 91–92
tissue placement, 89–90
Histopathology, 79
Histotechnology, 79
HL7. *See* Health Level Seven (HL7)
HL7's Version 2.x (V2), 222
HL7 Version 3 Clinical Document Architecture (CDA), 222
Homologous recombinant deficiency (HRD), 148–149
Hybrid DP workflows, 63–64
Hyperplane, 245*f*

I

ICA. *See* Industry consulting agreement (ICA)
Identifier standards, 223
IHC. *See* Immunohistochemical stains (IHC)
IHC. *See* Immunohistochemistry (IHC)
IHE pathology. *See* Integrating healthcare enterprise (IHE) pathology
Image-based prognosis, 149–151
Image management software (IMS), 235
Image quality/computational pathology tools, 119
Imaging Work Order Steps (IWOS), 237
Immunohistochemical markers, 144–145
Immunohistochemical stains (IHC), 309–310
Immunohistochemistry (IHC), 91–92, 233–234
IMPACT. *See* Innovative Models Promoting Access to Care Transformation (IMPACT)
IMS. *See* Image management software (IMS)
Industry consulting agreement (ICA), 259
Industry partner (licensor), 259
Information Object Definitions (IODs), 223–224
Innovation cycle, 260–261, 261*f*
Innovative Models Promoting Access to Care Transformation (IMPACT), 269–270
Input devices, for digital pathology, 13
augmented reality and virtual reality, 14
mobile devices, microscope attachments for, 14
mouse and keyboard, 13
3D input devices, 13–14
touchscreen displays, 13
voice and gesture recognition, 14
Integrating healthcare enterprise (IHE) pathology, 42–43, 119, 231–239
assessment of PD-L1, 146–147
Integrating the Healthcare Enterprise (IHE), 220
Intellectual property (IP), 257
International, and Digital Imaging and Communication in Medicine (DICOM), 220
International Color Consortium (ICC) standards, 110–111
International Organization for Standardization (ISO), 121–122, 220
Intraoperative consultation, 137
In vivo microscopy, 280
IODs. *See* Information Object Definitions (IODs)
IP. *See* Intellectual property (IP)
ISO. *See* International Organization for Standardization (ISO)
IWOS. *See* Imaging Work Order Steps (IWOS)

J

Joint Commission, 80

L

Laboratory-developed tests (LDTs), 187–188, 210, 214–216
Laboratory information system (LIS), 80, 105, 135, 228–230, 235–237
Laboratory testing workflow (LTW), 235–236
LDTs. *See* Laboratory-developed tests (LDTs)
Leadership sponsorship, 25
Lightsheet microscopy (LSM), 298–301, 302*f*
Light transmission microscopy, 278
LIS. *See* Laboratory information system (LIS)
Logical observation identifier names and codes (LOINC), 221
Logistical challenges, 92–94
LOINC. *See* Logical observation identifier names and codes (LOINC)
Long-short-term memory (LSTM), 250–251
LSM. *See* Lightsheet microscopy (LSM)
LTW. *See* Laboratory testing workflow (LTW)
Lymph nodes
metastases detection in, 143–144
micrometastases detection in, 144*f*

M

Machine learning (ML) algorithms, 243, 258, 267–269, 309
basics of, 243–244
ChatGPT, 253–254
convolutional neural networks, 248–250
deep learning, 243, 246–248, 249*f*
generative models, 251–252
autoencoders, 251–252
generative adversarial networks, 252
neuromorphic computing, 254–255
quantum computing, 255
recurrent neural networks, 250–251
reinforcement learning, 253
shallow learning, 244–246
Machine learning systems, 287
Maintenance costs, 53
Material transfer agreement (MTA), 259–260

MBSA. *See* Microscopy Bulk Simple Annotations (MBSA)
Memorial Sloan Kettering Cancer Center (MSK), 48, 57, 105, 107*f*
Messaging standards, 221–222
Microcomputed CT (microCT), 282–283, 295*f*
Microscopy Bulk Simple Annotations (MBSA), 228
Microscopy with ultraviolet surface excitation (MUSE), 282–283, 288–294, 291*f*, 298
Microtomy, 85–86, 86*t*
  artifacts at, 91
ML. *See* Machine learning (ML) algorithms
Modern cytology, 178
Modern cytopathology, 179
Modern deep learning, overview of, 254
Modern slide scanners, 172
Monitors devices, in digital pathology, 11–12
MSK. *See* Memorial Sloan Kettering Cancer Center (MSK)
MTA. *See* Material transfer agreement (MTA)
Multicheckpoint process, 104–106
Multiphoton microscopy, 302–305, 303*f*
MUSE. *See* Microscopy with ultraviolet surface excitation (MUSE)

N

National Electrical Manufacturers Association (NEMA), 224
National Provider ID (NPI), 223
National Society of Histotechnology (NSH), 79
Natural language processing (NLP), 250–251
NC. *See* Neuromorphic computing (NC)
NDA. *See* Non-disclosure agreement (NDA)
NEMA. *See* National Electrical Manufacturers Association (NEMA)
Neural network (NN), 246, 247*f*
Neuromorphic computing (NC), 254–255
Next generation imaging technologies, 277–278
  clearing techniques, 279
  confocal microscopy, 282–283
  digital pathology, 280
  ex vivo microscopy, 278
  multiphoton microscopy, 283
  MUSE/FIBI, 288–290
  new technology, description of, 281–294
  nonslide pathology techniques, 308
  pathology, new imaging methods for, 277
  photoacoustic microscopy, 288–290
  rapid onsite evaluation, 277
  selective plane illumination microscopy, 283
  stimulated Raman histology, 282–288
  superresolution, 308–310
  3D imaging, 283
  whole slide imaging, 278–280
NLM. *See* Nonlinear microscopy (NLM)
NLP. *See* Natural language processing (NLP)
NN. *See* Neural network (NN)
Non-disclosure agreement (NDA), 259–260
Nonlinear microscopy (NLM), 292
NPI. *See* National Provider ID (NPI)

O

Object ID (OID), 223
Open-top light sheet (OTLS) microscopy, 300–301
Opportunity costs, 124
Optical coherence microscopy (OCM), 282–283, 294–310
Optical coherence tomography (OCT), 294–310
Optical resolution photoacoustic microscopy (OR-PAM)., 290
Organic light-emitting diode (OLED) displays, 12
Osteosarcoma, 143*f*

P

PACS. *See* Picture archiving and communication system (PACS)
PaLM. *See* Pathology and laboratory medicine (PaLM)
Pathology
  AI/ML algorithm, 272
  artificial intelligence, 268–274
  bias and fairness, 269, 271*f*
  digital and computational, 268–272
  digital imaging standard in, 223
  equity in healthcare, 269–270
  ethical aspects of, 267–268
  machine learning, 272–273
  multimodal data integration, 273
  ownership and control, 272
  privacy and choice, 270–272
  prostate cancer detection, Augmenting pathologists in, 273–274
  stratifying prostate cancer risk, 273
Pathology and laboratory medicine (PaLM), 42–43, 231–232
Pathophysiologic process, 280–281
Philips Intellisite Pathology Solution (PIPS), 189, 257–258
Photoacoustic microscopy, 288–290, 289*f*
Photomicroscopy, digital, 6–7
Picture archiving and communication system (PACS), 15, 235–237
PIPS. *See* Philips Intellisite Pathology Solution (PIPS)
PI. *See* Principal investigator (PI)
Pixel pipeline, 189–190

PMA. *See* Premarket approval (PMA) application
Point of Use Quality Assurance Remote Image (POUQA) tool, 116
Postanalytic step, 104
automated digital slide quality tools, 117–119
display monitors, 115–116
QC at, 113
quality factors, 112–119
WSI viewer software, 114–115
Pramana 2-head scanning system, 121
Preanalytical step, 104
Preanalytic factors, in DP implementation, 79
histology laboratory, 88–92
artifacts, 90–91
coverslipping method, 89
immunohistochemistry, 91–92
microscope slide labels, 90
special stains, 91–92
tissue placement, 89–90
Premarket approval (PMA) application, 188
Primary diagnosis, 192, 211–212, 232
AI applications for, 174
development and deployment of AI solutions, 174–175
whole slide imaging for, 173
Principal investigator (PI), 260–261
Privacy and security standards, 222–223
Progesterone receptor (PR), 144–146
Programmed death-1 (PD-1)/death ligand-1(PD-L1), 146–147
Prostate cancer, 139–140
Pseudocoloring technique, 292–293
Public health emergency (PHE), 187, 190–191, 196–197

Q

Quality control/quality assurance (QC/QA), 103–104
Quality error rates, 119–121
Quality management systems (QMS), 88–89, 103
analytic step, 104, 106–112
cost of quality, 124–125
digital pathology, 121–123, 123*f*
ensuring image quality, 104–106
framework for, 122
image quality/computational pathology tools, 119
key practices, 122
multicheckpoint process, 104–106
postanalytic step, 104
preanalytical step, 104
process, 104, 105*f*
quality error rates, 119–121
quality/performance metrics, 122–123, 124*t*
slide and image quality checkpoints, 106*f*
Quality/performance metrics, 122–123, 124*t*
Quantitative biomarker detection, 144–147, 145*f*
Quantum computing, 255
Qubit-based algorithms, 255

R

Raman scattering, 284–285
Raman spectroscopy, 284–285
Rapid onsite evaluation (ROSE), 178–179, 277
AI applications in, 178–179
primary diagnosis and expert consultation, 179
reimbursement, 178
RCM. *See* Reflectance confocal microscopy (RCM)
ReceptorNet, 145–146
Recurrent neural networks (RNNs), 250–251
Reflectance confocal microscopy (RCM), 297–298
Region of interest (ROI), 227
Reinforcement learning (RL), 253
Remote whole slide imaging sign-out
billing, 212–214
and CLIA'88, 192–194
definition, 192
diagnostic procedures, 211–212
enforcement discretion "waiver," ramifications and requirements for, 194–197
Food and Drug Administration role on, 210
hardware and software specifications, 210
HIPAA/privacy-related considerations, 207
hospital IT, role of, 207–208
medicolegal considerations, 210–211
risks of using digital pathology for, 205
standard operating procedures (SOPs), 205–207, 206*t*
validating/verifying monitors/laptops from remote locations, 209
validation of digital pathology platforms, 208–209
RL. *See* Reinforcement learning (RL)
RNNs. *See* Recurrent neural networks (RNNs)
ROI. *See* Region of interest (ROI)
ROSE. *See* Rapid onsite evaluation (ROSE)

S

SaMD. *See* Software as a medical device (SaMD)
SBIR. *See* Small Business Innovation Research (SBIR)
SDAM. *See* Specimen domain analysis model (SDAM)
SDM. *See* SNOMED-DICOM microglossary (SDM)

SDO. *See* Standards development organization (SDO)
Seed/angel investment, 260–261
Selective plane illumination microscopy (SPIM), 283, 298–302, 300*f*
Serum protein electrophoresis (SPEPs), 18
Service costs, 53
Service object pair (SOP), 41–42
Seven S's for success
  digital imaging and communications in medicine, 41–42
  imaging modes, 32
  implementing digital pathology, 23*t*
  lab space/countertop weight loads, 36
  leadership sponsorship, 25
  live viewing capabilities, 34–35
  magnification/resolution, 34
  multi z-plane support, 35
  rescan rates, 35
  scan area, flexible selection of, 34
  scanners, 31–32
  scan speeds, 33
  service and maintenance, 38–40
  slide sizes and formats, 33
  slide throughput/continuous load, 33
  software, 37–38
  space allocation, 26
    current space utilization, assessment of, 26–27
    ergonomics and user experience, 27–29
  staffing resources, 29–31
  storage, 36–37
  user experience, 36
  whole slide image file formats, 36
Shallow learning, 244–246
Skin cancer, 141–142
Slide-free pathology, 283
Slide-free techniques, 282, 284*t*
Small Business Innovation Research (SBIR), 259
Small Business Technology Transfer (STTR), 259
SME. *See* Subject-matter experts (SME)
SNN. *See* Spiking neural networks (SNN)
SNOMED CT. *See* Systematized nomenclature of medicine-clinical terms (SNOMED CT)
SNOMED-DICOM microglossary (SDM), 231
SNOMED. *See* Systematized Nomenclature of Medicine (SNOMED)
Software as a medical device (SaMD), 273
Solvent-assisted clearing, 306–307
SOP. *See* Service object pair (SOP)
Space allocation, 26, 50
Spaghetti diagram, 26–27
Special stains, 91–92
Specimen domain analysis model (SDAM), 239
Specimen grossing, 80–83
Specimen label identifiers, 223
Speed *versus* quality balance, 282
SPEPs. *See* Serum protein electrophoresis (SPEPs)
Spiking neural networks (SNN), 254–255
SPIM. *See* Selective plane illumination microscopy (SPIM)
Sponsored program office (SPO), 259–260
Sponsored research agreement (SRA), 259
SPO. *See* Sponsored program office (SPO)
SRA. *See* Sponsored research agreement (SRA)
SRH. *See* Stimulated Raman histology (SRH)
Staffing
  in digital pathology, 54–56
  resources, 29–31
Standards development organization (SDO), 220
Standards, digital pathology, 219–220
  DICOM. *See* Digital Imaging and Communication in Medicine (DICOM)
  in healthcare, 220–223
  identifier, 223
  messaging and transport, 221–222
  primary digital imaging standard, 223
  privacy and security, 222–223
  vocabulary/terminology/ontology, 221
Stimulated Raman histology (SRH), 284–288, 285*f*, 286*f*, 288*f*, 295, 298
Stimulated Raman scattering (SRS), 282–288
STTR. *See* Small Business Technology Transfer (STTR)
Subject-matter experts (SME), 262
Superresolution techniques, 303, 308–310
Support Vector Machine (SVM), 244–245
Surgical pathology, use cases in, 135–152
SVM. *See* Support Vector Machine (SVM)
Systematized Nomenclature of Medicine (SNOMED), 220
Systematized nomenclature of medicine-clinical terms (SNOMED CT), 221
Systematized Nomenclature of Medicine (SNOMED) International, 220

T

Tag image file format (TIFF), 2–3
Technology licensing office (TLO), 262
Technology transfer agreement (TTA), 259
Technology transfer office (TTO), 259
Telecytology, 176–178
  configurations, 176–178
  dynamic system, 177
  image capture, 176–177
  image transmission, 177
  for primary diagnosis and expert consultation, 179

rapid onsite evaluation, 178–179
remote, 179
security, 177
static system, 177
Telepathology, 136–139
diagnostic accuracy, 136
feasibility, 136
intraoperative consultation, 137
primary diagnosis, 137–139
TeleROSE systems, 177–179
3D imaging, 305–308
TIFF. *See* Tag image file format (TIFF)
Tissue embedding, 85
Tissue fixation, 82
3D input devices, 13–14
Tissue processing, 83–85, 84*t*
TLO. *See* Technology licensing office (TLO)
Total product lifecycle (TPLC), 269–270
Transfer learning, 249–250
Transport standards, 221–222
TTA. *See* Technology transfer agreement (TTA)
TTO. *See* Technology transfer office (TTO)
Tumor detection/classification, 139–142
breast cancer, 140–141
prostate cancer, 139–140
skin cancer, 141–142
Tumor mutation, 148–149

U

Unified code for units of measure (UCUM), 221
U.S. Food and Drug Administration (FDA), 5
UV photoacoustic effect and reflective microscopy (UV-PARS), 290

V

Vendor neutral archive (VNA), 235
Verifying Accurate Leading edge IVCT, and Development (VALID) Act, 214–216
Video streaming technology, 8–9
VivaScope, 296–297
Vocabulary standards, 221

W

WG13, 224. *See also* Working group 13 (WG13)
WG26. *See* Working Group 26 (WG26)
Whole slide imaging (WSI), 1, 9–10, 14–15, 32*t*, 88–89, 103, 135, 171–176, 267–268, 278–279. *See also* Digital pathology
food and drug administration regulations, 187–188
CAP Laboratory Accreditation Checklists, 197
CAP Laboratory General Checklist, 197–204
Class II, 189
Class III, 188
general guiding principles, 198–204
good practice statements (GPSs), 198–202, 199*t*
history of, 188–189
legislation regulating LDTs, 214–216
pixel pipeline, 189–190
remote sign-out. *See* Remote whole slide imaging sign-out
subsystems, 189
three-tier medical device regulatory classification system, 188
user training, 197
validation/verification process, 198–204
genetic mutation on H&E, 147–149, 150*t*
histology slide racks, 93*f*
HPF in, 113–114, 115*f*
for primary diagnosis, 173
quality assessment, 118–119
scanners, 92–93, 95, 110–111
technical features, 110
validation, 95
viewer software, 114–115
workflow adoption, 94–95
Whole slide scanning, 9–10
Working group 13 (WG13), 224
Working Group 26 (WG26), 223
World Health Organization (WHO), 221

Z

Z-stack multi plane scanning, 35